Speech Processing for IP Networks

Speech Processing for IP Networks

Media Resource Control Protocol (MRCP)

Dave Burke

John Wiley & Sons, Ltd

Other Wiley Editorial Offices

John Wiley & Sons Inc., 111 River Street, Hoboken, NJ 07030, USA

Jossey-Bass, 989 Market Street, San Francisco, CA 94103-1741, USA

Wiley-VCH Verlag GmbH, Boschstr. 12, D-69469 Weinheim, Germany

John Wiley & Sons Australia Ltd, 42 McDougall Street, Milton, Queensland 4064, Australia

John Wiley & Sons (Asia) Pte Ltd, 2 Clementi Loop #02-01, Jin Xing Distripark, Singapore 129809

John Wiley & Sons Canada Ltd, 6045 Freemont Blvd, Mississauga, ONT, L5R 4J3, Canada

Anniversary Logo Design: Richard J. Pacifico

Library of Congress Cataloging-in-Publication Data

Burke, Dave.
    Speech processing for ip networks : media resource control protocol (mrcp) / Dave Burke.
      p. cm.
    Includes bibliographical references and index.
    ISBN 978-0-470-02834-6 (cloth : alk. paper)
    1. Speech processing systems.   2. Automatic speech recognition.   3. MRCP (Computer network protocol)
    I. Title.
    TK7882. S65B87 2007
    006.4′54—dc22

                                                                                       2006036199

British Library Cataloguing in Publication Data

A catalogue record for this book is available from the British Library

ISBN 978-0-470-02834-6 (HB)

Typeset in 9/11pt Times by Integra Software Services Pvt. Ltd, Pondicherry, India
Printed and bound in Great Britain by Antony Rowe Ltd, Chippenham, Wiltshire
This book is printed on acid-free paper responsibly manufactured from sustainable forestry in which
at least two trees are planted for each one used for paper production.

Contents

About the author

Dave Burke is Chief Technology Officer at Voxpilot, a leading company in next-generation interactive voice and video response systems. Dave is an active member of the IETF SpeechSC Working Group that developed the MRCPv2 standard. Dave has also been a member of the W3C's Voice Browser Working Group since 2002 and is an editor of the VoiceXML 2.1 and Semantic Interpretation for Speech Recognition specifications. Dave has written several papers in leading academic journals and conferences and is a regular speaker at international speech conferences. Dave holds a Ph.D., M.Eng.Sc., and B.E. in Electronic Engineering, all from University College Dublin, Belfield, Ireland.

Preface

The Media Resource Control Protocol (MRCP) is a key enabling technology delivering standardised access to advanced media processing resources including speech recognisers and speech synthesisers over IP networks. MRCP leverages Internet and Web technologies such as SIP, HTTP, and XML to deliver an open standard, vendor-independent, and versatile interface to speech engines. MRCP significantly eases the integration of speech technologies into network equipment and accelerates their adoption, enabling operators and service providers to deliver, efficiently and effectively, exciting and compelling interactive services over the telephone.

MRCP is not a stand-alone protocol but is based on several technologies that span both the Internet and Web worlds. This book brings these technologies together into a single volume and provides a thorough introduction to the fundamentals underlying MRCP. The content is targeted at software engineers, network administrators, technical managers, product managers, and technical marketing professionals. The material is covered in such a way as to be approachable for readers new to both IP telecommunications and speech processing. The treatment of each subject commences with introductory level material and progresses with a 'deep dive' to cover important, advanced material. Both the MRCP protocol itself and the data representation formats it leverages are text based. This not only makes debugging and troubleshooting easier, it also helps newcomers to understand quickly how the protocol operates. In this book, we have taken every opportunity to supply complete protocol exchange examples, depicting messages exactly as they appear 'on the wire'.

Part I of the book is aimed at providing introductory background material. Chapter 1 begins with an introduction to speech applications and discusses the benefits that speech technologies bring to the realm of interactive voice response. The value proposition of MRCP is discussed in the context of problems previously faced with integrating speech technologies into network equipment. The chapter concludes by providing a background on how MRCP came to be and important information relating to its evolution in the relevant standard bodies. For readers new to speech processing, Chapter 2 introduces the principles of speech processing by describing the basic underlying techniques employed by modern speech synthesisers, speech recognisers, and speaker verification and identification engines. Chapter 3 provides an introduction to the basics of MRCP. The different media processing resource types are discussed and several typical network architectures are presented. This chapter serves to provide a general context for more detailed information to follow.

Part II of the book is concerned with the media and control sessions used to communicate with a media resource server. Chapter 4 introduces the Session Initiation Protocol (SIP), which provides session establishment and management functions within MRCP, and Chapter 5 discusses the specifics of how SIP is used in MRCP. Chapter 6 is concerned with the media session and in particular how the media is encoded and transported to and from the media resource server. Finally, Chapter 7 covers the control session and introduces the general principles of the MRCP protocol itself.

Part III of the book focuses on the data representation formats used by the different MRCP media resource types. These formats convey information to and from the media resources and are either referenced from or carried in the payload of MRCP messages. Chapter 8 introduces the W3C Speech Synthesis Markup Language (SSML), which is used to provide marked up content to speech synthesiser resources to render to audio. Chapter 9 discusses the W3C Speech Recognition Grammar Specification (SRGS) and its role in specifying and constraining the words and phrases a recogniser looks for when analysing an audio stream. Chapter 10 is concerned with the Natural Language Semantics Markup Language (NLSML) – a data representation format used for encapsulating recognition results from a speech recogniser resource as well as results from speaker verification and identification resources. Chapter 11 introduces the W3C Pronunciation Lexicon Specification (PLS) – a standard syntax for specifying pronunciation lexicons for use with speech synthesisers and speech recognisers.

Part IV of the book provides in-depth details on each of the four MRCP media resource types. Chapters 12 to 15 cover the speech synthesiser, speech recogniser, speech recorder, and speaker verification and identification resources respectively. Each method and event is presented along with a description of the header fields used to parameterise them.

Part V of the book discusses a popular approach for programming speech applications known as VoiceXML. This Web-based paradigm was in many ways a key driver to the development of several MRCP features and is introduced in detail in Chapter 16. Chapter 17 studies some fundamental VoiceXML-MRCP interworking patterns and presents an example of a complete VoiceXML application with associated MRCP flows.

While this book focuses on the MRCPv2 standard developed by the IETF's SpeechSC Working Group, much of the content is also relevant to the previous, non-standard but widely implemented version of MRCP, now known as MRCPv1. Loosely speaking, MRCPv1 is a limited subset of MRCPv2 that employs a different protocol for managing media sessions and transporting control messages, namely the Real Time Streaming Protocol (RTSP).

Appendices A, B and C (located on the web at http://www.daveburke.org/speechprocessing/) describe MRCPv1 and how it differs from MRCPv2, and offer short introductions to the Hyper Text Transfer Protocol (HTTP) and eXtensible Markup Language (XML) respectively for readers new to these technologies.

After reading this book, you should have a solid technical understanding of the principles of MRCP, how it leverages other protocols and specifications for its operation, and how it is applied in modern telecommunication networks to deliver advanced speech capabilities. Readers wishing to further their knowledge in any particular area will find useful references to the relevant IETF and W3C specifications in the references section at the end of the book.

Acknowledgements

Writing this book has been both a rewarding experience and a privilege. Through my involvement in Voxpilot, I have been afforded the opportunity to work on both the "academic" and "industrial" aspects of the technologies covered in this book. On the academic side, I have been fortunate to be able to work with a wonderful set of brains and personalities at the W3C's Voice Browser Working Group and IETF's SpeechSC Working Group. On the industry side, I have had the chance to work with a group of world-class engineers and architects at Voxpilot. In particular, Andrew Fuller, James Bailey, and Stephen Breslin deserve a special mention as the team who helped develop early designs and implementations of the technologies described herein and enabled me to garner a really in-depth understanding of speech processing technologies.

Of course, no acknowledgement section would be complete without an expression of gratitude to the Wiley team encompassing commissioning editors, copy editors, typesetters, and project managers. Your positive encouragement and professionalism was much appreciated throughout this project!

Part I

Background

1

Introduction

The Media Resource Control Protocol (MRCP) is a new protocol designed to provide a standardised, uniform, and flexible interface to network-based media processing resources such as speech recognisers and speech synthesisers. The standard MRCP specification is being developed by the Internet Engineering Task Force (IETF) and has been designed to dovetail with existing IETF protocols and Web standards created by the World Wide Web Consortium (W3C). This chapter provides a background to MRCP by introducing some of the benefits of using speech technologies and the motivations behind MRCP. A brief history of the standardisation efforts that led to the development of MRCP is also covered.

1.1 Introduction to speech applications

Speech processing technologies facilitate conversational dialogues between human and machine by exploiting the most natural and intuitive communication modality available to man, namely speech. One powerful application of speech, and the focus of this book, is its use in interactive services delivered over telephony devices. Speech processing technologies can breathe new life into rigid interactive voice response (IVR) systems resulting in a more user friendly voice-user interface (VUI) that understands natural, conversational language. Speech allows the user to circumvent complex and confusing dual tone multifrequency (DTMF) touch-tone menu structures and instead navigate quickly and fluidly to the particular service or option they seek. The result is shortened call times, increased customer satisfaction, and enhanced automation levels. Speech-enabled VUIs possess several advantages over other human–computer interaction paradigms. For a start, conversational dialogues do not interfere with visual or manual tasks such as driving a car. Secondly, dynamic, timely information may be synthesised to the user through the use of text-to-speech (TTS) technologies. Finally, audio-based media can be easily incorporated, for example music or radio broadcasts.

Telephony applications incorporating speech typically rely on *network-based* speech processing technologies. Free from the constraints of limited processing power, memory, and finite battery life, network-based media processing resources can deliver high accuracy, large vocabulary, speaker-independent speech recognition services in conjunction with natural sounding speech

Speech Processing for IP Networks   Dave Burke
© 2007 John Wiley & Sons, Ltd

synthesis. A further important advantage of network-based speech processing is that it obviates the need for specialist client-side software tailored for a limited set of devices; network-based speech services may be delivered uniformly to any telephony device without limitation. Given the ubiquity of telephones in comparison with desktop computers, and coupled with the attractiveness of providing a natural and intuitive interface, it is obvious why many companies are deploying telephony speech applications to deliver premium services to their customers. Network-based speech processing permits a plethora of advanced interactive services to be delivered by standard telephony, which are equally applicable across industries ranging from financial services to healthcare. Popular examples include automated directory enquiries, store locator, bill payment, account activation, password reset, flight information, ticket reservation and purchasing, biometric applications, and a wide variety of information services.

The speech-enabled IVR market is growing and expected to continue to grow. *DataMonitor* [1] reports that the total value of the North American IVR market (the largest IVR market in the world) will grow to $709 million in 2009. Within that statistic there is some further interesting information. The revenue from standards-based IVR built on technologies such as VoiceXML and MRCP will more than double from $90 million in 2004 to $196 million in 2009, at the expense of traditional, proprietary IVR. Further, by 2009, speech-enabled IVR port shipments will account for more than 50 % of total IVR port shipments.

1.2 The MRCP value proposition

The underlying technology of speech processing has not changed significantly in the last ten years but rather has undergone a steady evolution with increasing accuracy, performance, and scalability available year on year. The increase in scalability can be largely attributed to Moore's law – indeed most modern network-based speech processors run on general purpose hardware and operating systems. There is, however, a revolution taking place in the speech industry, and it is centred on the new and standardised mechanisms by which one can integrate and use speech technologies. Until recently, speech technologies remained costly and difficult to integrate into other network equipment. Once integrated, vendor lock-in was somewhat inevitable due to the complex and proprietary interfaces exposed by a particular speech media resource. In addition, maintenance costs were incurred as a result of the need to frequently update integrations as the speech technology evolved. Prior to MRCP, speech media resources typically exposed a client–server style interface. The client was shipped as a library exposing an interface implemented in low-level languages such as C, C++, and Java. The interfaces differed significantly across different vendors, and even versions from the same vendor, thus making support of multiple speech resources very difficult and costly to maintain. Compounding this is the fact that 'one size fits all' does not apply for speech technologies since different vendors' strengths lie in different spoken languages and dialects, for example.

MRCP helps alleviate much of the headache and cost associated with the integration of speech processing technologies by delivering a standard, uniform, vendor-independent network protocol to control speech processing resources. MRCP itself is based on modern IP telephony and Web technologies, making it ideal for deployment in next generation IP networks. For example, MRCP leverages the Session Initiation Protocol (SIP), which is the core signalling protocol chosen by the 3GPP for its popular IP Multimedia Subsystem (IMS) architecture aimed at both mobile and fixed networks. As a standard, MRCP brings with it vendor independence, a proven interface, durability, and assurances of an evolution path for the protocol. Indeed, MRCP offers a win–win situation for network equipment providers and speech resource providers alike, since both parties may concentrate on their core business without expending substantial resources on designing, developing, and evolving different interfaces and APIs. Furthermore, by removing many of the integration burdens, MRCP serves to accelerate the adoption of speech technologies. MRCP allows network operators to procure and deploy network equipment from separate providers to those providing speech processing equipment and yet still retain assurances of interoperability.

1.3 History of MRCP standardisation

Before delving into the details of how MRCP came about, it is instructive to introduce briefly two of the most important standards organisations behind the Internet and Web, namely the IETF and W3C.

1.3.1 Internet Engineering Task Force

The IETF is an open international community of network designers, operators, vendors and researchers concerned with the evolution of the Internet architecture and the smooth operation of the Internet. Founded in 1986, the IETF has grown to become the principal body engaged in the development of new Internet standard specifications.

Membership to the IETF is free and open to any interested person. The technical work within the IETF is done within working groups, which are focused on a particular topic or set of topics defined by their charter. Each working group has one or more working group chairs whose responsibility it is to ensure forward progress is made with a view to fulfilling the group's charter. Working groups maintain their own mailing list where the majority of the work is carried out. In addition, the IETF meets several times a year, enabling working groups to discuss topics face-to-face. Working groups belong to a particular Area (e.g. transport, network management, routing, etc). Each Area is overseen by its corresponding Area Director. The set of Area Directors collectively make up the Internet Engineering Steering Group (IESG). The IESG is responsible for the technical management of the IETF such as approving new working groups and giving final approval to Internet standards.

Early versions of an IETF specification are made available through the IETF website and mailing lists as an Internet-Draft. An Internet-Draft is released with the intention of receiving informal review and comment from a wide audience. Usually, during the development of a specification, the document is revised several times and republished. Internet-Drafts expire 6 months after publication at which time they are immediately removed from IETF repositories (though archives of older drafts can be readily found on the Web on so-called mirror websites). At any time, the current Internet-Drafts may be obtained via the Web from:

```
http://www.ietf.org/internet-drafts/
```

After a specification is deemed stable and has reached a sufficient level of technical quality, an area director may take it to the IESG for consideration for publication as a Request For Comments (RFC). Each RFC is given a unique number. A particular RFC is never updated but rather a new number is assigned for a revision. Many IETF protocols rely on centrally agreed identifiers for their operation. The Internet Assigned Numbers Authority (IANA) has been designated by the IETF to make assignments for its RFC document series. RFCs pertaining to Internet standards go through stages of development denoted by maturity levels. The entry level is called a Proposed Standard. Specifications that have demonstrated at least two independent and interoperable implementations from different code bases, and for which sufficient successful operational experience has been obtained, may be elevated to the Draft Standard level. Finally, a specification for which significant implementation and successful operational experience has been obtained may be elevated to the Internet Standard level. RFCs can be obtained via the Web from:

```
http://www.ietf.org/rfc.html
```

1.3.2 World Wide Web Consortium

The W3C is an international consortium focusing on its mission to lead the World Wide Web to its full potential by developing protocols and guidelines that ensure long-term growth of the web. Tim Berners-Lee, who invented the World Wide Web in 1989 by providing the initial specifications for URIs, HTML, and the HTTP (a protocol that runs over the Internet[1]), formed the W3C in 1994. Today, the W3C consists of a large number of member organisations and a full-time staff, which collectively publish open, non-proprietary standards for Web languages and protocols.

The technical work in the W3C is carried out by working groups consisting of participants from member organisations, W3C staff, and invited experts. Working groups conduct their work through mailing lists, regular teleconferences, and face-to-face meetings. Each working group has one or more working group chairs and is categorised into a particular activity (e.g. HTML Activity or Voice Browser Activity). Each activity falls within a domain (e.g. Architecture Domain or Interaction Domain). The W3C Team manages both the technical activities and operations of the consortium by providing overall direction, coordinating activities, promoting cooperation between members, and communicating W3C results to the members and press.

W3C technical reports or specifications follow a development process designed to maximise consensus about the content and to ensure high technical and editorial quality. Specifications start out as a Working Draft, which is a document published for review by the community, including W3C members, the public, and other technical organisations. A specification advances to the Candidate Recommendation level after it has received wide review and is believed to satisfy the working group's technical requirements. Next, a working group will usually seek to show two interoperable implementations of each feature at which time the specification may be elevated to the Proposed Recommendation level. Finally, a specification is published as a W3C Recommendation when the W3C believes that it is appropriate for widespread deployment and that it promotes the W3C's mission. W3C Technical Reports and publications may be obtained via the Web from:

```
http://www.w3.org/TR/
```

1.3.3 MRCP: from humble beginnings toward IETF standard

Cisco, Nuance, and SpeechWorks jointly developed MRCP version 1 (MRCPv1). MRCPv1 was first made publicly available through the IETF as an Internet-Draft in 2001 and was later published as an 'Informational' document under RFC 4463 [13]. Although it has enjoyed wide implementation – a testament to the popularity and benefit of the MRCP approach – this version of the protocol has not been, and will not be, developed into an actual IETF standard. MRCPv1 has several restrictions and deficiencies that ultimately prevented it from becoming a standard. The protocol leverages the Real Time Streaming Protocol (RTSP) for both setting up media streams and for transport of the

---
[1] The HTTP protocol is now maintained by the IETF while HTML continues to be evolved by the W3C.

MRCP messages. SIP has since become the preferred mechanism for media session initiation in MRCP architectures, and the tunnelling mechanism that leverages RTSP as a transport protocol is largely regarded as an inappropriate use of RTSP. MRCPv1 also suffers from interoperability problems due to a weakly defined data representation for recognition results returned from speech recognisers. This, coupled with many vendor-specific extensions that were required (because they were not included in MRCPv1 in the first place) meant that true interoperability was not achieved and platform vendors were often still forced to perform some level of specific integration work for each speech technology vendor. Finally, the scope of MRCPv1 is limited and there is a lack of support for speaker verification and identification engines and speech recording.

In March 2002, a formal IETF Birds-of-a-Feather (BOF) was held at the 53rd IETF meeting where like-minded vendors came together to express their interest in developing a new and significantly improved version of MRCP. Later that year, the IESG chartered the Speech Services Control (SpeechSC) Working Group comprising leading speech technology and protocol experts. The core of the charter was to develop a standard protocol to support distributed media processing of audio streams with the focus on speech recognition, speech synthesis, and speaker verification and identification. The SpeechSC Working Group's first deliverable was a requirements document [10] outlining the specific needs of a standardised successor to MRCPv1. Particularly salient requirements for the new protocol were the proposed reuse of existing protocols while, at the same time, the avoidance of redefining the semantics of an existing protocol (something which MRCPv1 did with its particular reuse of RTSP). The SpeechSC Working Group has subsequently developed MRCP version 2 (MRCPv2), a significantly improved, extended, and standardised version of MRCPv1. At the time of going to press, the MRCPv2 specification [3] had undergone wide peer review and was at 'Last-Call' – a stage in the standardisation process used to permit a final review by the general Internet community prior to publishing the specification as an RFC.

MRCPv2 leverages SIP for establishing independent media and control sessions to speech media resources, adds support for speaker verification and identification engines, and includes many welcomed extensions and clarifications that improve flexibility and interoperability. The use of SIP allows MRCPv2 to leverage the many benefits that SIP offers, including resource discovery and reuse of existing network infrastructure such as proxies, location servers, and registrars. Significantly, MRCPv2 does not tunnel the control messages through SIP but rather uses SIP to establish a dedicated connection. MRCPv1 may be thought of as an early, proprietary version of the standard MRCPv2. MRCPv1 shares many similarities to MRCPv2 and is outlined in more detail in Appendix A. Throughout the rest of the book, unless otherwise stated, the term MRCP is used to refer to MRCPv2.

Independent of the IETF standardisation work, the W3C established the Voice Browser Working Group (VBWG) in 1999, with a charter to apply Web technology to enable users to access services from their telephone via a combination of speech and DTMF. The VBWG's deliverable is the W3C Speech Interface Framework, a collection of markup languages for developing Web-based speech applications. VoiceXML [4, 5] is a core component of that framework and is a standard, easy-to-learn, Web-based technology for authoring telephony applications that employ DTMF, speech recognition, and speech synthesis. VoiceXML depends on its sibling languages for specifying speech recognition and speech synthesis behaviours. The W3C Speech Recognition Grammar Specification (SRGS) [6] is a standard, XML-based markup approach for specifying speech grammars (the set of words and phrases a speech recogniser may recognise at a given point in a dialogue). A closely related language is the W3C Semantic Interpretation for Speech Recognition (SISR) [7]. This specification enables 'tagging' of semantic information to speech grammars to facilitate a basic form of natural language understanding. The W3C Speech Synthesis Markup Language (SSML) [8] is a standard, XML-based markup approach for specifying content for speech synthesis together with a mechanism for controlling aspects of speech production such as pronunciation, volume, pitch, and rate. Both SRGS and SSML can leverage the W3C Pronunciation Lexicon Specification (PLS) [9], which allows pronunciations for words or phrases to be specified using a standard pronunciation alphabet.

The VoiceXML language is a common 'user' of the MRCP protocol, that is, many VoiceXML platforms employ the MRCP protocol to interact with third party speech recognisers and speech synthesisers. In many ways (though not formally called out anywhere), VoiceXML capabilities were a primary driver behind a large number of functional additions to MRCP. Historically, the VBWG and SpeechSC Working Group have shared several participants, which has served to provide a healthy amount of cross-pollination. One obvious result of this is that MRCP leverages several W3C Speech Interface Framework specifications for specifying speech recogniser and speech synthesiser behaviours including SRGS, SISR, and SSML.

1.4 Summary

In this chapter, we have provided a brief introduction to speech applications by focusing particularly on network-based speech processing technologies and the many benefits that they bring to the world of IVR. The business case for MRCP was presented, including how it significantly helps to alleviate the burden and cost of integrating speech technologies, helps to open the market by allowing network operators to 'mix and match' the best technology for their particular purpose, and assists in accelerating the adoption of speech technologies for commercial applications. Finally, we presented a short history of MRCP standardisation by introducing the two standard bodies that were key to its development and discussing how MRCP and its related specifications came about.

In Chapter 2, we provide a background on how modern speech processing technologies function. This chapter is recommended for readers new to speech technologies; experienced readers may prefer to skip directly to Chapter 3, which introduces the basic principles of MRCP.

2

Basic principles of speech processing

The speech processing research community is a large and active one with an amassed set of results and techniques. The most effective, robust, and practical of these techniques have naturally filtered through to the commercial domain and are employed in leading speech processing products. The goal of this chapter is to introduce some of the fundamental principles underlying speech processing by focusing on the core techniques used in modern speech systems. This material is highly relevant to understanding the underlying principles of the different MRCP media resources. As we will see in Chapter 3, MRCP provides specialised media resources to cover four specific speech processing functions, namely speech synthesis, speech recognition, speech recording, and speaker verification and identification.

We commence the chapter with a background study of the principles of the human speech production system. We then discuss core techniques used in contemporary speech recognition products. Several of these techniques are also relevant to the following discussion on speaker verification and identification, and, to a lesser extent, the subsequent one on speech synthesis. The treatment of speech processing in this chapter is more descriptive than mathematical, and is intended to impart an intuitive understanding of the basic concepts. For more detailed information, the interested reader is referred to a number of recommended books and tutorial papers on speech recognition (Deller *et al.* [60]; Rabiner and Juang [61]; Makhoul and Schwartz [73]), speaker verification and identification (Campbell [71]; Bimbot *et al.* [72]), and speech synthesis (Dutoit [70]; Schroeter [74]).

2.1 Human speech production

Many of the speech processing techniques used in practice are derived from or strongly influenced by principles of the human speech production system. It is therefore prudent to undertake a study of how human speech is produced before delving into the details of modern speech processing.

Figure 2.1 illustrates a schematic of the human speech production system. The lungs expel air upwards through the *trachea*, passing through the *larynx*. The tensed *vocal cords* in the larynx

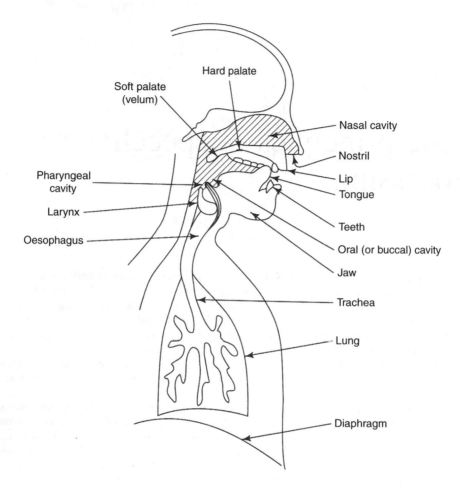

Figure 2.1 Human speech production system (reproduced by permission of John Wiley & Sons, Inc. from [60]).

vibrate with the passing air and cause the resulting air flow to adopt a roughly periodic flow that is subsequently modulated in frequency as it passes through the *pharyngeal cavity* (throat), the mouth cavity, and possibly the nasal cavity. The resultant sound is radiated primarily from the lips and, to a lesser extent, the nose. The combination of the pharyngeal cavity and mouth cavity is known as the *vocal tract*. The *nasal tract*, on the other hand, begins at the velum and ends at the nostrils. The *velum*, whose soft tip is seen to hang down at the back of the mouth when it is wide open, acts as a trap door to couple acoustically the nasal tract for certain 'nasal' sounds.

Figure 2.2 illustrates a simplified 'engineer's view' of the human speech production system. The three main cavities act as an acoustic filter whose properties are determined by the changing position of the articulators (i.e. jaw, tongue, velum, lips, mouth, etc.). Each vocal tract shape is characterised by a set of resonant frequencies called *formants*. The formant central frequencies are usually denoted by $F_1, F_2, F_3, \ldots$ with lower numbers corresponding to lower frequencies. The first 3–5 formant

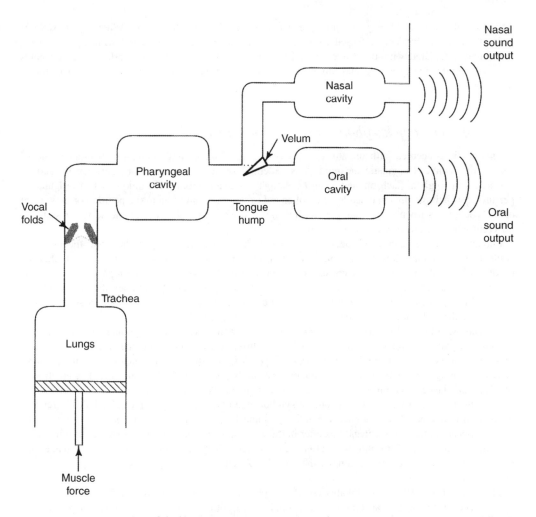

Figure 2.2 Engineer's view of the human speech production system (reproduced by permission of John Wiley & Sons, Inc. from [60]).

frequencies play a significant part in shaping the overall spectrum by amplifying certain frequencies and attenuating others.

When the vocal cords are tensed they vibrate and the air passing through the *glottis* (an opening between the vocal folds) becomes quasi-periodic, resulting in the production of *voiced* sounds (roughly corresponding to the vowel sounds in English). For men, the pitch range for voiced sounds varies approximately within the range of 50–250 Hz, while for women the range is roughly 120–500 Hz. When the vocal cords are relaxed, sound can still be produced. In this case, the airflow must pass through a restriction in the vocal tract and become turbulent to produce the *unvoiced* sounds (for example, the sound associated with 's' in six). If there is a point of total closure in the vocal tract (usually near the front), air pressure can build up behind this point. When the closure is opened, the air is suddenly released causing a brief transient sound associated with a *plosive* (e.g. the sound associated with the letter 't' in English). It is also possible to produce mixed sounds that are simultaneously voiced and unvoiced (e.g. the sound associated with 's' in 'is', pronounced like a 'z').

2.1.1 Speech sounds: phonemics and phonetics

Phonemics is concerned with the study of phonemes. A *phoneme* is the smallest theoretical unit of speech that conveys linguistic meaning. It is generally accepted that there are between 40 and 45 phonemes in English. Each phoneme can be thought of as a symbol for a unique set of articulatory gestures consisting of the type and location of sound excitation in addition to the position or movement of the vocal tract articulators. Changing a phoneme in a word to a different phoneme will result in a different word (or possibly nonsense).

Phonetics, on the other hand, deals with the speech sounds themselves rather than the language context of the sound. Phonetics is concerned with the study of *phones* – the actual sounds that are produced in speaking. A phoneme consists of a family of related phones called *allophones* – variations on phones that represent slight acoustic variations of the basic unit. Allophones thus represent the permissible freedom in a language in producing a phoneme. The number of allophones associated with a phoneme depends on the particular phoneme and its position within speech.

We can use minimal pairs to show that two phones constitute two different phonemes in a language. A minimal pair is a pair of words or phrases that differ in only one phoneme and have a distinct meaning. For example, the English words *pet* and *bet* differ by the first consonant or phone and have distinct meanings. Hence the phones /p/ and /b/ are actually phonemes for the language. The notation of two forward slashes is commonly used to denote a phoneme. Allophones are often denoted by the use of diacritical[1] or other marks to the phoneme symbols and placed in square brackets [ ] to differentiate them from phonemes. For example, in English, [p] and [pʰ] are allophones of the /p/ phoneme. The former appears in the word *spit* and the latter in the word *pit* – the inclusion of a superscript *h* indicates aspiration (loosely, an extra puff of air after the consonant). Note that phonemes in one language may be allophones in another. For example, the sounds /z/ and /s/ are distinct phonemes in English but allophones in Spanish.

The International Phonetic Alphabet (IPA) is a phonetic notation (mostly) expressible in Unicode and devised by linguists to accurately and uniquely represent each of the wide variety of phones and phonemes used in all spoken languages. IPA aims to provide a separate unambiguous symbol for each contrastive (i.e. phonemic) sound occurring in human language. IPA supports an exhaustive list of vowel and consonant symbols, numerous diacritics, stress symbols, intonational markers and more. The IPA symbols used for English[2] are illustrated in Table 2.1 (consonants) and Table 2.2 (vowels). In some of our examples, we will also see *suprasegmental* symbols so called because they apply to more than one segment (vowel or consonant). Primary stress is indicated by the ' symbol and secondary stress via the ˌ symbol.

[1] A diacritical mark or diacritic (sometimes called an accent mark) is a mark added to a letter to alter the containing word's pronunciation.
[2] The pronunciation given here corresponds to that of standard British English (the pronunciation associated with southern England, which is often called Received Pronunciation).

Table 2.1	List of IPA symbols for consonants in English

Consonants
[tʃ]: **chat**
[x]: **loch**
[g]: **get**
[dʒ]: **jingle**
[ŋ]: **ring**
[ʃ]: **she**
[ʒ]: **measure**
[θ]: **thin**
[ð]: **then**
[j]: **yes**

The remaining constants use the following symbols and correspond to their usual English usage [b], [d], [f], [h], [k], [l], [m], [n], [p], [r], [s], [t], [v], [w], [z].

Table 2.2	List of IPA symbols for vowels in English

Short vowels	Long vowels	Diphthongs[a]	Triphthongs[b]
[a]: **mat**	[ɑ:]: **father**	[ʌɪ]: **my**	[ʌɪə]: **fire**
[ə]: **ago**	[ə:]: **her**	[əʊ]: **how**	[aʊə]: **sour**
[ɛ]: **red**	[ɛ:]: **hair**	[eɪ]: **say**	
[ɪ]: **pit**	[i:]: **see**	[əʊ]: **toe**	
[ɪ]: **cosy**	[ɔ:]: **saw**	[ɪə]: **beer**	
[ɒ]: **pot**	[u:]: **too**	[ɔɪ]: **toy**	
[ʌ]: **bun**		[ʊə]: **poor**	
[ʊ]: **put**			

[a] A diphthong is a two-vowel combination with a quick but smooth movement between the vowels.
[b] A triphthong is a variation of a diphthong with a three-vowel combination.

2.2 Speech recognition

Our focus in this section is on task-oriented speech recognition technology of the sort applied in IVR scenarios. This type of technology can be typically characterised as speaker independent and potentially large vocabulary (upwards of hundreds of thousands of words), albeit constrained by a grammar to a specific domain (e.g. flight booking service, parcel tracking application, etc.). The speech recognition problem can be posed as one of converting a continuous-time speech signal into a sequence of discrete words or subunits such as phonemes. The conversion process is not trivial, however, primarily because of the variability in speech signals both within and across different speakers. Speech signals do not comprise discrete well-formed sounds but rather a series of 'target' sounds (some quite short) with intermediate transitions. Compounding this is the fact that preceding sounds and following sounds can significantly affect whether the target sound is reached and for how long. This interplay of sounds

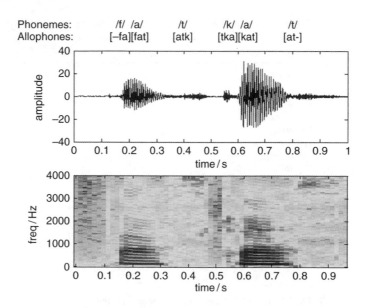

Figure 2.3 Time-series (top) and time-frequency (spectrogram) (below) plot of the utterance 'fat cat'.

within an utterance is called *coarticulation* and is a natural consequence of physical constraints due to motion and continuity of the human articulatory apparatus. Figure 2.3 shows a time-series plot of the utterance 'fat cat' with the corresponding phonemes and allophones. The lower plot illustrates a spectrogram with the time on the abscissa and frequency on the ordinate. Darker shading corresponds to increased energy – resonances in the vocal and/or nasal tracts. The two words in this utterance are minimal pairs – differing only in one phoneme. Take a closer look at the /a/ phoneme appearing in both words. While the spectrogram reveals some similarities, it also elucidates some differences: how a particular phoneme is spoken is contingent on its neighbouring phonemes. Notice also that while the phonemes themselves are considered discrete entities, there is no physical separation between the different sounds in the speech signal.

In addition to the complexities of coarticulation where the spectral characteristics of a phoneme or word vary depending on the surrounding phonemes or words, several other factors contribute to the considerable variability in speech signals, notably:

- *Speaker differences*: Each person's voice is unique and voices can differ significantly from dialect to dialect.
- *Speaking style*: People speak at different rates and with different pronunciations at different times. For example, a stressed person may articulate a sentence differently than if they were relaxed.
- *Channel and environment*: Different channel characteristics can modulate the original speech signal making it difficult to recognise. A noisy environment can further mask the speech signal of interest.

It is crucial, therefore, that speech recognisers work to mitigate some of these sources of variability if robust recognition performance is to be achieved.

Figure 2.4 illustrates the basic components of a speech recogniser. The first stage in the speech recognition process is concerned with feature extraction. While one could in principle operate on

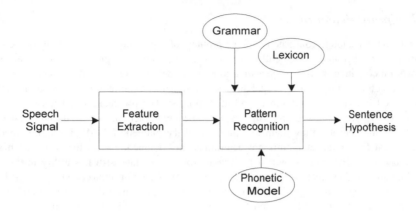

Figure 2.4 Basic components of a speech recogniser.

raw sample values of the audio signal, performing judicious feature extraction allows some of the sources of variability in the input signal to be reduced and leads to more robust and discerning features for subsequent processing. The feature extraction stage in a speech recognition system accesses the spectral envelope of the signal and discards unimportant information pertaining to whether the speech is voiced or unvoiced and, in the case of voiced sounds, also removes the speaker dependent pitch information. The feature extraction stage usually operates on about 20 ms quasi-stationary overlapping frames of speech, which are updated every 10 ms (resulting in 100 frames per second). Each frame results in a set of feature values describing it and is called a *feature vector*. The feature extraction stage is discussed in more detail in Section 2.2.2 in the context of the mel-cepstrum.

The goal of the pattern recognition stage is to match a sequence of feature vectors from the input utterance to a set of previously trained phonetic models. Modern speech recognisers usually employ Hidden Markov Models (HMM) to represent phonetic speech events statistically. A common approach is to use a HMM to model a phoneme with a separately parameterised HMM for each phonetic context of interest. A lexicon supplies a mapping of orthographic (written) words to their corresponding phonetic transcription (or transcriptions in the case a word has multiple pronunciations). Usually, the speech recogniser will incorporate its own comprehensive lexicon and offer the user the ability to augment or extend[3] that lexicon, for example by adding external lexicon documents such as the one described in Chapter 11. The user-supplied grammar applies syntactic (word order) constraints on what can be recognised by defining the allowable set of words and phrases. This top down, hierarchal, collection of the grammar, lexicon, and phonetic models can be composed as a set of concatenated HMMs to which the sequence of feature vectors is matched. A likelihood estimate is associated with each path and a search algorithm finds the string of phonetic models (hence string of words) that gives the highest likelihood. The string of words from the optimum path yields the sentence hypothesis returned in the recognition results.[4] The HMM approach is discussed further in Section 2.2.3.

---

[3] Modern recognisers also include the capability to derive a phonetic transcription from the orthographic form for novel words although this process can be error prone.
[4] Many recognisers can return the N-best results allowing the application to provide further logic to prune unlikely results based on specific domain knowledge.

2.2.1 Endpoint detection

Speech endpoint detection is concerned with the problem of separating speech portions of a continuous signal from the background noise. Older speech recognition technologies, such as those limited to isolated-word recognition, were very sensitive to accurate determination of speech endpoints. Modern continuous speech recognisers often model the acoustic signal as a continuum of silence (or background noise), followed by the actual utterance, and then more silence (or background noise). This can be achieved either by providing training samples with silence at each end or by training separate non-speech HMMs and concatenating them to either end of the phonetic HMMs. Using this mechanism, the precise locations of the endpoints are determined in conjunction with the recognition strategy. A separate endpoint detection as part of the feature extraction stage still has utility in this case for (a) providing an initial estimate of possible endpoints, (b) offloading unnecessary processing burden from the pattern recognition stage (i.e. when only non-speech portions are present), and (c) potentially providing early speech onset information for the purposes of enabling barge-in functions in spoken dialogue systems (i.e. to stop prompt playback if a user starts speaking).

Speech endpoint detection in itself is a non-trivial task to get right. Endpoint algorithms need to take special care of certain speech sounds such as weak unvoiced sounds (e.g. /f/ and /h/), weak plosives such as /p/, /t/, and /k/, final nasals (e.g. 'gone'), trailing off sounds (e.g. 'three'), and voiced fricatives[5] becoming unvoiced (e.g. 'has'). A widely used algorithm for endpoint detection is that of Rabiner and Sambur [69]. This algorithm uses two measurements: the short-term energy of the signal (obtained in [69] by summing the absolute value of the signal over a 10-ms window) and the number of zero crossings measured over a frame of 10 ms in duration. Figure 2.5 illustrates an example for the word 'four'.

Two thresholds, an upper E_{tu} and lower E_{tl}, for the energy are determined by analysing the first 100 ms of the signal, which is assumed to be silence. A single threshold, Z_t, for the zero crossing rate is also determined during this period. The algorithm first operates on the energy by looking for the point at which the energy goes above E_{tu} and then back tracks to find the point at which the energy crossed the lower threshold, E_{tl}. This point is designated N_1. The same process is applied to determine N_2 starting from the other end. Using this information only, the weak fricative at the beginning of the word 'four' in Figure 2.5 would be erroneously truncated. The second part of the algorithm uses the zero crossing measure. The algorithm searches backwards from N_1 over 25 frames (250 ms), counting the number of intervals greater than Z_t. If this count is greater than 3, the point at which the Z_t threshold is first surpassed is found, designated by N'_1. This constitutes the first endpoint. A similar approach is used to determine N'_2 to give the final second endpoint.

2.2.2 Mel-cepstrum

The most popular features employed in today's speech recognisers are derived from the so-called *mel-cepstrum*. Cepstral analysis is motivated by the problem of separating the excitation component from the impulse response of the vocal system model. In particular, it has been found that speech recognition accuracy is greatly enhanced by discarding the excitation component. Cepstral analysis was initially proposed by Bogart *et al.* [68] – the word 'cepstrum' is an anagram of 'spectrum' chosen because of the technique's unusual treatment of the frequency domain data as though it were time domain data. Figure 2.6 illustrates the block diagram for the real cepstrum. The real cepstrum is obtained by taking

---

[5] Fricatives are sounds associated with exciting the vocal tract with a steady air stream that becomes turbulent at some point of constriction. Voiced fricatives have a simultaneous voicing component while unvoiced fricatives have no voicing component.

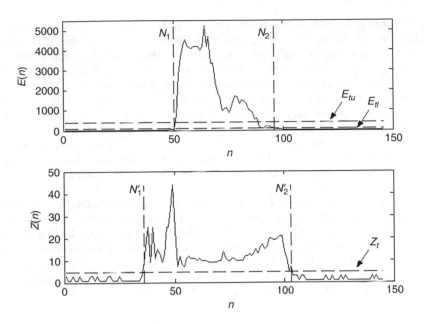

Figure 2.5 Endpoint detection technique using the algorithm from [69] for the utterance 'four'.

Figure 2.6 Block diagram of the real cepstrum.

a Discrete Time Fourier Transform (DTFT) of the speech signal, applying an absolute log function, and supplying the result to an inverse DTFT (IDTFT). We discuss the thought process behind the cepstrum next.

The observable human speech signal is a convolution of the voiced excitation sequence and the impulse response due to the vocal system (see Figure 2.2). Since convolution in the time domain is the equivalent to multiplication in the frequency domain, we can write:

$$S(\omega) = E(\omega)\Theta(\omega)$$

where $S(\omega)$ denotes the DTFT of the speech signal, $E(\omega)$ represents the DTFT of the excitation, and $\Theta(\omega)$ is the DTFT of the vocal system impulse response. By taking the logarithm of $|S(\omega)|$, one can separate the spectrum into two parts:

$$\log|S(\omega)| = \log|E(\omega)| + \log|\Theta(\omega)|$$

Now, by taking the unusual step of interpreting this log transformed frequency domain as a new, linear time domain, the $\log|S(\omega)|$ signal may be viewed as a periodic signal made up of two linearly combined parts. Heuristically, we can view the $\log|E(\omega)|$ part as a high frequency component (originally consisting of impulses corresponding to voiced excitation), and the $\log|\Theta(\omega)|$ part as a low frequency component (originally consisting of the spectral envelope resulting from the vocal system spectral shaping characteristics). Bogart *et al.* reasoned that by performing a frequency domain transform of this signal, the two parts could be separated – the $\log|E(\omega)|$ part occupying high 'frequency' values and the $\log|\Theta(\omega)|$ part occupying low 'frequency' values. Of course, this is not the conventional frequency domain and hence the 'frequencies' are usually referred to as *quefrencies*. Since the $\log|S(\omega)|$ signal is periodic, we can determine its Fourier series coefficients corresponding to the harmonics of the signal. Now, noting that $\log|S(\omega)|$ is a real, even, function of ω, we can equivalently use the IDTFT[6] in place of the usual equation for determining Fourier series coefficients. This brings us to the standard definition of the real cepstrum, as shown in Figure 2.6.

Figure 2.7 illustrates an example of the real cepstrum for the phoneme /i:/ for the author's voice. The vocal system response manifests as a rapidly decaying portion of the cepstrum at low quefrency. The voiced excitation appears as a clearly separated train of impulses separated by the pitch period located at higher quefrencies. For the purposes of feature extraction applied to speech recognition applications, we are only interested in the low quefrency components corresponding to the voice system response. These coefficients are easily extracted by a rectangular window that omits the excitation impulses or equivalently we can just take the first N cepstrum coefficients and discard the rest.

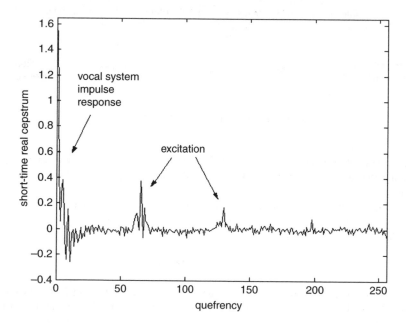

Figure 2.7 Real cepstrum for the phoneme /i:/. The 8-bit speech signal was sampled at 10 kHz and windowed using a 512 point Hamming window.

---

[6] In practice, the evenness of the signal is exploited and the Discrete Cosine Transform (DCT) is equivalently used.

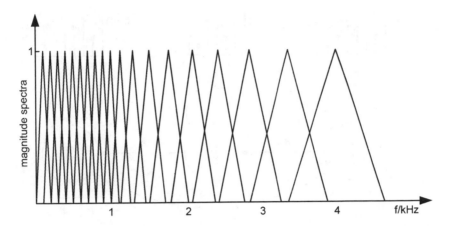

Figure 2.8 Critical band filters used to generate the mel-cepstrum.

One of the reasons why spectral features work so well in speech recognition is due at least in part to the fact that the cochlea in the human ear itself employs a quasi-frequency analysis. The frequency scale employed by the cochlea is non-linear and is known as the *mel* scale. Up to about 1 kHz, the scale is linear and thereafter it is logarithmic. In the logarithmic range, if a person is to perceive a doubling in frequency, the actual physical frequency of the signal must significantly more than double. This warping of the frequency axis is commonly applied to the cepstrum – specifically by warping the speech signal spectrum prior to applying the logarithm function. Often a second psychoacoustic principle is also applied, based on the fact that the perception of a particular frequency in the auditory system is influenced by energy in a critical band of frequencies around that frequency. This critical band begins at about 100 Hz for frequencies under 1kHz and increases logarithmically above that. A common approach is to use the log total energy in critical bands around the mel frequencies as the input to the final IDTFT by applying a half-overlapped triangular window of increasing base length centred on the mel frequencies prior to applying the log transformation; see Figure 2.8.

The first 12 or so mel-cepstrum coefficients[7] form a set of features for input to the pattern recognition stage. In practice, it is common also to include the delta cepstrum – the difference in the spectral coefficients from the previous frame. Some implementations also provide the second derivative (i.e. the delta of the delta features). The delta cepstrum directly provides the pattern recognition stage with information related to the time evolution of the feature vectors and can significantly improve recognition accuracy. In addition, the energy measure over the frame and its delta is often included in the set of features. The collection of features are assembled into a single feature vector and supplied as input to the pattern recognition stage, discussed next.

2.2.3 Hidden Markov models

Most modern speech recognisers employ HMMs in the pattern recognition stage. The HMM is a mathematical construct that enables us to model speech as a two-part probabilistic process. The first

---

[7] Commonly referred to as the Mel Frequency Cepstral Coefficients (MFCC).

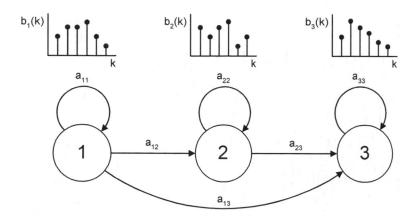

Figure 2.9   Example of a three-state HMM.

part models the speech as a sequence of transitions (the changing sounds over time) through a directed graph. The second part models the features in a given state (the individual sounds) as a probability density function over the space of features. This doubly stochastic nature of the HMM is well suited to the task of modelling speech variability in time and feature space. Moreover, efficient algorithms exist to train HMMs automatically on a give corpus of speech data.

A HMM consists of a set of states connected by a number of possible transition paths (see Figure 2.9). Each transition path has an associated transition probability labelled here by a_{ij}. A symbol is emitted on arrival (or indeed re-entry) into a state, assuming for now we are working with a discrete observation HMM. Each state in the HMM has associated with it a set of output symbols with its own probability distribution. The notation $b_i(k)$ indicates the observation probability that symbol k is emitted given state i.

The HMM derives its name from the fact that given a series of observations from the model, the exact sequence of states is hidden from the observer. However, it is possible to determine the sequence of states *most likely* to have produced the observation and with it a measure of likelihood for that path through the HMM. It is this capability that is exploited by speech recognisers. For example, a simple isolated word recogniser might employ a HMM for each word in its vocabulary. Given an input of a sequence of feature vectors, each HMM can be tested to find the best path through it and a likelihood measure for that path. The word associated with the HMM giving the highest likelihood is the speech recogniser's output hypothesis for the analysed speech.

Large vocabulary, continuous, speech recognisers usually model speech on a sub-word basis. That is to say, the acoustic unit modelled by the HMM corresponds to some subunit of a word. HMMs can be simply concatenated to form full words or sentences. This important step obviates the need for every word in the vocabulary to appear in the training data. By employing a lexicon, new words can be specified as compositions of their acoustic units even when those words do not appear in the training data. Different speech recognisers use different acoustic units. One common approach is to model a phoneme using a three-state model of the form illustrated in Figure 2.9. The motivation for choosing three states can be thought of in terms of providing a state for each part of the phoneme – the first corresponding roughly to the left part of the phoneme, the middle state corresponding to the middle, and the last state corresponding to the right part of the phoneme. Notice that the transitions are from left to right only, thus encapsulating the causal nature of the speech signal. Transitions to the same state allow the natural variability in duration of different instantiations of phonemes to be accounted for. Indeed, more states could be used to model a phoneme but this brings with it the burden of more data required to train it and increased computational overhead.

As Figure 2.3 illustrates, coarticulation effects can result in a different allophone being uttered for the same phoneme dependent on its context, i.e. depending on what its neighbouring phonemes are. A popular approach to capture this context dependency is to use a triphone acoustic unit which uses a different model for each permutation of left and right neighbour. An immediate problem resulting from this strategy is the sheer number of triphones. For example, assuming 45 phonemes in a language, a total of $45^3 = 91\,125$ HMM models would result and with it a requirement for a prohibitively large amount of data to train those models sufficiently. One solution for reducing the number of triphones is to cluster similar triphones together, based on phonetic knowledge, or to use some automatic means. This can dramatically reduce the number of triphones, and hence the amount of training data required, yet still admit phonetic context information to be used.

The HMM example presented in Figure 2.9 is a discrete observation HMM – on entry to each state, one symbol from a set of possible symbols is emitted according to some probability distribution associated with that state. However, the feature vectors produced by the feature extraction stage are both continuously varying and non-scalar. To reconcile this difference, feature vectors can be discretised into one of a set of output symbols in a codebook of templates using a process called vector quantisation. There are typically in the order of 256 entries in a codebook. Given an input feature vector, we find the template closest to it in a previously trained codebook and its index serves as the HMM output symbol. The Euclidean distance measure can be readily used with features derived from cepstral analysis to find its closest matching template in the codebook.

An alternative approach to performing vector quantisation of the feature vectors is to use a continuous observation HMM. Instead of a discrete set of output symbols at each state, we now have a continuous set of symbols comprising the uncountable possible values of the feature vector. We can write the observation probability as $b_i(y(t))$, i.e. the probability that the feature vector $y(t)$ will be emitted on entering state i. A multivariate probability distribution function is associated with each state. Typically, a Gaussian mixture model is applied for the probability density function. A Gaussian mixture model consists of a weighted sum of multivariate Gaussian distributions. The model is parameterised by the weights in addition to the mean vector and covariance matrix associated with each Gaussian component. Techniques for training HMMs incorporate mechanisms for determining the parameters of the Gaussian mixture model (see below).

2.2.3.1 Recognition

The problem of recognition can be stated thus: given a particular observation sequence of feature vectors and a particular HMM, how do we calculate the likelihood that the HMM produced that sequence? There are two approaches to calculating the likelihood typically used, each leading to its own algorithm. The first approach derives a likelihood measure based on the probability that the observations were produced using *any* state sequence of paths through the model. The associated algorithm is called the Forward–Backward (F–B) or Baum–Welch algorithm. The second approach employs a likelihood measure based on the *best* path through the model. The associated algorithm is called the Viterbi algorithm and, due to its superior efficiency and suitability for language modelling (see next section), is the more popular of the two in practice.

The Viterbi algorithm proceeds by laying out the states in time to form a state lattice as illustrated in Figure 2.10. For each feature vector $y(t)$ in the observation sequence, we make a corresponding state transition to one of the states on its right. The Viterbi algorithm recasts the problem as one of sequential optimisation. A cost is associated with each transition and is taken as the negative logarithm of the transition probability a_{ij}. A cost is also associated with arrival

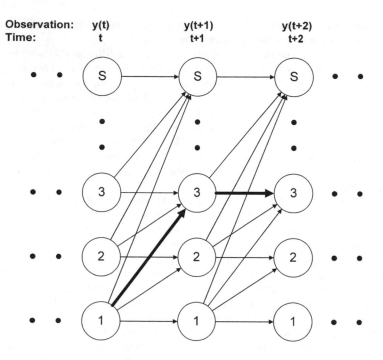

Figure 2.10   State lattice representation of a HMM with S states.

in each state – the negative logarithm of $b_i(y(t))$. The two costs are summated. Taking the negative logarithm has two effects. The first is to turn higher probabilities into lower costs. The second is to turn numerically problematic multiplications[8] into sums of numerically stable numbers. The best path through the lattice is therefore the path of least cost. The Viterbi algorithm applies techniques from dynamic programming to find the least cost path. All paths start at the bottom left of the lattice. The dynamic programming approach applies sequentially in time across the lattice from left to right. At each point in time, a minimum distance cost from the origin, labelled D_{min}, is updated for each state. This can be calculated incrementally, e.g. for state 3 at time $t+1$ we choose the path from the particular state at time t that yields the lowest value for its associated D_{min} value plus the cost associated with the transition from that state to the current state. The algorithm thus records the lowest cost path to each state. In addition, each state records the previous state that was selected by least cost. The algorithm terminates when all paths have reached a final state or are pruned (see below). If there is more than one final state, we choose the path with the minimum cost. Since each state knows its previous state in the path, we can backtrack from the final state to determine the path taken.

The Viterbi algorithm is quite efficient, requiring computations proportional to the number of states in the model and the length of the observation sequence. Contemporary implementations include various optimisations to maximise the density of concurrent speech recognition jobs that can be handled in real-time on a single server. The beam search is one such popular optimisation commonly applied. In

---

[8] Sequentially multiplying probabilities results in very small numbers that may not be easily represented on a particular computer.

practice, as one executes the Viterbi algorithm, relatively few of the partial paths will maintain low enough distance costs and can be safely pruned. The beam search operates by pruning any paths whose D_{min} exceeds that of the best path in the lattice column given at time t by a certain amount. In this fashion, only those paths that remain in a certain acceptable beam of likelihoods are retained.

2.2.3.2 Training

Training a HMM involves finding the optimum values of the transition probabilities a_{ij} and observation probabilities $b_i(y(t))$ given a corpus of training data. Three main inputs are required for training a speech recogniser:

  (i) a corpus of sample audio utterances;
 (ii) a written transcription of the utterances, and
(iii) a lexicon giving the phonetic representation of each word.

The training of a speech recogniser is labour intensive because a human has to listen to each utterance and write down the sequence of words that makes up its transcription. In addition, a person must write the one or more phonetic representations for each word and add them to a lexicon (although dictionaries do exist to assist this stage).

HMMs are usually trained using the so-called F–B (Baum–Welch) reestimation algorithm. The algorithm is applied iteratively using the same observation sequence and each iteration results in a modified HMM (i.e. new values for a_{ij} and $b_i(y(t))$ that gives an improved likelihood that the HMM produced the observation sequence). The iteration continues until such time as the incremental improvement falls below a threshold. The algorithm is guaranteed to converge to a local maximum. How good that local maximum is depends on whether the model is seeded with good initial conditions. One of the remarkable properties of the HMM is that we do not need to worry about the exact temporal alignment of the speech utterance and the corresponding words and phonemes. Given good initial conditions, the F–B algorithm will perform an implicit alignment[9] of the input feature vector sequence and the states of the HMM as it iterates. In particular, this removes the time-consuming step of having to mark-up the training database temporally. In practice, seed models are available for previously trained word or phone models, or are derived from an available marked database.

2.2.4 Language modelling

While we have already discussed how lexicons can be used to construct word models based on constituent phone models, we have so far sidestepped the issue of how words are assembled to form sentences. Specification of super-word knowledge can significantly improve speech recognition performance. For task-oriented speech recognition problems such as the ones considered in this book, so-called *finite state* or *regular grammars* are most commonly used to specify the combinations of words and phrases that comprise a sentence. Formally, a grammar is a set of rules by which symbols or terminals (i.e. words) may be properly combined in a language. The language is the set of all possible combinations of symbols. A finite-state grammar is a particular type of grammar that consists of rules

---

[9] Recall the earlier discussion about endpoint detection. Given a crude estimate for the endpoints and optionally a HMM modelling noise concatenated to either end of the word / phone HMM sequence, the F–B algorithm will automatically find the best alignment.

that map a non-terminal into either another non-terminal and a terminal, or just a terminal. In Chapter 9 we will study a standard grammar format used with MRCP, called the Speech Recognition Grammar Specification (SRGS), which has the expressive power of a finite state grammar.

Using finite state grammars in speech recognition problems brings with it two advantages:

(i) The grammars are straightforward to understand and write (e.g. see Chapter 9).
(ii) The grammars can be modelled by sentence-level, discrete observation HMMs.

HMMs are essentially finite state machines – an automaton capable of generating language produced by a finite state grammar. Consider a simple SRGS grammar rule that accepts the input utterances 'a sunny day' or 'a rainy day':

```
<rule id="weather">
    a
    <one-of>
        <item>sunny</item>
        <item>rainy</item>
    </one-of>
    day
</rule>
```

This may be mapped to a simple HMM as shown in Figure 2.11. Here we are using the Mealy form of a HMM instead of the Moore form previously discussed. The Mealy form is identical to the Moore form in that it has transition probabilities, a_{ij}, but differs from it in that the observations are emitted during the transitions instead of on entry into a state. The numbers in parenthesis are the observation probabilities for emitting the particular word given the transition they label. In this example, state I corresponds to the initial state and state F to the final state. The other states correspond to non-terminals and each transition produces a terminal. For example, after the initial ('a'), the probability that a 'sunny' or 'rainy' will follow is given by the transition probability (1.0) multiplied by the observation probability (0.5). Since we can construct sentence-level HMMs with a one-to-one correspondence to a given finite state grammar, we can use the HMM essentially as a parser for that grammar. Further, since the parsing problem is recast as a HMM problem, it is amenable to the techniques used in the acoustic-level HMMs described earlier.

Speech recognisers frequently combine the sentence-level HMM with acoustic-level HMMs in a top-down approach. The linguistic decoder (LD) manages the sentence-level HMM while the acoustic decoder (AD) manages the acoustic-level HMMs. For the present discussion, assume for simplicity that the acoustic-level HMMs model complete words. The LD propagates word hypotheses down to the AD, which returns a likelihood for each hypothesis based on phonetic analysis of the input feature vector sequence. Embedded in each transition in the sentence-level HMM (the linguistic search) is an entirely parallel search at the acoustic level for each hypothesised word. When an acoustic-level HMM

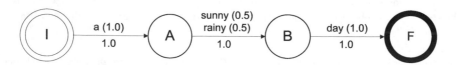

Figure 2.11 Sentence-level HMM for a simple grammar rule.

reaches a final state, the AD can return the likelihood result upwards. A Viterbi algorithm executes at the sentence-level HMM in an attempt to find the least cost path through the word lattice. The cost associated with each node in the sentence-level HMM is calculated using a combination of the transition and observation probabilities as before, but this time augmented by the cost (i.e. negative logarithm of the likelihood) returned from the acoustic-level search. If the transition cost is excessive, this means that the transition does not explain the feature vector observations sufficiently well and the path is not extended. The sentence-level HMM and acoustic-level HMMs typically operate in a frame synchronous manner. The LD requests from the AD the likelihood values for different word hypotheses given a specific start time. Each time a new start time is specified (i.e. after the sentence-level HMM makes a successful transition), the AD locates a new HMM for each word hypothesis. The AD 'remembers' its current position within the Viterbi algorithms running in parallel within each acoustic-level HMM, updating each one when a new observation feature vector becomes available. The AD reports the likelihood for a word hypothesis when the corresponding acoustic-level HMM reaches a final state.

For large vocabulary speech recognisers, the acoustic-unit modelled by the acoustic-level HMM will be word subunits such as context dependent phones. This requires an extra layer in the LD to provide word models usually as a concatenation of phone models based on the lexicon (a separate word-level HMM is also possible). The LD requests likelihood values from the AD for each phone model comprising its lexical representation of a word.

We can remove the complexity of communicating between the layers by compiling the recogniser into a single, composite HMM. For example, assuming each word is represented by the appropriate concatenation of phone models, we could insert that string of phonetic HMMs directly into the transitions in the sentence-level HMM as illustrated in Figure 2.12.

The resulting HMM can be processed to find the maximum likelihood path and with it the most likely sequence of words. One problem faced by large vocabulary speech recognisers is the potentially very large size of the search space. Beyond the beam search technique described earlier, many algorithms have been conceived for improving the performance of HMMs with respect to searching. Most algorithms attempt to exploit the redundancy within the network. For example several alternate words may commence with the same phonemes allowing the network to be organised into a tree structure (where redundancies are exploited as common ancestor nodes) for more efficient processing.

Another popular grammar format is the so-called n-gram Markov grammar – also known as a Statistical Language Model (SLM). These grammars specify the conditional probability of a word appearing in a sentence as a function of the previous $n - 1$ words. The probability of the sentence is then the product of these conditional probabilities. For example, trigrams give the probabilities of triplets of words in the lexicon and perform well in capturing the natural constraints imposed on the sequences of words in a language. However, one of the drawbacks of n-grams is that large amounts of

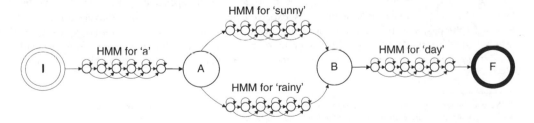

Figure 2.12 Acoustic-level HMMs inserted into the transitions of a sentence-level HMM.

written text are required for training data to compute the probability of observed n-grams as well as to estimate probabilities for unseen word subsequences. The considerable expense of gathering relevant data has impeded the adoption of n-gram grammars for speech recognition used with typical IVR applications. It is usually considerably more cost effective and easier to write a good quality finite-state grammar. One class of application that does benefit from n-gram grammars is those services where the user is greeted with open-ended prompts such as 'How may I help you today?'

2.3 Speaker verification and identification

Speaker recognition is a close cousin of speech recognition that is concerned with the problem of recognising the speaker from a spoken utterance. As we will see, many of the techniques discussed in the context of speech recognition also apply for speaker recognition. Speaker recognition encompasses the two tasks of *speaker verification* and *speaker identification*. Speaker verification is concerned with the problem of verifying that the speaker is the person they purport to be (their claimed identity is usually asserted via a customer code, pin number, etc.). Speaker identification deals with the problem of determining who is speaking from a set of known speakers. While the speaker makes no claim as to their identity, usually the speaker identification system is aided with the knowledge that the speaker comes from a fixed group of known speakers. For speaker identification, the output result is the identity of the person most likely to have spoken the analysed utterance (or possibly a 'none-of-the-above')[10] while for speaker verification the output result is ultimately an accept/reject decision.

Several useful biometric signals exist today including retina, face, fingerprint, and voice. Different biometrics have different strengths and weakness. A primary strength of applying a voice signal to biometrics is derived from the fact that it is a natural modality for communication and users are generally comfortable with allowing entities – both humans and machines – hear and analyse their voice. A secondary strength comes from the fact that thanks to the ubiquity of the telephone, no specialised equipment is required to gain access to the voice. Indeed, speaker recognition is a natural 'add-on' to speech-enabled IVR applications. Combining speaker verification with speaker recognition delivers twofold security through knowledge verification and speaker verification. Speaker verification technologies are used in IVR applications where it is desirable to secure access to information or to secure specific actions. Common applications include financial and retail transactions and account management, personal information access, and password reset. Speaker identification has arguably fewer use-cases and is principally used in IVR to allow a group of users to share the same account and to personalise access based on the particular user. For example, an individual speaker from a family of speakers could be identified using their voice and the closed set of speakers comprising the family could be indicated by the caller ID of their home phone.

Speaker recognition applications can be further partitioned into those that operate in a text-dependent mode and those that work in a text-independent mode. For text-dependent applications, the system knows what text is to be spoken and it is expected that the user will cooperatively speak that text. Many deployed application use this mode of operation since it is more robust. Typically the user is asked to speak a string of digits or to repeat a random[11] phrase of words from a small vocabulary. Text-independent applications on the other hand are less intrusive by obviating the need

---

[10] We use the term *open set* identification when there is a possibility the speaker is not known to the system. *Closed set* identification is where all the speakers are assumed to be known to the system.
[11] Requesting that the user speak a random string of words helps reduce the possibility of spoof attacks where a recording of the target speaker is used.

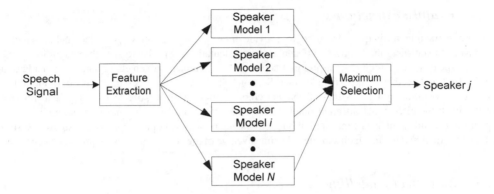

Figure 2.13 Block diagram of a speaker identification system.

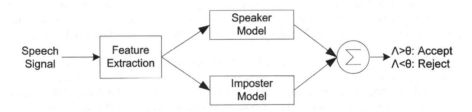

Figure 2.14 Block diagram of a speaker verification system.

for the user to speak specific phrases and hence offer greater flexibility. For example, a speaker verification system could perform background verification while the user interacts with the system and the speaker can be automatically granted access to sensitive information if their identity is correctly verified.

Figures 2.13 and 2.14 illustrate a block diagram for a speaker identification and speaker verification system respectively. The feature extraction stage extracts speech signal features that convey speaker-dependent information. While speech recognition and speaker recognition use the same feature extraction strategies, they differ in how they use those features. The combination of the speaker's vocal tract and speaking style is what characterises his or her speech making it distinctive. In the training or enrolment phase, the features are used to create a statistical representation of the person's voice called a speaker model or voiceprint. During the test phase, the acquired sequence of feature vectors is compared to one or more statistical speaker models to determine the likelihood of a match.

In the speaker identification case, the features are compared to a set of speaker models in a group to generate a set of likelihood ratios – one for each model comparison. The speaker model whose likelihood ratio statistic indicates the closest match identifies the speaker.

In the speaker verification case, the claimed identity is used to select a speaker model. The likelihood that that speaker model produced the features is compared with an imposter model (also called a background model) that represents the set of speakers that is not the claimed speaker. The comparative likelihood value is subjected to a threshold that determines the output accept / reject decision.

2.3.1 Feature extraction

Feature extraction techniques are used to provide more compact, robust and discernable features amenable to statistical modelling and distance score calculations. Similar to speech recognition, most modern speaker recognition systems use mel-cepstral features along with an energy measure in addition to the first and second time derivatives (see Section 2.2.2). A final processing step often applied in speaker recognition systems is that of channel compensation. The purpose of this step is to remove or diminish the effects that different input devices (i.e. telephone handsets) have on the audio signal spectrum (i.e. different band-limits and frequency characteristics). A popular approach is to use Cepstral Mean Subtraction (CMS), which subtracts the time average of the cepstrum from the produced cepstrum.

2.3.2 Statistical modelling

There is a variety of different statistical modelling techniques that has been used for speaker models with different approaches being applied depending on whether a text-dependent or text-independent mode of operation is sought. For text-dependent systems, speaker-specific HMMs are most often applied. The associated likelihood score therefore corresponds to the likelihood that the observed sequence of feature vectors was spoken by the speaker modelled by the corresponding HMM. Just as for speech recognition, HMMs can be used to model whole phrases, individual words, or individual phonemes.

For text-independent systems, where there is no prior knowledge of what the speaker will say, a statistical model that is independent of temporal aspects of the speech is employed. The most successful statistical model for text-independent applications is the Gaussian Mixture Model (GMM). A GMM[12] is a linear combination of N unimodal Gaussian distributions, where the probability is given by:

$$p(y|\lambda) = \sum_{i=1}^{M} a_i p_i(y)$$

where M is the number of components in the mixture model, a_i is the mixture proportion of component i, and p_i is the probability density function of component i. Each component is an N-dimensional Gaussian distribution parameterised by its $N \times 1$ mean vector μ_i and its $N \times N$ covariance matrix Σ_i:

$$p_i = \frac{1}{(2\pi)^{N/2} |\Sigma_i|^{1/2}} \exp\left(-\frac{1}{2}(x - \mu_i)^T \Sigma_i^{-1}(x - \mu_i)\right)$$

We collect the parameters of the density model together as $\lambda = (a_i, \mu_i, \Sigma i)$, $i = 1 \ldots M$. Figure 2.15 illustrates a three-component GMM where $N = 1$. Typically, 64 to 256 mixture components are used for a speaker model. In practice, a diagonal covariance matrix is usually employed, implying no correlation between the components of the feature vectors. Diagonal covariance matrices have been found to give better results and are more efficient to invert.

Given a collection of training feature vectors, $Y = y_1, y_2, \ldots, y_T$, we can use the maximum likelihood method to find the model parameters which maximise the likelihood of the GMM given by

$$p(Y|\lambda) = \prod_{i=1}^{T} p(y_i|\lambda)$$

---

[12] A GMM can be thought of as a single state continuous observation HMM.

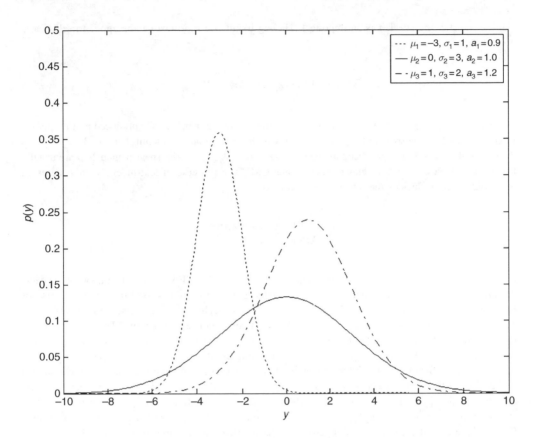

Figure 2.15 GMM with three components where $N = 1$.

The parameter estimates can be found using the iterative expectation–maximisation (EM) algorithm (this is the same technique used for estimating HMM parameters for the Baum–Welch re-estimation algorithm). The basic idea of the EM algorithm is to start with an initial model λ and estimate a new model $\hat{\lambda}$ such that $p(Y|\hat{\lambda}) \geq p(Y|\lambda)$. The new model is subsequently used as the initial model for the next iteration (typically about five to ten iterations are involved for parameter convergence).

During the test phase, given a sequence of feature vectors of length $Y = y_1, y_2, \ldots, y_T$, the log-likelihood that model λ produced the features can be given by

$$\log p(Y|\lambda) = \frac{1}{T} \sum_{i=1}^{T} \log p(y_i|\lambda)$$

Duration effects are normalised by scaling by the length of feature vector sequence, T.

We conclude this section by focusing on the problem of speaker verification. The output of a speaker verification system is a binary result of 'accept' or 'reject'. This amounts to binary hypothesis testing – that the speaker is the person they claim to be or that the speaker is not the person they claim to be (i.e. the speaker is an imposter). The log likelihood ratio test can be applied in this case, given by

$$\Lambda(Y) = \log p\left(Y \mid \lambda_{speaker}\right) - \log p\left(Y \mid \lambda_{imposter}\right)$$

where $p\left(Y \mid \lambda_{speaker}\right)$ is the probability that the feature vector sequence Y was produced by the speaker model $\lambda_{speaker}$ (i.e. the speaker model associated with the user's claimed identity) and $p\left(Y \mid \lambda_{imposter}\right)$ is the probability that Y was produced by the imposter model $\lambda_{imposter}$. The speaker models p are usually HMMs in the case of text-dependent systems and GMMs in the case of text-independent systems. A threshold function decides the outcome according to:

$$\Lambda(Y) \geq \theta \Rightarrow accept$$
$$\Lambda(Y) < \theta \Rightarrow reject$$

If the threshold, θ, is too low, there is a danger of False Accepts (FA) – i.e. when a system accepts the identity claim of an imposter. On the other hand, if the threshold is too high, there is a risk of False Rejects (FR), that is, a valid identity claim is rejected. Thus, there is a trade-off between false accepts and false rejects and hence a trade-off between convenience and security. The combination of the FA rate and FR rate is called the operating point of the system. Given a corpus of test data, the θ threshold value can be tuned to give the desired operating point.

As with speech recognition, the main enemy of speaker verification and identification technologies is the variability in the speech signal. In particular, the possible mismatch of training data and test data is a major problem. The differences stem primarily either from intraspeaker factors or from environmental factors. For the former, a speaker's voice might vary due to health, emotion, or age. For the latter, a different phone may have been used to capture training data than the device used for the test data. There may also be different acoustic environments used in the training data and test data. Speaker verification systems employ the concept of imposter modelling to perform normalisation in an attempt to minimise non-speaker related variability (e.g. device type, noise, and text in the case of text-independent applications). There are two main approaches to constructing an imposter or background model. The first approach uses a collection of other speaker models to compute an imposter likelihood match. The set may comprise speaker models from other speakers trained on the system or may be obtained from some other source. Usually the set needs to be speaker specific in order to produce good results. The likelihood match is often calculated from a maximum or average over the set. The second, and more popular, approach of the two uses a single, speaker-independent model trained on a large number of speakers. This approach is sometimes called a world model or Universal Background Model (UBM). The advantage of this approach is that only a single imposter model needs to be trained and scored. For text-independent applications, GMMs between 512 and 2048 components are typically used. Some newer speaker verification systems use the imposter model as an initial model for new speakers in the training phase. The imposter model is adapted (via the EM algorithm) to the new speaker's training data. This strategy provides a tighter coupling between the speaker's model and the imposter model and results in better performance. One reason for improved performance is that a new speaker's training data may not be sufficiently representative of the speaker's voice and hence test data may exhibit mixture components not seen in the training data. If the training data originates from an adaption of the imposter model whose mixture components represent the broad space of speaker-independent acoustic classes of speech sounds, then any novel mixture components in the speaker's test data will provide little discrimination from the training data and hence not adversely affect verification.

2.4 Speech synthesis

The term Text-to-Speech (TTS) synthesis refers to the process of automatically converting any given input text into audible speech for the purpose of conveying information from a machine to a human in a natural manner. For a machine to convincingly mimic a human speaking, the resultant speech must exhibit certain key characteristics including *accuracy*, *intelligibility*, and *naturalness*. Accuracy refers to the machine's ability to understand the various nuances of the languages in the same way a knowledgeable human is expected to (e.g. speaking abbreviations, common names, dates, times, ordinals, etc.). Intelligibility is a basic characteristic of rendered speech that determines how easily the average listener can understand it. Naturalness is a somewhat more subjective measure concerned with how close the system sounds to a recording of a human speaking. Intelligibility was a prime challenge in earlier TTS engines but has now been largely surmounted in contemporary systems. However, and although systems have made considerable advances in the last decade, naturalness of synthetic speech over a wide range of input text remains a significant challenge.

TTS is hugely valuable in IVR systems for rendering information stored or received in textual form into speech. For example, applications often use TTS to render any source of immediate or dynamic information such as news reports, traffic reports, driving directions, e-mails, bank account transactions, and directory listings. Outside of IVR, TTS has utility in embedded devices to provide more natural user interfaces (particularly when visual modalities are otherwise occupied such as when driving), in communication devices for the visually and speech impaired, in computer games for increased interactivity, and in the output component of language translation systems.

Broadly speaking, there are three main approaches to speech synthesis:

- articulatory synthesis;
- formant synthesis;
- concatenative synthesis.

Articulatory synthesis employs models of the articulators (tongue, jaw, lips, etc.) in conjunction with a model of the glottis to synthesise speech. The models used may vary from crude abstractions of the human speech production system through to complex, biologically realistic constructs. Today the interest in articulatory synthesis is due mostly to its utility in exploring speech production mechanisms rather than its ability to provide high quality, natural sounding speech efficiently.

Formant synthesis uses a model of speech comprising a source excitation – either periodic glottal pulses for voiced sounds or noise for unvoiced sounds such as fricatives – and a cascade of filters to model the resonances (formants) in the vocal and nasal tracts. For sounds that involve the vocal tract, a series of filters usually suffices; for sounds that involve the nasal tract or combine voiced/unvoiced sounds, a parallel arrangement of filters is preferred. A formant synthesis model can be parameterised on a given input segment of speech by using an analysis-by-synthesis approach, whereby the parameters are modified such that the synthesised speech matches the input segment as closely as possible. Formant synthesisers are computationally efficient and often feature in embedded devices where processing power and battery life is constrained. While formant synthesisers provide high intelligibility, they do not provide stellar naturalness.

Concatenative synthesis is a data driven approach that employs short segments of speech recorded from a human speaker. The segments are efficiently recombined in a specified order at synthesis time, with possible modifications to prosody, to produce arbitrary speech. The majority of contemporary speech synthesisers used today in IVR applications are based on concatenative synthesis – a consequence of the fact that this approach can deliver speech that is both highly intelligible and natural. For this reason, the remainder of this chapter will focus on the concatenative synthesis approach.

Figure 2.16 illustrates a block diagram of a concatenative speech synthesiser. The architecture can be divided into a front-end section and a back-end section. The front-end of the system presented here is quite general and is equally applicable in different types of speech synthesiser (formant, articulatory, concatenative) while the back-end is specific to a concatenative synthesiser.

The purpose of the front-end is to accept text input and produce a phonetic transcription with the associated prosody targets (duration, pitch, amplitude) suitable for consumption by the back-end. The input text may be in raw form or may be marked up with tags[13] to provide 'hints' to the system about how to pronounce certain parts of the text. The back-end of the system is concerned with finding the optimum segments of speech (units) given the phonetic transcription and accompanying prosody. The speech segments may then undergo some modification necessary to reach the desired pitch, duration, and amplitude and are subsequently rendered (played) resulting in audio output. We drill down on the front-end processing and back-end processing steps in the following sections.

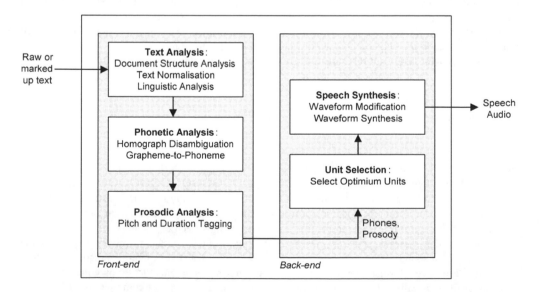

Figure 2.16   Block diagram of a concatenative text-to-speech system.

---

[13] One such markup format is the W3C Speech Synthesis Markup Language (SSML) discussed in Chapter 8.

2.4.1 Front-end processing

The front-end section accepts text as input and produces a sequence of phones and associated prosody at its output. The front-end section can be subdivided into three distinct sub-blocks: text analysis, phonetic analysis, and prosodic analysis – we discuss each block next.

The text analysis block first performs a preprocessing step to analyse the document structure and organise the input sentences into manageable lists of words. In particular, punctuation must be correctly handled. For example, the text analysis block must understand that the colon in '23:45' indicates a time, and to disambiguate between an end of sentence period and decimal point such as in the sentence 'It is 3.14 miles to the city'. Text normalisation deals with transforming abbreviations, acronyms, numbers, dates, and times into full text. This requires careful processing. For example, '20/08/1976' must be transformed into 'twentieth of August nineteen seventy six' and not erroneously as 'twenty forward slash zero eight forward slash one thousand nine hundred and seventy six', for example. As another example, '$123' should be pronounced 'one hundred and twenty-three dollars' and not 'dollar sign, one, two, three'. Again, ambiguities arise in text normalisation. Take for example the sentence 'Dr. Ryan lives on Reynolds Dr.' where 'Dr.' can be interpreted as 'doctor' or 'drive'. It should be clear from these examples that the performance of the document structure and text normalisation tasks are critical for ensuring accuracy of the TTS system. The text analysis block also performs some linguistic analysis. The part of speech category (e.g. noun, verb, adjective, etc.) for each word is determined based on its spelling. Contextual information may also be used to reduce the possible part of speech categories for a word. Word sense must also be taken into account – for example consider the different pronunciations for the word 'bass' in 'I like to fish for sea bass' and 'I like to play bass guitar'. Finally, words may be decomposed into their elementary units or *morphemes* to facilitate pronunciation lookup later (e.g. the word 'unbelievable' comprises three morphemes 'un-' (negative), '-believe-', and '-able').

The phonetic analysis block is concerned with grapheme-to-phoneme[14] conversion (also called letter-to-sound conversion). Pronunciation dictionaries are employed at word level to provide the phonetic transcriptions. In order to keep the size of the dictionary manageable, words are generally restricted to morphemes. A set of morphophonemic rules is applied to determine how the phonetic transcription of a target word's morphemic constituents is modified when they are combined to form that word. Automatic grapheme-to-phoneme conversion based on rules is used for words not found in the dictionary as a fallback, though this approach is often error prone. The phonetic analysis block must also provide homographic[15] disambiguation – for example 'how much produce do they produce?' Contextual information can aid in selecting the right pronunciation – a popular approach is to use a trained decision tree called a Classification and Regression Tree (CART) that captures the probabilities of specific conversions given the context.

The prosodic analysis block deals with determining how a sentence should be spoken in terms of melody, phrasing, rhythm, and accent locations – factors critical to ensure both intelligibility and naturalness of the resultant speech. From the perspective of the speech signal, prosody manifests as dynamic pitch changes, amplitude and duration of phones, and the presence of pauses. Prosodic features have specific functions in speech communication. Human readers vary prosody for two main reasons:

(i) To partition the utterance into smaller parts to help the listener to 'digest' the utterance.
(ii) To set the focus through contrastive means. For example, certain pitch changes can make a syllable stand out in an utterance and as a result the corresponding word it belongs to will be highlighted.

---

[14] A grapheme is an atomic unit of a written language – for English, letters are graphemes.
[15] A homograph is one of two or more words that have the same spelling but different origin, meaning, and sometimes pronunciation.

Prosodic analysis that takes into account the first point is commonly implemented in modern TTS synthesisers. The latter point requires that the system has some understanding of meaning (known as semantics and pragmatics) – a difficult challenge – and is often sidestepped by using neutral prosody resulting in speech that, while not always containing the most appropriate intonation, is still plausible.

2.4.2 Back-end processing

The back-end stage of a concatenative TTS synthesiser consists of storing, selecting, and smoothly concatenating prerecorded segments of speech (units) in addition to modifying prosodic attributes such as pitch and duration of the segments (i.e. subject to the target prosody supplied by the front-end). Some of the key design questions relating to back-end processing include what unit of speech to use in the database, how the optimum speech units are chosen given phonetic and prosodic targets, how the speech signal segments are represented or encoded, and how prosodic modifications can be made to the speech units. We discuss theses questions in the following two sections.

2.4.2.1 Unit selection

Different types of speech unit may be stored in the database of a concatenative TTS system. At the most obvious, whole words may be stored (this is the approach taken by the MRCP basic synthesiser resource as we will see in Chapter 12). However, whole word units are impractical for general TTS due to the prohibitively large number of words that would need to be recorded for sufficient coverage of a given language. Also, the lack of coarticulation at word boundaries results in unnatural sounding speech. When choosing an optimum unit of speech for a TTS system, the following considerations need to be taken into account:

- The units should account for coarticulatory effects as much as possible.
- The smaller the units, the more concatenations are required to make up a word or sentence, adversely affecting quality.
- The larger the units, the more of them that are needed to cover an application domain.

Up until the mid-90s, designers of TTS synthesisers tried to strike a balance by typically employing diphones. A diphone is the portion of speech corresponding to the middle of one phone to the middle of the next. The middle of a phone is the most stable part and hence a diphone captures the coarticulatory effects resulting from the transition between phones. The number of diphones required to synthesise unrestricted speech is quite manageable (in English there are about 1000 needed, for example) thereby facilitating moderate memory requirements for the database. Diphone units are usually obtained from recordings of a speaker speaking 'diphone-rich' text. However, since one diphone unit is used for all possible contexts, the resultant speech does not sound particularly natural.

Modern speech synthesisers have evolved away from using databases with a single, 'ideal' diphone for a given context to databases containing thousands of examples of a specific diphone. By selecting the most suitable diphone example at runtime, and in many cases avoiding making quality-affecting prosodic adjustments to the segment, significant improvements in the naturalness of the speech can be obtained. This is a fundamental paradigm change. Formerly a speech expert had to choose the optimum single diphone from a corpus of recordings for the database. However, the new paradigm takes a data driven approach and leaves it to the system to decide the best unit to select from a set. The approach for

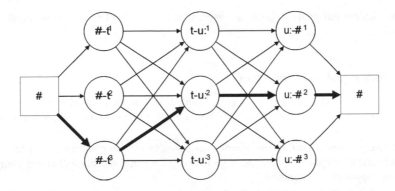

Figure 2.17   Unit selection example for the word 'two' [tuː].

selecting the optimum unit is called unit selection [75]. The process involves treating the database as a state transition network as outlined in Figure 2.17. Each column contains multiple examples (units) of the same diphone. A selection of a particular unit incurs a *concatenation cost* and a *target cost*. The concatenation cost depends on the spectral similarity at unit boundaries; a suitable cost function can be derived from a combination of cepstral distance, amplitude difference, and pitch difference. The target cost depends on the phonetic context and the distance of the prosodic targets (amplitude, duration and pitch) set by the front-end and the unit in the database. Given multiple examples for each diphone in each column, the problem is to find the optimum path (lowest cost) through the network. The Viterbi search algorithm described in Section 2.2.3.1 in the context of HMMs[16] is usually applied and the beam search approach is used to optimise the search. Note that one consequence of employing the unit selection approach is that the memory requirements for the database can be significant.

There are two reasons why unit selection leads to significant improvements in the naturalness of the produced speech. The first relates to the fact that the optimum path may find whole words and sequences of words that were actually present in the training data. In particular, high quality speech can be obtained from modest sized databases where the training data is targeted at a specific domain. Second, since there are multiple diphones in the database for different linguistic and prosodic contexts, the need to employ prosody modifying techniques that adversely affects quality is reduced (i.e. the unit may already have the desired prosody applied).

To create the speech segments used in concatenative TTS synthesisers, recorded speech must be captured and annotated with the position of phones, diphones, word labels, tone labels, etc. The databases used in conjunction with unit selection can contain several dozen hours of speech, thus making manual labelling by speech experts an intractable task. Fortunately, automatic labelling tools are available. Automatic phonetic labelling is achieved using a speech recogniser trained specifically to the speakers voice (speaker dependent) operating in forced alignment mode (i.e. the text of the utterance is supplied to the recogniser *a priori*). This allows the recogniser to focus on locating the phone boundaries rather than having to generate different recognition hypotheses. Automatic prosodic labelling tools work by searching for linguistically relevant acoustic features such as normalised durations, maximum and average pitch ratios, etc., in addition to using phonetic labelling information

---

[16] Note the unit selection state transition network uses cost functions instead of probabilistic costs as in the case of HMMs but this does not affect the functioning of the Viterbi algorithm in any way.

and possibly lexical information (such as whether there is word-final or word-initial stress, for example).

2.4.2.2 Speech signal representation

So far, we have not discussed how the speech segments or units are encoded in the database. The approach used to encode the segments needs to take a number of important factors into account:

(i)   Memory requirements for the unit selection database can be improved if the encoder compresses the audio. The compression rate must be traded off against the computational complexity of decoding the speech for synthesis.
(ii) There should be no perceived loss in speech quality as a result of applying the encode/decode process.
(iii) It should be straightforward to make natural sounding prosodic modifications such as duration, pitch, and amplitude.

Modern, high-quality, TTS synthesisers typically employ either time domain approaches such as the Time Domain Pitch-Synchronous Overlap Add (TD-PSOLA) approach [76] or frequency domain approaches such as the Harmonic Plus Noise Model (HNM) approach [77] for encoding the speech signal. We introduce the TS-PSOLA and HNM approaches next.

The TD-PSOLA technique operates by windowing the speech segment with a series of Hanning windows centred on periodic glottal closures (the point of maximum excitation). The windowing operation is thus 'pitch-synchronous'. A Hanning window with a width of two periods is normally applied. This windowing procedure results in a series of overlapping short-term signals. The short-term signals may be recombined by simply summing them (overlap and add). The pitch can be increased by reducing the distance between the short-term signals before adding them. Conversely, the pitch can be reduced by increasing the distance between the short-term signals (see Figure 2.18 for an illustration).

The duration of a speech segment can be increased by inserting duplicates of the short-term signals according to some time warping function. Conversely, the duration of a speech segment can be reduced by omitting short-term signals according to some time warping function. The TD-PSOLA is a simple and efficient technique that operates in the time domain allowing both time-scale modifications and pitch-scale modifications to be made to the original signal. The TD-PSOLA can be used in conjunction with any speech compression technique (e.g. Differential Pulse Code Modulation or DPCM is one such approach). One problem with the TD-PSOLA approach is that it can introduce audio glitches at the concatenation points due to pitch, phase, or spectral envelope mismatch.

The alternative HNM approach uses a hybrid approach to model the speech signal as being composed of a harmonic part and a noise part. Harmonically related sinusoids account for the quasiperiodic component of the speech signal while the noise part accounts for its non-periodic components such as fricative or aspiration noise, period-to-period variations of the glottal excitation, etc. The harmonic part is parameterised by i sinusoidal components each with its own amplitude α_i and phase ϕ_i. The noise part consists of an autoregressive filtered Gaussian noise source. The two parts are linearly summed to produce synthesised speech. With the HNM approach, prosodic features (specifically, fundamental frequency, duration and amplitude) may be altered easily by adjusting model parameters. Another significant advantage of the HNM technique is that sophisticated smoothing can be performed at inter-unit concatenation points. Since discontinuities in the noise part are not perceived, the HNM

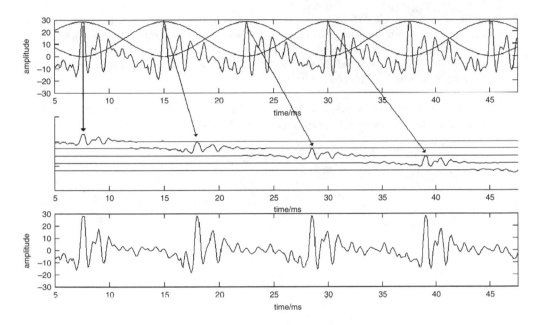

Figure 2.18 Pitch reduction using the TD-PSOLA technique. The top plot consists of a voiced portion of speech with superimposed, pitch-synchronised Hanning windows of width two pitch periods. The middle plot shows the result after windowing. The bottom plot recombines (overlap and add) the segments stretched out to give a lower pitch.

only operates on the harmonic part to smooth pitch discontinuities, harmonic amplitude discontinuities and phase mismatches. A disadvantage of the HNM approach compared with the TD-PSOLA approach is that it is computationally intensive.

2.5 Summary

We commenced this chapter with an overview of the human speech production system and touched on categorisations of speech sounds by discussing aspects of phonemics and phonetics including the IPA. We will come across the IPA again in the context of the Pronunciation Lexicon Specification (PLS) in Chapter 11 – a specification used in conjunction with both speech recognisers and speech synthesisers. The remainder of the chapter provided an introduction to many of the core techniques commonly applied in the different speech processing tasks starting with speech recognition, followed by speaker recognition (encompassing both speaker verification and speaker identification), and finally speech synthesis. The material in this chapter is intended to give the reader an intuitive feel and understanding for how the different MRCP resources operate 'under the hood'. Speech processing technology is a broad and fascinating field that progresses at a rapid rate. More detailed information can be found in the references cited at the beginning of the chapter. New results in the field are published in a number of journals and conferences proceedings including *IEEE Transactions on Speech and Audio Processing*, *IEEE Transactions on Signal Processing*, *Speech Communications Journal*, *IEEE Conference on Acoustics, Speech and Signal Processing* (ICASSP), *Eurospeech*, and the *International Conference on Speech and Language Processing*.

The next chapter provides an overview of MRCP by discussing the different media resource types supported, some network scenarios, and introduces some of the key features of the MRCP protocol.

3

Overview of MRCP

In this chapter, we provide a high-level overview of MRCP, review some network scenarios employing MRCP, and introduce protocol operation fundamentals with the help of some simple examples. After reading this chapter, you should have a good understanding of the basic principles, which should prove useful as a solid basis for more detailed material to follow in later chapters.

3.1 Architecture

MRCP is both a framework and a protocol. The framework defines network elements and their relationship to each other, and also specifies how existing protocols such as the Session Initiation Protocol (SIP) and Real-Time Protocol (RTP) may be used to provide session management and media transport functions within this framework. The MRCP protocol itself provides the mechanism to control media resources such as speech recognisers and speech synthesisers. For brevity, and as is convention, we will often use the term 'MRCP' when we mean 'MRCP framework'.

MRCP is designed to allow clients to invoke and control specialised media processing resources deployed on a network. MRCP is targeted at speech processing resources including speech synthesisers, speech recognisers, speech recorders, and speaker verification and identification engines. The speech processing resources are deployed remotely from the MRCP client on dedicated speech processing servers referred to as media resource servers. Different media resource types may be combined on the same server, e.g. a speech recogniser may be co-located with a speech synthesiser. An MRCP client, on the other hand, is any media processing device that requires speech processing capabilities. Examples of common media processing devices that use MRCP include:

- Interactive Voice Response (IVR) platforms;
- advanced media gateways;
- IP media servers.

Speech Processing for IP Networks   Dave Burke
© 2007 John Wiley & Sons, Ltd

Distributing the media resource server remotely from the MRCP client leads to several technical hurdles that must be surmounted. In particular, one needs the ability to:

  (i) locate and discover a media resource server;
 (ii) establish communication channels with the media resource server, and
(iii) remotely control the media resources.

MRCP leverages the Session Initiation Protocol (SIP) [2] to address requirements (i) and (ii). SIP enables an MRCP client to locate a resource type on the network using a SIP Uniform Resource Indicator (URI) and also enables a media resource server's capabilities to be queried. Once a suitable media resource server is located, SIP is used to establish two communication conduits with it. One conduit is required to send/receive an audio stream to/from the media resource server (called the media session). A second conduit is needed to satisfy requirement (iii) above, namely to communicate control requests from the MRCP client to the media resource server and to return responses and events from the media resource server to the MRCP client (called the control session). The MRCP protocol runs over the control session.

Figure 3.1 illustrates the generalised MRCP architecture. The MRCP client includes a SIP stack and an MRCP stack – the latter performing the role of a client. The platform-specific media resource API mediates between the lower level protocol machinery layers and the platform-specific application layer. When the application layer requires a speech resource, it invokes the media resource API, which creates a SIP dialog with the media resource server via the SIP stack. SIP initiates a media session running over the Real-Time Protocol (RTP) [11] and a control session running over the MRCP protocol with the media resource server. The session initiation process also includes the implicit reservation of one or more dedicated media resources on the server on behalf of the client. Subsequently, the MRCP client can directly invoke the reserved media resource through the MRCP control session. The MRCP client may include the media source/sink within the same device or alternatively this may be part of another entity. SIP facilitates this flexibility by virtue of the fact that the signalling (SIP) and media

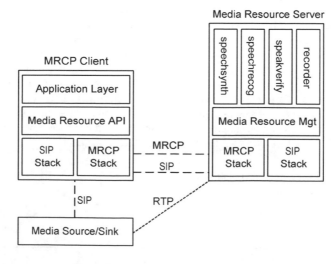

Figure 3.1   Generalised MRCP architecture.

(RTP) may follow different paths. Often, a distributed media source/sink will use SIP to communicate with the MRCP client as shown in Figure 3.1.

The media resource server similarly includes a SIP stack and an MRCP stack. The MRCP stack in this case performs the role of a server — responding to requests from the MRCP client and generating events in the direction of the client. The media resource server includes one or more media resources types such as a speech recogniser, speech synthesiser, speaker verification and identification engine, or speech recorder. When the MRCP client invokes multiple resources, these resources may share the same media session or there may be a dedicated session per media resource. Co-location of multiple resources can be useful in certain scenarios, for example, a barge-in detected event by a speech recogniser resource may be relayed directly to a co-located speech synthesiser resource to stop audio promptly.

Prior to MRCP, many speech server vendors supplied their own proprietary client–server APIs to speech resources using low-level programming languages such as C, C++, or Java. With the advent of MRCP, speech server vendors are moving away from providing proprietary client–server APIs and instead just offering an MRCP protocol interface. Use of standard APIs written in low-level programming languages for speech application has been largely eclipsed by the development of VoiceXML [4, 5], which has become the preferred approach for developing speech applications. VoiceXML is a markup-based Web programming language developed by the W3C for creating advanced speech telephony applications. VoiceXML exploits many of MRCP's features and is discussed in Part V of this book.

MRCP's usage of SIP facilitates an elegant solution to the problem of managing media and control sessions. From the perspective of SIP, the properties of the sessions it manages are not important: SIP is just responsible for locating the media resource server and providing rendezvous functions. SIP enables basic resource discovery so that MRCP clients can learn about specific capabilities of a media resource server. SIP allows the media resource server to be addressed using standard SIP URIs and thus SIP location services may be used to configure pools of media resource servers centrally (see Chapter 5). Finally, note that the MRCP protocol itself is designed in such a way that it is independent of the session establishment and management mechanisms, and hence it is theoretically possible to use a different session management approach in place of SIP, although this is not commonplace.

3.2 Media resource types

MRCP defines a total of six media resource types, which are summarised in Table 3.1. One or more media resources may be co-located in the same media resource server.

Table 3.1   Media resource types

Resource type	Description
basicsynth	Basic synthesiser
speechsynth	Speech synthesiser
dtmfrecog	DTMF recogniser
speechrecog	Speech recogniser
recorder	Speech recorder
speakverify	Speaker verification

Speech synthesisers come in two variants: a basic synthesiser resource (basicsynth) and a speech synthesiser resource (speechsynth). A basic synthesiser is a media resource that operates on concatenated audio clips and provides limited synthesis capabilities. More specifically, a basicsynth only supports a subset of the SSML standard [8] – specifically the `<speak>`, `<audio>`, `<mark>`, and `<say-as>` elements (see Chapter 8 for more information on SSML). A speech synthesiser resource, on the other hand, is a full-featured resource capable of rendering text into speech and supports the entire SSML standard.

Recogniser resources come in two flavours: those that support DTMF digit recognition – the DTMF recogniser resource (dtmfrecog), and those that support speech recognition (and usually DTMF recognition) – the speech recogniser resource (speechrecog). Both resources leverage the W3C Speech Recognition Grammar Specification (SRGS) [6] to specify the words and phrases (or DTMF digit sequences) that the recogniser can accept as input (see Chapter 9 for more information on SRGS). The speech recognisers used in conjunction with MRCP are usually large vocabulary, speaker-independent, recognisers that support many languages and dialects. The speech recogniser resource usually includes the capability to perform natural language understanding by post-processing the recognition results and applying semantic interpretation mechanisms such as those specified in the W3C Semantic Interpretation for Speech Recognition (SISR) [7] (see Chapter 9 for more information on SISR). In addition, speech recogniser resources may support the ability to handle voice enrolled grammars (grammars created at runtime from utterances spoken by the user).

The speech recorder (recorder) resource provides the capability to save audio to a specified URI, usually including the ability to endpoint the audio for suppressing silence at the beginning and end of the recording (and automatically terminating recording after the user has stopped speaking for a certain amount of time).

Finally, the speaker verification resource (speakverify) provides biometric authentication functionality typically used to authenticate the speaker in order to permit access to sensitive information or services. Speaker verification operates by matching a media stream containing spoken input to a pre-existing voiceprint. The caller establishes a claimed identity, e.g. by speaking a pass phrase. The collected audio may then be used to verify the authenticity by comparing the audio to the voiceprint for that identity. Speaker identification involves matching against more than one voiceprint to establish the identity of the speaker within a group of known speakers.

3.3 Network scenarios

In the following sections, we look at some network scenarios involving different kinds of media processing devices leveraging MRCP for distributed speech processing.

3.3.1 VoiceXML IVR service node

Figure 3.2 illustrates an architecture incorporating a VoiceXML Interactive Voice Response (IVR) service node for the Public Switched Telephony Network (PSTN). The VoiceXML IVR service node performs the role of the MRCP client in this architecture and comprises a VoiceXML interpreter, a SIP stack, and an MRCP stack. The VoiceXML IVR service node requires several speech processing services including speech recognition (and DTMF recognition), speech synthesis and speech recording. These services are made available on the media resource server through the speechrecog, speechsynth, and recorder resources respectively. The media gateway bridges the traditional PSTN network with the SIP network by mapping PSTN signalling such as Integrated Services Digital Network (ISDN)

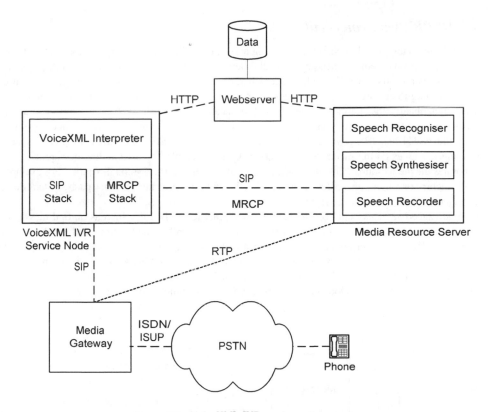

Figure 3.2   VoiceXML IVR service node example.

or ISDN User Part (ISUP) to SIP and converting circuit-switched audio to packet-based audio (RTP). The media gateway provides the media source/sink functionality in this architecture and connects to the VoiceXML IVR service node via SIP (see [15] for a specification of that interface). VoiceXML IVR service nodes are commonly deployed standalone or as part of a bigger system such as in a call centre. In the latter case, the VoiceXML IVR service node is often invoked prior to putting the call into a queue to be handled by a human agent and the information gleaned from the VoiceXML application may be used to perform optimum routing and to also to display a 'screen pop' on the agent's desktop.

The VoiceXML application resides on a Web server in the form of one or more VoiceXML documents, which are fetched by the VoiceXML interpreter using the Hyper Text Transfer Protocol (HTTP) [12]. VoiceXML pages may be created dynamically or 'on-the-fly' by using data from external sources such as a database. The VoiceXML interpreter processes the VoiceXML documents and interprets instructions to perform speech synthesis, speech recognition and audio recording. These instructions are used to drive different media resources hosted on the media resource server via the MRCP protocol. The VoiceXML document may reference other resources on the Web server such as prerecorded audio files or speech grammars. These references, in the form of URIs, may be passed over within the MRCP protocol allowing the media resource server subsequently to fetch those resources directly from the Web server via HTTP.

3.3.2 IP PBX with voicemail

Figure 3.3 illustrates an architecture incorporating an IP Public Branch Exchange (PBX) (also referred to as a softswitch) with built-in messaging capabilities providing voicemail services to both IP phones and phones residing on the PSTN. The voicemail service running on the messaging device requires speech recording to capture callers' messages and a speech synthesiser resource to playback messages, indicate dynamically the number of messages, and speak other details such as the time a message was received and the originating phone number. Similarly to the previous example, a media gateway bridges the PSTN and IP networks. The messaging device uses SIP and MRCP to invoke speech processing resources on the media resource server. In this example, we assume the messaging device incorporates the media source/sink: this design choice affords the IP PBX some flexibility. For example, it could incorporate a circuit switch interface directly without requiring a media gateway and do so independently of the messaging system.

More advanced services could be provided as an adjunct to the IP PBX such as unified messaging (for example, e-mails could be stored as part of the user's mailbox and synthesised by the speech synthesiser). Auto-attendant applications are also possible with the help of MRCP. For example, the caller could say the name of the person they wish to speak to and an MRCP-driven speech recogniser would return the name of the callee to the IP PBX, which would then route the calls accordingly (by using the name of the callee to look up the corresponding phone number up in a database, for example).

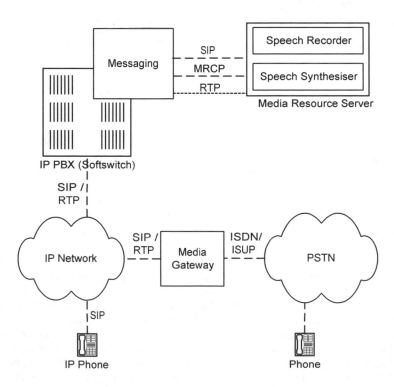

Figure 3.3   IP PBX with voicemail example.

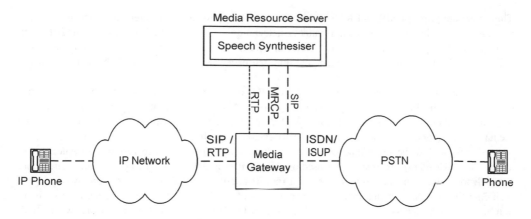

Figure 3.4 Media gateway example.

3.3.3 Advanced media gateway

Figure 3.4 illustrates an architecture consisting of a media gateway with advanced, media-rich capabilities. As in previous examples, the media gateway bridges the PSTN and IP network. In this scenario, the media gateway may directly invoke a media resource server to provide speech synthesis services to the caller. The media gateway can use this capability to relay information to the caller. For example, consider the case where a caller on the PSTN tries to contact a user's IP phone but that user's phone returns a SIP error response such as 480 Temporarily Unavailable with additional information (specified with the Retry-After SIP header field) indicating a better time in the future to reattempt the call. This could occur if the SIP phone had a 'do not disturb' option selected. Without the speech synthesiser resource, the media gateway may simply have to return a busy signal to the PSTN phone. However, by invoking the speech synthesiser via MRCP, the media gateway can deliver an informational message to the caller, e.g. 'The person at extension 4123 is temporarily unavailable to take your call. Please call back in 1 hour'. The speech synthesiser resource can be used for other features, e.g. customised 'ringback' audio streamed to the caller while the callee's phone is in the ringing state, etc. In this architecture, the media source/sink is integrated into the media gateway – this allows the speech synthesiser to be uniformly invoked from calls originating from the PSTN or IP network.

3.4 Protocol operation

In this section, we take an introductory look at the protocol operation details of SIP and MRCP using some simple examples as a guide.

3.4.1 Establishing communication channels

MRCP does not define its own mechanism for resource discovery and session establishment but rather engages SIP for these functions. SIP is used in the IP telephony world to establish media sessions between user agents. The media session itself is transported independently of SIP via RTP, which is a lightweight protocol for transporting time-sensitive data such as audio and video media.

This loose coupling of SIP and RTP allows the signalling path to be different from the media path (separating the signalling and media paths is a technique which the PSTN world learned through evolution and is applied in IP telephony architectures). SIP further uses the Session Description Protocol (SDP) [14] during session establishment and mid-call exchanges to describe and negotiate the media session.

Figure 3.5 illustrates a simple example of how SIP is used to establish a call between two IP phones or user agents. In this case, Alice places a call to Bob. Alice's user agent sends an INVITE request to Bob's user agent. SDP is included in the INVITE message, describing the characteristics of the media stream Alice's user agent would like to establish (called the SDP offer). Bob's user agent starts to ring and at the same time returns an informational 180 Ringing response to Alice's user agent. The 180 Ringing response is used by Alice's user agent to trigger ringback tone generation. When Bob answers the call, Bob's user agent returns a 200 OK final response to Alice's user agent including the SDP answer that indicates which aspects of the offered SDP to use for the media stream and also includes details of where to send the media to. This negotiation of the media, where both participants arrive at a common view of the media session between them, is called the offer/answer model. Finally, Alice's user agent sends an ACK, which completes the so-called three-way handshake and a bi-directional media stream is established between the two user agents. At the end of the call Bob chooses to terminate the call – Bob's user agent issues a BYE message and Alice's user agent responds with a 200 OK response. More detailed information on SIP is provided in Chapter 4.

In the context of MRCP, SIP is used to establish a media session between the media sink/source (which may or may not be directly co-located with the MRCP client) and the media resource server.

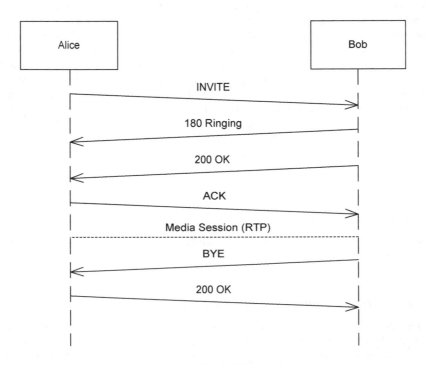

Figure 3.5 Simple SIP example.

In the simple case, where a single media resource type on the media resource server is used at a time, the media session will be unidirectional (e.g. audio flows from a speech synthesiser but to a speech recogniser). When multiple media resource types are used at the same time a bi-directional media stream may be established. Unlike regular IP telephony, MRCP includes the establishment of a control session in addition to the media session. This is achieved by adding further information into the SDP description and also by including the type of media resource that the MRCP client requires. Note that while the media session is transported using RTP running over UDP, the control session is transported using the MRCP protocol running over a lower-level connection-oriented protocol such as TCP or SCTP. The format of control messages and the message exchange patterns employed in the MRCP protocol forms the bulk of the MRCP specification. Figure 3.6 illustrates an example of an MRCP client connecting to a media resource server.

The flow is similar to a regular IP telephony call but the 180 Ringing response is not required and both a media session and a control session are established as a result of the three-way INVITE handshake. The SDP descriptions used in the INVITE and 200 OK messages as part of the offer/answer model share many similarities to the ones used in the previous example but there are differences on account of the need to establish the additional control session. Chapter 5 discusses MRCP's use of SDP and the offer/answer model in detail for establishing control sessions.

3.4.2 *Controlling a media resource*

The MRCP protocol itself is text-based. Text-based protocols such as HTTP, SIP, and MRCP are especially useful when it comes to debugging problems in deployments. At first glance, MRCP messages appear similar in structure to HTTP messages and indeed many of the definitions and syntax are identical to HTTP 1.1 (see Appendix C for an introduction to HTTP, located on the Web at http://www.daveburke.org/speechprocessing/). However, there are substantial differences too. MRCP

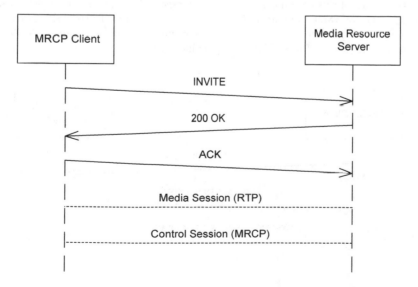

Figure 3.6 MRCP session initiation example.

messages may be one of three types: a request, a response or an event. A request is sent from the MRCP client to the media resource server. After receiving and interpreting a request (though not necessarily processing it), the server returns a response. The response includes a three-digit status code indicating success or failure (e.g. a response code of 200 indicates success). In addition, the response includes the current request state, which may be either PENDING, IN-PROGRESS or COMPLETE. The PENDING state indicates that the request is in a queue and will be processed in a first-in-first-out manner. The IN-PROGRESS state indicates that the request is currently being processed (the more common case). The COMPLETE response indicates that the request is complete and that no further messages are expected in connection with the request. Both the PENDING and IN-PROGRESS states indicate that the request is not yet complete and further event messages are expected. Events allow the media resource server to communicate a change in state or the occurrence of a certain event to the MRCP client. Event messages include the event name (for example, a START-OF-INPUT event is generated by a speech recogniser when it first detects speech or DTMF) and also includes the request state.

3.4.3 Walkthrough examples

In this section, we describe the salient features of the MRCP protocol using simple examples – the first example illustrates the basic message flow for a speech synthesiser and the second example shows the basic message flow for a speech recogniser.

3.4.3.1 Speech synthesiser

Figure 3.7 illustrates a typical MRCP message exchange between an MRCP client and media resource server that incorporates a speech synthesiser resource. We assume the control and media sessions have already been established via SIP as described above. The messages illustrated in Figure 3.7 are carried over the control session using TCP or SCTP transport. The current request state is shown in parentheses. Audio is flowing from the media resource server to the media source/sink (not shown).

The MRCP client requires text to be synthesised and begins the process by issuing a SPEAK request to the media resource server. The SPEAK message is illustrated below:

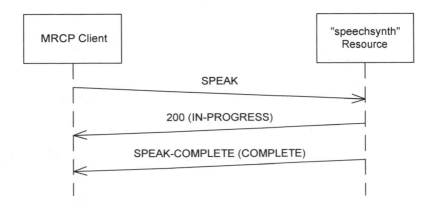

Figure 3.7   Speech synthesiser example.

```
MRCP/2.0 380 SPEAK 14321
Channel-Identifier: 43b9ae17@speechsynth
Content-Type: application/ssml+xml
Content-Length: 253

<?xml version="1.0" encoding="UTF-8"?>
<speak version="1.0" xmlns="http://www.w3.org/2001/10/synthesis">
    <emphasis>Good afternoon</emphasis> Anne. <break/>
    You have one voice message, two e-mails, and three faxes
    waiting for you.
</speak>
```

Since MRCP is text-based, the messages appear on the network exactly as depicted above. MRCP request messages can be broadly divided into three sections: the request line, the header fields, and the optional message body. The request line commences with the protocol name and version separated by a forward slash character, in this case MRCP/2.0. The next number refers to the length of the message in bytes (380). The method name of the request follows (in this case SPEAK), and finally a unique request ID (14321) is included for the purposes of correlating the subsequent response and any event(s) to this request. The request line is terminated by a carriage return and line feed (CRLF). The next lines contain header fields in the format of header-name: header-value CRLF. All MRCP messages must include the Channel-Identifier (obtained in the SDP answer as mentioned in Section 3.4.1). MRCP messages that include a message body must specify the Content-Type and Content-Length headers. The Content-Type specifies the kind of message (an IANA registered type) and the Content-Length specifies the length of the message body in bytes. A blank line separates the header fields and message body (a blank line is present even if there is no message body). In this example, the message body includes 253 bytes of SSML identified by the application/ssml+xml content type. The SSML describes the document to be synthesised using an XML representation. This SSML example applies special emphasis to the phrase 'Good afternoon' and inserts a pause between the two sentences. SSML is described further in Chapter 8. The response to the SPEAK request is illustrated below:

```
MRCP/2.0 119 14321 200 IN-PROGRESS
Channel-Identifier: 43b9ae17@speechsynth
Speech-Marker: timestamp=857206027059
```

The response line looks similar to the request line – the message is 119 bytes long and is in response to the request identified by request ID 14321. The 200 status code indicates success. MRCP uses response codes with similar semantics to those defined in HTTP. Table 3.2 summarises the MRCP response codes and their meaning.

Table 3.2   MRCP response status codes.

Status code	Meaning
200–299	Success
400–499	Client error
500–599	Server error

The request is in the IN-PROGRESS state, which most likely means that audio is currently being streamed to the client. As usual, the corresponding Channel-Identifier is included to identify the particular channel. The Speech-Marker header specifies the time the SPEAK request started (the value is an NTP timestamp [35]).

Subsequently, a SPEAK-COMPLETE event from the media resource server is generated and the request transitions to the COMPLETE state indicating that no more events will be sent for this particular request:

```
MRCP/2.0 157 SPEAK-COMPLETE 14321 COMPLETE
Channel-Identifier: 43b9ae17@speechsynth
Speech-Marker: timestamp=861500994355
Completion-Cause: 000 normal
```

The event line starts with the protocol identifier and message length (157 bytes). The event name follows (SPEAK-COMPLETE) and the remainder of the line contains the request ID (14321) and request state (COMPLETE). The Speech-Marker header specifies the time the SPEAK request completed. The Completion-Cause header field is included in this event. In this case, the value for the Completion-Cause indicates that the SPEAK request terminated normally (i.e. ran to completion). The speech synthesiser resource is described in detail in Chapter 12.

3.4.3.2 Speech recogniser

Figure 3.8 illustrates a typical MRCP message exchange between an MRCP client and media resource server that incorporates a speech recogniser resource. We assume the control and media sessions have already been established via SIP as described above. The messages illustrated in Figure 3.8 are carried over the control session. The current request state is shown in parentheses. Audio is flowing from the media source/sink to the media resource server (not shown).

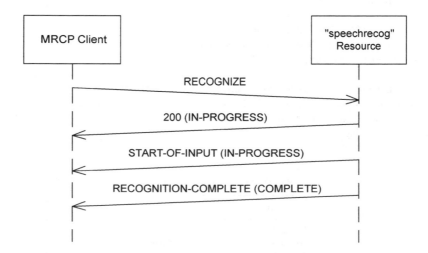

Figure 3.8   Speech recogniser example.

The MRCP client requires speech recognition to be performed on an audio stream directed to the media resource server and begins the process by issuing a RECOGNIZE request to the media resource server. The RECOGNIZE message is illustrated below:

```
MRCP/2.0 461 RECOGNIZE 32121
Channel-Identifier: 23af1e13@speechrecog
Content-ID: <grammar1@form-level.store>
Content-Type: application/srgs+xml
Content-Length: 289

<?xml version="1.0" encoding="UTF-8"?>
<grammar version="1.0" xmlns="http://www.w3.org/2001/06/grammar"
        xml:lang="en-GB">
    <rule id="yesno">
        <one-of>
            <item>yes</item>
            <item>no</item>
        </one-of>
    </rule>
</grammar>
```

The message format is similar to the SPEAK message format. The channel identifier includes the suffix speechrecog to indicate that the channel is controlling a speech recognition resource. The body of the RECOGNIZE request, identified by the application/srgs+xml content type, includes a speech grammar using the XML form of SRGS to define the words and phrases that may be recognised as a result of this recognition request. In this example, the recogniser may recognise just two alternatives, 'yes', or 'no' (this simple grammar is employed for pedagogical reasons – a realistic grammar would contain many more alternatives and sequences). The grammar is identified by a client-supplied value in the Content-ID header field. This allows the grammar to be referenced later (e.g. in recognition results or reused in subsequent recognition actions). The response from the media resource server is illustrated below:

```
MRCP/2.0 79 32121 200 IN-PROGRESS
Channel-Identifier: 23af1e13@speechrecog
```

This response is similar to the example given above for the SPEAK request: it is a successful response (200) informing the client that the request is currently in progress and most likely means that the speech recogniser is currently collecting audio and will soon generate a hypothesis of what is being said. When the speech recogniser detects speech, it generates the START-OF-INPUT event:

```
MRCP/2.0 111 START-OF-INPUT 32121 IN-PROGRESS
Channel-Identifier: 23af1e13@speechrecog
Input-Type: speech
```

The client may use this event to terminate a speech synthesiser (also known as 'barge-in'), for example. Subsequently, when the speech recogniser has finalised its result, it generates the RECOGNITION-COMPLETE event and includes the recognition results in that event's message body:

```
MRCP/2.0 472 RECOGNITION-COMPLETE 32121 COMPLETE
Channel-Identifier: 23af1e13@speechrecog
Completion-Cause: 000 success
Content-Type: application/nlsml+xml
Content-Length: 289

<?xml version="1.0" encoding="UTF-8"?>
<result grammar="session:grammar1@form-level.store"
        xmlns="http://www.ietf.org/xml/ns/mrcpv2">
   <interpretation confidence="0.9">
      <instance>
         yes
      </instance>
      <input>yes</input>
   </interpretation>
</result>
```

The `Completion-Cause: 000 success` header field indicates a successful recognition result. Other common alternatives include `001 nomatch` if the speech input did not match an active grammar and `002 no-input-timeout` if no speech was heard before the no input time expired. The recognition results are contained in the RECOGNITION-COMPLETE body using the Natural Language Semantic Markup Language (NLSML) indicated by the type `application/nlsml+xml`. NLSML was originally created by the W3C but was superseded by work on a new W3C standard called Extensible Multimodal Annotation Markup Language (EMMA) before work on NLSML was completed. The MRCPv2 specification adopts the original, unfinished version of NLSML and provides a complete specification of it sufficient for its purposes. A future version of MRCP will probably adopt EMMA. The NLSML result in this example above indicates the word 'yes' was recognised with a confidence of 90%. For this simple grammar, the raw input, which is contained within the `<input>` element, is the same as the semantic interpretation, contained in the `<instance>` element (see Chapter 10). The speech recogniser resource is described in detail in Chapter 13.

3.5 Security

Security is an important consideration for speech applications. For example, it is common for speech synthesisers to render financial information to the user or for a speech recogniser to accept sensitive information such as credit card numbers, etc. Speech applications must ensure confidentiality and integrity of the data and its transformations to and from spoken form. The use of speaker verification, a biometric technique, requires extra considerations. Voiceprint data used in conjunction with speaker verification is extremely sensitive and can present substantial privacy and impersonation risks if stolen.

Broadly speaking, there are five areas within the MRCP framework where security mechanisms need to be employed:

  (i)   session establishment;
 (ii)   control session protection;
(iii)   media session protection;
 (iv)   indirect content access;
  (v)   protection of stored media.

Since MRCP employs SIP for session establishment, it is able to leverage security mechanisms already provided in SIP. This is introduced generally in Chapter 4 and applied specifically to MRCP in Chapter 5. Control session protection and media session protection is discussed in Chapters 6 and 7 respectively.

MRCP uses content indirection frequently. Content may be fetched or stored on systems identified by URIs that reside outside of the MRCP client or media resource server. For example, sensitive text to be synthesised might be stored on a remote web server in the form of an SSML document that is fetched using HTTP. Voice recordings containing sensitive information may be sent to a remote Web server over HTTP. MRCP clients and media resource servers are specifically required to support HTTP running over Secure Sockets Layer (SSL) or Transport Layer Security (TLS). Resources protected by SSL or TLS are identified by HTTPS URIs.

Media resource servers often make use of stored audio. Voice recordings are both stored (as part of an application such as voicemail, for example) and fetched (for replaying audio files to the user, for example). It is also very common to store user utterances captured from speech recognisers for diagnosis and tuning purposes. This information represents a potential security vulnerability because utterances could be used to reveal sensitive information. Speaker verification and identification technologies pose additional security problems. Theft of voiceprints or a corpus of recorded utterances sufficient to derive a voiceprint could be used by an attacker to compromise services and systems protected by biometric authentication. Further, and as with other biometric technologies, once voiceprint data has been compromised, revocation is not possible (for the obvious reason that one cannot change their voice). It is therefore imperative that centralised repositories of sensitive media are deployed and managed in a robust and secure fashion to minimise risk of identity theft or the theft of sensitive information. Various strategies exist for protecting sensitive data, for example the use of encrypted databases with strong access control and network security techniques ranging from technical (e.g. use of firewalls) to the obvious (protected server rooms with limited, secure access).

3.6 Summary

In this chapter, we introduced the general MRCP architecture and gave some examples of how different media processing devices can leverage MRCP to offer enriched, speech-enabled, services. The basic operation of MRCP and its use of SIP were introduced with examples given for two media resource types. A discussion on security implications of using MRCP was provided.

This concludes Part I of the book. With this high-level overview complete, we are ready to delve into more detailed material. In Part II, 'Media and Control Sessions', we focus on SIP and session management in MRCP, and provide a detailed look at the media and control sessions themselves.

Part II

Media and Control Sessions

4

Session Initiation Protocol (SIP)

This chapter focuses on the Session Initiation Protocol (SIP) defined in RFC 3261 [2]. SIP is the key mechanism used by MRCP to establish and manage media and control sessions. SIP's flexibility goes far beyond its use in MRCP – it is the lingua franca of modern day IP telecommunications. This chapter assumes no prior knowledge of SIP and is aimed at imparting a good grounding in the subject.

4.1 Introduction

SIP is an IETF signalling protocol designed for creating, modifying, and terminating sessions with one or more participants. Those sessions are commonly Internet telephony calls, conferencing media, and multimedia distribution. SIP's focus is on rendezvous and session management, and the protocol itself is independent of the session it manages. This fundamental design choice allows the media to be delivered via the most direct path while the signalling, which is not as sensitive to latencies, may be delivered through a series of routing elements called proxy servers. SIP is heavily inspired by HTTP in that it is a text-based, request–response protocol but they also differ in many aspects – a natural consequence of the increased complexity required to enable bi-directional signalling. Work on SIP began in the IETF Multiparty Multimedia Session Control (MMUSIC) working group, which published version 1.0 as an Internet-Draft in 1997. The protocol was significantly revised and version 2.0 was published as an Internet-Draft the following year. In March 1999, the protocol was published as RFC 2543 [30]. Work on SIP was subsequently taken over by the newly formed IETF SIP working group in 1999. The SIP working group published a significantly revised but mostly backwards compatible version of SIP under RFC 3261 [2] in 2002. This version renders the RFC 2543 version obsolete. In parallel, the IETF SIPPING working group was chartered to document the use of SIP for several applications related to telephony and multimedia, and to develop requirements for extensions to SIP needed for those applications. There are several extensions to SIP for different purposes; this chapter concentrates on the core SIP protocol details described in RFC 3261 [2], which includes the features commonly exploited by MRCP. A brief summary of several important SIP extensions is provided in Section 4.7.

Speech Processing for IP Networks   Dave Burke
© 2007 John Wiley & Sons, Ltd

4.2 Walkthrough example

In this section we introduce some of the major features of SIP by jumping straight into a walkthrough of a typical IP telephony call (illustrated in Figure 4.1). This is a convenient and fast way to introduce the salient points of SIP. The subsections to follow expand on particularly important points brought up by the example.

   In Figure 4.1, two participants Alice and Bob wish to establish an Internet telephone call via their respective SIP user agents (UA). A SIP user agent may take many forms, for example a hardware desk phone, a software program running on a computer (called a 'soft phone'), or may be embedded in a portable device such as a PDA or mobile phone. The process commences when Alice's user agent issues an INVITE message to Bob's user agent. Alice's user agent identifies Bob's user agent using a SIP URI sip:bob@gauss.com (see Section 4.3 for more information on SIP URIs). The INVITE message contains a description of the media session characteristics requested, which is encoded in the message body using the Session Description Protocol (SDP) [14]. The INVITE message is illustrated below:

```
INVITE sip:bob@gauss.com SIP/2.0
Via: SIP/2.0/UDP pc10.newton.com;branch=z9hG4bKqw1
Max-Forwards: 70
To: Bob <sip:bob@gauss.com>
From: Alice <sip:alice@newton.com>;tag=983174
Call-ID: b8a931af@pc10.newton.com
CSeq: 2314 INVITE
Contact: <sip:alice@pc10.newton.com>
Content-Type: application/sdp
Content-Length: 141

v=0
o=alice 2890844527 2890844527 IN IP4 pc10.newton.com
s=-
c=IN IP4 pc10.newton.com
t=0 0
m=audio 4122 RTP/AVP 0
a=rtpmap:0 PCMU/8000
```

SIP is a text-based, UTF-encoded protocol and hence the message is transmitted over the wire exactly as presented above. The SIP message consists of a request line, several header fields, and an optional message body. The request line and the header fields are terminated with a carriage return line feed (CRLF). The header fields (or 'headers' for short) employ the format of header-name: header-value CRLF. An additional CRLF delimits the header portion of the message from its optional body. In the example, the request line includes the method type (INVITE), the Request-URI (sip:bob@gauss.com), and the protocol version (SIP/2.0). All SIP requests have a similarly formatted request line. At a minimum, a valid SIP request must include the following six header fields: Via, Max-Forwards, To, From, Call-ID, and CSeq. The Via header field indicates the path taken so far by the request and is used to route a subsequent response message back over the same path. Each SIP entity that originates or forwards the request adds its host to the Via header, in effect creating a history of the path taken by the request. The Via header field also includes the protocol version and transport used and the branch parameter value uniquely identifies the request

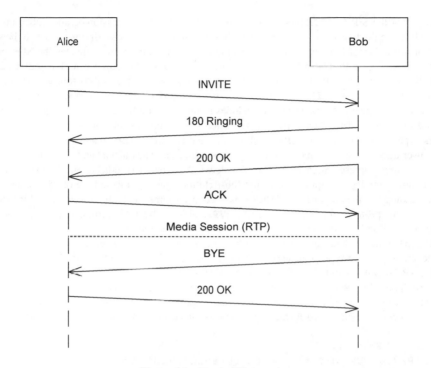

Figure 4.1   Simple SIP example.

transaction. A SIP transaction consists of a single request and the one or more responses to that request. The `Max-Forwards` header field is used to limit the number of hops a request traverses and is decremented by one for each hop. The `To` header field indicates to whom the request is directed. It includes an optional display name and a SIP or SIPS URI enclosed in angle brackets. The display name can be used to show the name of the caller on the device's screen during alerting and progress of the call, for example. The `From` header field indicates the originator of the request and includes a display name and SIP URI. The `From` header includes a `tag` parameter for identification purposes (note that Bob's user agent will add it's own `tag` parameter to the `To` header in the response). The `Call-ID` header field contains a globally unique identifier for this call, generated by the user agent using a combination of a random string and the user agent's host name or IP address. The combination of the `To` tag, `From` tag, and `Call-ID` completely defines a peer-to-peer SIP relationship between Alice and Bob and is referred to as a 'SIP dialog'. The `CSeq` header field is an abbreviation of 'command sequence' and consists of a sequence number and method name. The sequence number is incremented for each new request within a dialog and serves to differentiate retransmissions of requests from new requests. The `Contact` header field contains a SIP URI that represents a direct route to contact Alice and is called the *device URI* or *contact URI*. This is in contrast to the address in the `From` header field which uses an *address-of-record* URI that typically does not provide a direct route to Alice (see Section 4.3 for more information on SIP URIs). The `Content-Type` header provides a description of the message body, in this case indicating that the body contains SDP. The `Content-Length` is the length of the body in bytes and is used to facilitate efficient parsing.

The body of the SIP `INVITE` message contains SDP describing the session details (called the offer), which Alice would like to use in the call with Bob. SDP session descriptions are text based and consist of a number of lines identified by a single letter field. The v-line indicates the SDP protocol version – in this case 0. The o-line contains origin information including a user name (`alice`), a session ID and version (both use a 64-bit Network Time Protocol (NTP) [35] timestamp), a network type and address type (`IN IP4`, indicating Internet using IP4 addressing), and an address for the phone (`pc10.newton.com`). The s-line includes a subject name for the session – usually left blank, indicated by a single dash. The c-line contains connection data information with three sub-fields for network type (`IN`), address type (`IP4`), and the connection address (`pc10.newton.com`) for where Alice's user agent would like media to be sent. The t-line indicates a start and stop time respectively (using a decimal representation of NTP time values). A value of 0 indicates no bounded time. The m-line includes media descriptions. The first sub-field indicates the media type (`audio`). The second sub-field indicates the port for where Alice's user agent requires media to be sent. The third sub-field is the transport protocol for the media: `RTP/AVP` indicates the RTP Profile For Audio and Video Conferences defined in RFC 3551 [17]. The fourth sub-field is the media payload format also defined in RFC 3551. Finally, the a-line contains attribute information for the preceding m-line. The `rtpmap` attribute maps payload type 0 appearing on the m-line to additional information, namely the audio encoding (PCM μ-law) and clock frequency (8000 Hz).

Prior to the final response to the `INVITE` request, one or more provisional responses (also known as informational responses) may be made. Provisional responses are identified by $1xx$ responses and are optional – in our example a `180 Ringing` response indicates that Bob's user agent is alerting the user:

```
SIP/2.0 180 Ringing
Via: SIP/2.0/UDP pc10.newton.com;branch=z9hG4bKqw1
 ;received=10.0.0.1
Max-Forwards: 70
To: Bob <sip:bob@gauss.com>;tag=32512
From: Alice <sip:alice@newton.com>;tag=983174
Call-ID: b8a931af@pc10.newton.com
CSeq: 2314 INVITE
Contact: <sip:bob@host2.gauss.com>
Content-Length: 0
```

SIP responses commence with a status line consisting of the SIP version, the response code, and the reason phrase. SIP's response codes are based around HTTP's [12] with several additions and extensions. Table 4.1 summarises the SIP response codes and their meanings.

Table 4.1 SIP response status codes

Status code	Meaning
100–199	Provisional (informational)
200–299	Success
300–399	Redirect
400–499	Client error
500–599	Server error
600–699	General error

The reason phrase beside a response code is used solely for the purpose of making the protocol human readable and is not acted upon by a user agent. Alice's user agent can use the `180 Ringing` response to start playing ringback tone. The rest of the headers are copied from the `INVITE` request with the exception of the `Content-Length`, which this time indicates that no message body is present. Note that the `To` and `From` header fields are not reversed even though the message is going in the reverse direction. This is because SIP defines the `To` and `From` header field values to be set according to the direction of the original request (which in this case was the `INVITE`). Bob's user agent adds a `tag` parameter to the `To` field value thus allowing the SIP dialog established by the `INVITE` request to be uniquely identified by the `tag` parameter values on the `To` and `From` header fields along with the `Call-ID` header field value. It is important that Bob's user agent contributes to building the unique identifier for the SIP dialog since it is possible that an intermediate proxy could fork the original request and, as a result, establish more than one SIP dialog with different user agents. The provisional response message is part of the `INVITE` transaction as is evident by the same `branch` parameter value on the `Via` header. A new parameter, `received`, is added to the `Via` header field by Bob's user agent and is used to indicate the source IP address from which the packet was received and serves to assist in routing the response.

When Bob answers the phone, his user agent generates a `200 OK` final response message:

```
SIP/2.0 200 OK
Via: SIP/2.0/UDP pc10.newton.com;branch=z9hG4bKqw1
 ;received=10.0.0.1
Max-Forwards: 70
To: Bob <sip:bob@gauss.com>;tag=32512
From: Alice <sip:alice@newton.com>;tag=983174
Call-ID: b8a931af@pc10.newton.com
CSeq: 2314 INVITE
Contact: <sip:bob@host2.gauss.com>
Content-Type: application/sdp
Content-Length: 139

v=0
o=bob 3812844327 3812844327 IN IP4 host2.gauss.com
s=-
c=IN IP4 host2.gauss.com
t=0 0
m=audio 8724 RTP/AVP 0
a=rtpmap:0 PCMU/8000
```

The `200 OK` response message header fields are similar to the `180 Ringing` provisional response's. The message body includes Bob's user agent's session description answer. The answer responds to the offer by accepting or refusing aspects of the media session described by the m-line (in this case it accepts the media session offer). The m-line in the answer indicates the port to send the media to (i.e. 8724). The c-line in the answer indicates the IP address to send the media to (`host2.gauss.com`). See Section 4.5.2 for more information on the SDP offer/answer model.

Alice's user agent now generates an acknowledgement or ACK message:

```
ACK sip:bob@host2.gauss.com SIP/2.0
Via: SIP/2.0/UDP pc10.newton.com;branch=z9hG4bKery
```

```
Max-Forwards: 70
To: Bob <sip:bob@gauss.com>;tag=32512
From: Alice <sip:alice@newton.com>;tag=983174
Call-ID: b8a931af@pc10.newton.com
CSeq: 2314 ACK
Content-Length: 0
```

Note that the ACK is routed directly to Bob's device URI (gleaned from the Contact header in the responses to the INVITE) and placed in the Request-URI. The three-way handshake is now complete and media may start flowing in both directions via RTP: the SIP dialog has established an associated media session. Note that there is a new branch parameter value on the Via header field value, signalling that the ACK is part of a new transaction.[1] Of course, this transaction is still within the same SIP dialog (as indicated by the 3-tuple consisting of the Call-ID, From tag and To tag).

Bob decides to terminate the call first and his user agent issues a BYE request to Alice's user agent:

```
BYE sip:alice@newton.com SIP/2.0
Via: SIP/2.0/UDP host2.gauss.com;branch=z9hG4bKqjd
Max-Forwards: 70
To: Alice <sip:alice@newton.com>;tag=983174
From: Bob <sip:bob@gauss.com>;tag=32512
Call-ID: b8a931af@pc10.newton.com
CSeq: 3254 BYE
Content-Length: 0
```

Notice that the To and From fields are reversed since this request is in the opposite direction to the INVITE and ACK requests. There is a new branch parameter value on the Via header field indicating this is another new transaction. The BYE request is of course still part of the original SIP dialog as is evident by the same Call-ID and tag values. Alice's user agent responds with a 200 OK response:

```
SIP/2.0 200 OK
Via: SIP/2.0/UDP host2.gauss.com;branch=z9hG4bKqjd
Max-Forwards: 70
To: Alice <sip:alice@newton.com>;tag=983174
From: Bob <sip:bob@gauss.com>;tag=32512
Call-ID: b8a931af@pc10.newton.com
CSeq: 3254 BYE
Content-Length: 0
```

Receipt of a BYE request terminates the SIP dialog and its associated media session(s) thus ending the call. The example of the BYE request reveals an important distinction between SIP and other protocols such as HTTP. While HTTP is a client–server protocol with requests permitted in one direction only (i.e. from the client to the server), SIP is a client–server protocol with requests permitted in both directions. In effect, a SIP user agent must behave as both a client and a server at the same time. When a user agent originates a request, we say it is taking on the role of a user agent client (UAC) and its

---

[1] Note that all branch parameter values commence with z9hG4bK as required by RFC 3261 to allow backward compatibility with an older transaction identification mechanism specified in RFC 2543.

peer is taking on the role of a user agent server (UAS). During the lifetime of a SIP dialog, a user agent changes roles between a UAC and UAS.

4.3 SIP URIs

A SIP URI traditionally identifies a user but may also be used to identify a communication resource such as a mailbox, an IVR service, or, in the context of MRCP, a media resource server. A SIPS URI specifies that the resource is contacted securely. The SIP and SIPS schemes follow the general syntax defined in RFC 2396 [22]. The basic[2] SIP URI format is:

```
sip:user@host:port;param-name=param-value
```

(SIPS URI schemes simply use `sips` in place of `sip`). The token before the '@' is termed the *user part* and may be omitted for certain kinds of resources. The host part consists of a fully qualified domain name or a numeric IP4 or IP6 address. The port number is optional (the default port number for SIP using TCP or UDP is 5060; the default port number for SIPS using TLS over TCP is 5061). URI parameters may be optionally appended to the URI delimited by semicolons. The following are valid SIP addresses:

```
sips:megan@192.168.1.2:5070
sip:mark@192.168.1.2;transport=tcp
sip:reg01.example.com
```

With SIP user agents, one usually associates two addresses. The first is called the address-of-record, which is the public, fixed address for a user. The second address is known as the contact URI or device URI and is the address of the physical device (this address appears in the `Contact` header field). In the example above, `sip:alice@newton.com` is Alice's address-of-record and `alice@pc10.newton.com` is Alice's device URI. Using two different address types is the key to how SIP implements mobility functionality. Alice can change the physical device address of the device she uses (for example, using it on a different network) or may change the device itself (for example, using a SIP mobile PDA when away from her desk instead of a desktop phone). SIP networks store the address-of-record and device URI mapping in a centralised database called a location service. A user agent registers its device URI via a SIP server called a registrar, which inserts the mapping into the database. By placing a specialised SIP server called a proxy server in front of Alice's user agent, for example, the proxy server may consult the location service and route inbound `INVITE` requests to the appropriately registered user agent or user agents. See Section 4.6 for more information on registrars and proxy servers.

4.4 Transport

SIP is independent of the underlying transport protocol (a feature that has inevitably increased the complexity of the protocol). SIP messages may be transported using an unreliable transport protocol such as the User Datagram Protocol (UDP) or a reliable, connection-oriented, transport protocol such

---

[2] More complex SIP URIs are possible but are somewhat less common, e.g. URIs specifying a password or additional SIP headers – see RFC 3261 [2] for full details.

as the Transmission Control Protocol (TCP) or the Stream Control Transport Protocol (SCTP). When secure SIP is required as indicated by a SIPS URI, the Transport Layer Security (TLS) protocol is used, which is based on the Secure Sockets Layer (SSL) protocol first used in web browsers. TLS uses TCP for transport.

All SIP entities must support both UDP and TCP. When SIP operates with UDP, the SIP protocol undertakes responsibility for implementing reliability functions for message delivery: SIP supports retransmissions if a message does not reach its destination and hence is not acknowledged within a designated time period. UDP offers a simple transport mechanism but does have some disadvantages. For a start, UDP does not contain any congestion control. For example, if a heavily loaded network starts to drop packets, retransmissions will generate more packets, which will load the network further and result in more dropped packets. A second problem with using SIP with UDP is that the message size is limited by the Maximum Transmission Unit (MTU) size of the network or networks, thus causing problems for large SIP messages (in such cases, a protocol such as TCP is recommended).

TCP, on the other hand, provides a more advanced transport mechanism that offers a faster retransmit if packets are lost, congestion control, and no MTU-enforced limitation on message size. Using TCP does incur some cost in the form of increased transmission delay and complexity.

SCTP shares the same advantages as TCP does over UDP but also has some unique advantages over TCP. SCTP does not suffer from 'head of the line' blocking: when multiple SIP transactions are being delivered over the same TCP connection and one message is lost, a retransmission will occur and the other transactions (such as those belonging to other SIP dialogs) must wait for the failed one to proceed. SCTP avoids this problem since it is message based. SCTP also allows connections to be associated with multiple IP addresses on the same host enabling the use of a set of fail-over addresses should the primary one fail.

4.5 Media negotiation

The types of devices that connect to the Internet vary widely in their capabilities to render media and perform user interaction. SIP is designed to support heterogeneous clients with highly variable capabilities by providing an extension and capability negotiation framework that covers many aspects of the protocol. The following subsections look at the media capability negotiation aspects of SIP.

4.5.1 Session description protocol

SIP user agents employ SDP to describe the desired multimedia session that each user agent would like to use. The term 'protocol' in the SDP acronym is somewhat of a misnomer since SDP is much more of a format than a protocol *per se*. SDP is defined in RFC2327 [14] and was developed by the IETF MMUSIC working group. SDP was originally intended for use on the Internet multicast backbone (Mbone), to advertise multimedia conferences and communicate the conference addresses and conference tool-specific information necessary for participation. SDP can also be used for unicast transmission and of course is used by SIP to provide session descriptions.

SDP session descriptions are text based and consist of a number of lines in the format of <type>=<value> where <type> is always one character in length and is case-sensitive. Each <type> is called a field and the <value> consists of one or more sub-fields. There is a total of 16 fields defined in RFC 2327 [14], of which five fields are mandatory. SDP fields are required to be in a specified order and unrecognised fields or out of order fields are simply ignored. Table 4.2 summarises

Table 4.2 SDP fields in the required order

Field	Required?	Level	Name
v=	Yes	Session-level	Protocol version
o=	Yes	Session-level	Owner/creator and session identifier
s=	Yes	Session-level	Session name
i=	No	Session-level	Session information
u=	No	Session-level	URI of description
e=	No	Session-level	E-mail address
p=	No	Session-level	Phone number
c=	Yes[a]	Session-level	Connection information
b=	No	Session level	Bandwidth information
t=	Yes[b]	Session-level	Time the session is active
r=	No[b]	Session-level	Zero or more repeat times
z=	No	Session-level	Time zone adjustments
k=	No	Session-level	Encryption key
a=	No	Session-level	Session attribute lines
m=	No	Media-level	Media name and transport address
i=	No	Media-level	Media title
c=	Yes[a]	Media-level	Connection information
b=	No	Media-level	Bandwidth information
k=	No	Media-level	Encryption key
a=	No	Media-level	Media attribute lines

[a] The c= field must be present at the session-level or at the media-level for all media descriptions.
[b] The t= and r= fields may appear multiple times.

the fields in their required order. A session description consists of a session-level description (details that apply to the whole session and all media streams) and optionally several media-level descriptions (details that apply to a single media stream). An example of using SDP within SIP is illustrated in the walkthrough example in Section 4.2 above.

The media-level descriptions play an important part in SIP's media negotiation mechanism. Each media-level description starts with an m-line in the format of:

```
m=<media>  <port>  <transport>  <fmt list>
```

The first sub-field is the media type. For MRCP, the relevant types are audio (for the media session) and application (for the control session). The second sub-field is the transport port to which the media stream will be sent. For RTP, this is always an even port. Note that the IP address to send the media to is specified in the c-line, which may be present in each media-level description or, as is more common, is present in the session-level description. The third sub-field refers to the transport protocol. Typical values are RTP/AVP (RTP Profile for Audio and Video Conferences, defined in RFC 3551 [17]), UDP, and TCP. MRCP uses RTP/AVP for its media sessions and TCP for its control sessions (see Chapter 5 for more information). The final sub-field, <fmt list>, consists of one or more media formats represented by a number. When a list of payload formats is given, this implies that all of these formats may be used in the session, but the first of these formats is the default format for the session. Payload numbers from 0 – 95 are reserved for static definitions (numbers are assigned to a fixed codec type in IETF standards) and numbers 96–127 are dynamic. When dynamically allocated payload types are used, additional encoding information is provided in `rtpmap` attributes in the form of:

```
a=rtpmap:<payload type>  <encoding name>/<clock rate>
```

The `<payload type>` sub-field corresponds to one of the payload formats in the `<fmt list>` in the m-line. The `<encoding name>` sub-field indicates an IANA registered type for the codec. The `<clock rate>` sub-field indicates the sampling rate measured in hertz. Additional parameters may be added by appending a single '/' character followed by the parameters. For example, a media-level description offering G.711 μ-law and linear audio would use the following media-level description:

```
m=audio 3241 RTP/AVP 0 96
a=rtpmap:0 PCMU/8000
a=rtpmap:96 L8/8000
```

Strictly speaking, the first `rtpmap` attribute is superfluous since the payload format 0 is statically defined by RFC 3551 [17]. There are many other attributes that may be present in the media-level description. Here we summarise some important ones that crop up regularly and that apply to MRCP. The directionality of a media stream may be specified with an a-line with the sub-field of `sendrecv`, `sendonly`, or `recvonly`. The default is `sendrecv`. The direction is taken in context of the entity that is supplying the SDP description. Example:

```
a=sendrecv
```

The `fmtp` attribute is used to specify format specific parameters. The syntax is

```
a=fmtp:<payload type>  <format specific parameters>
```

The `mid` attribute is used for identifying a specific media stream and is defined in RFC 3388 [26]. MRCP makes use of this attribute to facilitate identifying a mapping between a media session and one or more control sessions (see Chapter 5 for more information). In the following example, the media stream is given an identifier of '3':

```
m=audio 3241 RTP/AVP 0
a=rtpmap:0 PCMU/8000
a=mid:3
```

4.5.2 Offer/answer model

SDP on its own is not sufficient to allow two endpoints to arrive at a common view of a multimedia session between them. For this functionality, SIP makes use of the offer/answer model defined in RFC 3264 [27]. In this model, one participant (called the 'offerer') generates an SDP message that constitutes the offer – the set of media streams and codecs the offerer wishes to use along with the IP addresses and ports for those streams. In response, the other participant (called the 'answerer') responds with the answer – an SDP message that responds to the different streams in the offer indicating whether each is acceptable or not. This process allows an overlapping set of codecs to be chosen from the set of codecs supported by each participant. For example, the offer might contain the following media-level session descriptions describing an audio and video multimedia session:

```
m=audio 1232 RTP/AVP 8
a=rtpmap:8 PCMA/8000
m=video 1242 RTP/AVP 34
a=rtpmap:34 H263/90000
```

The answer could choose the audio stream only by responding to the video stream using port 0 (indicating a wish to disable it):

```
m=audio 1232 RTP/AVP 8
a=rtpmap:8 PCMA/8000
m=video 0 RTP/AVP 34
a=rtpmap:34 H263/90000
```

In the usual case, the offer is placed in the SIP INVITE message and the answer is returned in the 200 OK response message. This was the case in the example in Section 4.2 – Alice was the offerer and Bob was the answerer. Alternatively, the offer may be omitted in the INVITE, in which case the 200 OK response will contain the offer and the answer will be placed in the ACK message. It is a minimum requirement that SIP user agents must support the offer/answer model using the INVITE / 200 OK and 200 OK / ACK messages. An offer may be rejected outright if it is placed in the INVITE by responding with a 488 Not Acceptable Here message. More complex behaviours are possible if reliable provisional responses are used (specified in RFC 3262 [21]) or the UPDATE method supported (specified in RFC 3311 [24]), in particular by allowing completion of the offer/answer before the final response to the INVITE.

It is possible for a user agent to change the media session mid-call by issuing a re-INVITE message with a new media description (note that this media description indicates the full state of the session and not simply a delta to it). A re-INVITE message is the same as an INVITE message except that it is issued within a previously established SIP dialog. A user agent can differentiate a re-INVITE from a new INVITE because a re-INVITE will have the same Call-ID and To and From tags as an existing dialog. The re-INVITE will also have a new CSeq value. A user agent can decline a re-INVITE request that attempts to change the characteristics of the media session by responding with a 488 Not Acceptable Here message. In this case, the session continues using the previously negotiated characteristics.

4.6 SIP servers

In a SIP network, the user agents are the endpoints that originate or terminate SIP requests. The example in Section 4.2 illustrated a simple peer-to-peer SIP call that operated without the need of intermediate servers. In practice, SIP networks are often more complex and include several SIP servers, principally to aid routing. The following subsections introduce the different types of SIP server. It is important to note that a SIP server is a logical concept and not a physical one: it is possible to co-locate more than one logical SIP server into a single physical server.

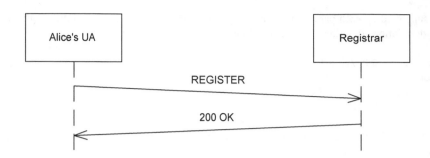

Figure 4.2 SIP registration.

4.6.1 Registrars

A registrar is a special type of server that accepts REGISTER requests from SIP user agents and stores information in a central database called a *location service*. The information stored represents a binding between the address-of-record and the device URI. Figure 4.2 depicts a REGISTER request and a successful response.

An example REGISTER message is illustrated below:

```
REGISTER sip:registrar.newton.com SIP/2.0
Via: SIP/2.0/UDP pc10.newton.com;branch=z9hG4bKqw2
Max-Forwards: 70
To: Alice <sip:alice@newton.com>
From: Alice <sip:alice@newton.com>;tag=456248
Call-ID: e23a31af@10.0.0.2
CSeq: 1826 REGISTER
Contact: <sip:alice@10.0.0.2>
Expires: 7200
Content-Length: 0
```

The address-of-record is contained in the To header field and the device URI is placed in the Contact header field. The From header field indicates who is performing the registration. The Expires header tells the registrar for how long the user agent would like the binding to exist in seconds. A successful response to the request is:

```
SIP/2.0 200 OK
Via: SIP/2.0/UDP pc10.newton.com;branch=z9hG4bKqw2
 ;received=10.0.0.2
Max-Forwards: 70
To: Alice <sip:alice@newton.com>;tag=23141
From: Alice <sip:alice@newton.com>;tag=456248
Call-ID: e23a31af@10.0.0.2
CSeq: 1826 REGISTER
Contact: <sip:alice@10.0.0.2>
Expires: 7200
Content-Length: 0
```

The `Contact` header in the response specifies the current bindings and the `Expires` header field indicates how long the binding will stay current in seconds (a registrar may choose an expiration value less than the value requested by the user agent). Note that there may be more than one contact address specified if multiple devices are registered for the same address-of-record. More than one contact address can be specified either with multiple `Contact` headers or a comma-delimited value specified in a single `Contact` header.

4.6.2 Proxy servers

A SIP proxy server primarily plays the role of routing. A request may traverse several proxies on its way to the UAS. Each proxy makes routing decisions, often modifying the request slightly before forwarding it on. Proxies are also useful for enforcing a policy or set of permissions (e.g. is the caller allowed place a call to a specific network?). Proxies come in two types: stateless and stateful. A stateless proxy processes each SIP request or response based on the contents of the particular message only – no additional state is maintained. A stateful proxy, on the other hand, does maintain state outside of the information contained in each request or response. A stateful proxy stores information related to the SIP transaction and can therefore deliver more advanced services. For example, a 'forking' proxy can forward an `INVITE` request from a user agent to several different target endpoints in parallel with the goal of setting up a SIP dialog between the user agent and the first SIP endpoint that answers. Stateful proxies are more common in practice than stateless ones.

Note that a proxy is a logical role, which is specifically defined to be restrictive. For example, proxies operate only on the Request-URI and certain header fields – they do not process message bodies. Proxies do not incorporate media capabilities and a proxy never initiates a request but rather only respond to requests. SIP is designed around the idea of 'smart endpoints', that is, the bulk of the intelligence is provided in the user agent. This allows innovation to take place at the user agent where it may be tailored more appropriately to the specific device while freeing up the 'in network' proxy servers from maintaining whole call state, and allowing them to scale better and provide more reliability (e.g. a transaction stateful proxy can fail mid-call without affecting ongoing calls).

Figure 4.3 extends the example in Figure 4.1 by introducing proxy servers acting on behalf of Alice and Bob. This typical arrangement is referred to as the 'SIP trapezoid' because of the shape of the

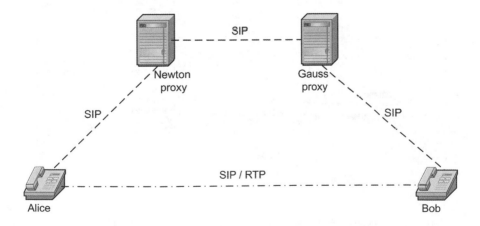

Figure 4.3 SIP trapezoid.

SIP signalling and media session routing. In this configuration, each user agent is configured with an outbound proxy server in its own domain to which it sends all its outbound requests. Requests that cross domains typically use DNS resolution to locate the next hop – in this case the other proxy server. Each proxy server also behaves as an inbound proxy for requests it receives from other domains and routes the request appropriately (e.g. by consulting a location service). A proxy is configured to be responsible for one or more domains. In the example, the Newton proxy is responsible for routing within the `newton.com` domain and the Gauss proxy is responsible for routing within the `gauss.com` domain.

Figure 4.4 illustrates the message flow (the `Via` headers are included in the diagram to show the routing process but the `branch` and `received` parameters have been omitted for clarity). Alice's user agent places an `INVITE` to Bob's user agent as before. However, in this example, Alice's user agent is configured to use the Newton proxy. The Newton proxy adds its address in an additional `Via` header and forwards the `INVITE` to `proxy.gauss.com` – the Gauss proxy (the address was returned from a DNS lookup of the host part of Bob's URI). The Gauss proxy recognises that it is responsible for the domain specified in the `INVITE`'s Request-URI and accesses a location service to obtain Bob's device URI, which Bob's user agent had previously registered. The Gauss proxy in turn adds its address in another `Via` header and forwards the `INVITE` to Bob. When Bob's user agent starts alerting, it responds with a `180 Ringing` message. Bob's user agent knows to send the provisional response message to the first `Via` entry (`proxy.gauss.com`). The Gauss proxy inspects the message and removes the `Via` header with its address and forwards the message to the next `Via` header address (`proxy.newton.com`). The Newton proxy, in turn, removes the `Via` header with its address and delivers the message to Alice's user agent (indicated in the last remaining `Via` header). A similar process occurs when Bob answers the call and the `200 OK` final response is sent. The `200 OK` response includes a `Contact` header field, thus allowing Alice's user agent to send the `ACK`

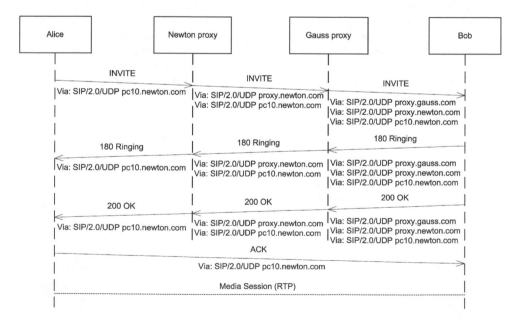

Figure 4.4 Session initiation example with the SIP trapezoid.

directly to that address (specified in the Request-URI of the ACK message) and thereby bypassing the proxy servers. This allows the proxy servers to 'drop out' of the call flow – in essence their work to provide routing to set up the call is complete at this point.

The example in Figure 4.4 illustrates how the Via header plays a crucial role in routing response messages. As the request makes its way to its final destination, each hop is recorded within the message thus providing sufficient information to allow the same return path to be used. In some cases, the proxy server may wish to stay in the signalling path for the duration of the call. The flow in Figure 4.4 assumes that the IP network provides end-to-end connectivity. This may not be case if there are firewalls or Network Address Translators (NAT) in place – the proxy may need to stay in the signalling path to facilitate end-to-end signalling. A proxy can insert itself into subsequent signalling flows by putting its address into a Record-Route header field. Endpoints inspect the Record-Route header(s) in both received requests and received responses to learn the ordered set of proxy addresses that must be traversed for new requests within the same dialog. The ordered set of proxies is called a 'route set'. The route set is placed in the Route header(s) for each new request within the dialog. Figure 4.5 illustrates the same example as before but this time both proxies remain in the signalling path (the Record-Route and Route headers are shown and the Via headers apply as before but are not shown). Each proxy adds its own address to a new Record-Route header field. The lr parameter indicates 'loose routing' and is used to differentiate the routing mechanism in RFC 3261 [2] from previous mechanisms. Through this mechanism, both Alice's and Bob's user agents learn of the route set comprising the two proxies.

Alice's user agent inserts the route set into the ACK message (a new request) using the Route header field. As the request propagates through each proxy, the proxy checks if a Route header field is present containing its own address, removes it, and forwards the request to the address in the next Route header. When the message reaches the Gauss proxy, it notes it is not responsible for the domain

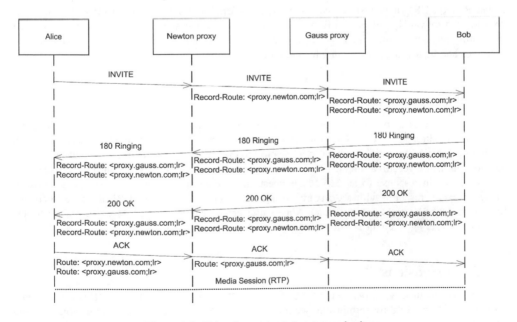

Figure 4.5 Using the Record-Route mechanism

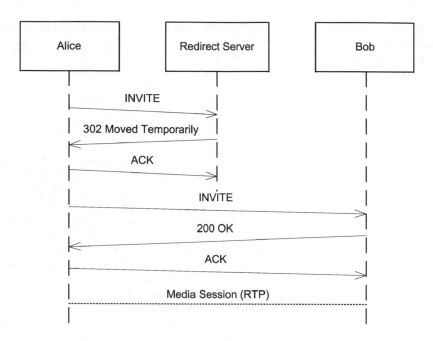

Figure 4.6 SIP redirection example.

in the Request-URI (it is responsible for the `gauss.com` domain not the `host2.gauss.com` domain) and forwards it directly to Bob's user agent. The `Route` header plays a similar role to `Via` for routing messages – the important distinction, however, is that `Route` is used for routing request messages and `Via` is used for routing response messages.

4.6.3 Redirect servers

A redirect server is a special kind of UAS that responds to requests with a $3xx$ redirect response containing an alternate set of URIs on which to contact the client. Redirection allows servers to push routing information for a request back in a response to the client, thereby taking itself out of the loop for this transaction yet still aiding in locating the target of the request. Figure 4.6 illustrates an example of redirection. In response to an `INVITE` request, the redirect server responds with a `302 Moved Temporarily` response with a `Contact` header field set to the Request-URI of where to contact Bob (`sip:bob@host2.gauss.com`). Alice's user agent generates a new `INVITE` request to this address (the user agent client is said to 'recurse' on the $3xx$ response).

4.7 SIP extensions

There are six method types defined in RFC 3261 [2]: `INVITE`, `ACK` and `CANCEL` associated with setting up sessions, `BYE` for terminating sessions, `REGISTER` for registering contact information, and `OPTIONS` for querying servers about their capabilities. Other RFCs specify common extensions to

SIP. Extensions for SIP event notifications are specified in RFC 3265 [18] (defines the SUBSCRIBE and NOTIFY methods). A mechanism for referring a user agent to another resource (e.g. to perform call transfer) is specified in RFC 3515 [19] (defines the REFER method). A mechanism for supporting instant messaging and presence is specified in RFC 3428 [20] and RFC 3903 [23] (defines the MESSAGE and PUBLISH methods). An extension for delivering provisional responses reliably is specified in RFC 3262 [21] (defines the PRACK method). A mechanism for updating parameters of a session before the INVITE has completed is specified in RFC 3311 [24] (defines the UPDATE method). An extension for carrying session related control information (e.g. ISDN or ISUP signalling messages) is specified in RFC 2976 [25] (defines the INFO method).

SIP supports general extensibility through the use of the Supported, Require, and Unsupported headers. A UAC advertises that it supports a SIP extension that can be applied in the response to a request by including a Supported header in the request. The Supported header specifies a list consisting of one or more 'option tags'. For example, the option tag for PRACK is 100rel and thus a user agent that supports this extension would include the following header in an INVITE request:

Supported: 100rel

If the UAC wants to insist that the UAS understands an extension, it may add the Require header, listing the option tags that must be supported to process that request. If the UAS does not support the extension, it can reply with a 420 Bad Extension response including an Unsupported header listing the option tag(s) not supported. Finally, the Supported header field may be returned in the response to an OPTIONS message to indicate supported SIP extensions (see next section).

4.7.1 Querying for capabilities

The standard SIP OPTIONS method is used for querying a user agent or proxy server for its capabilities. This allows the client to discover information related to supported extensions, codecs, etc., without actually placing a call. Figure 4.7 illustrates a typical successful flow.

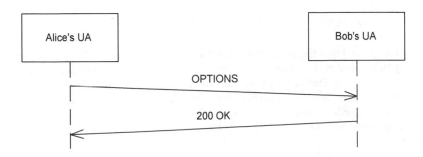

Figure 4.7   Querying for capabilities with SIP.

An example OPTIONS request is illustrated below:

```
OPTIONS sip:bob@gauss.com SIP/2.0
Via: SIP/2.0/UDP pc10.newton.com;branch=z9hG4bK1ea
Max-Forwards: 70
To: Bob <sip:bob@gauss.com>
From: Alice <sip:alice@newton.com>;tag=14124
Call-ID: a8a931aa@pc10.newton.com
CSeq: 1000 OPTIONS
Contact: <sip:alice@pc10.newton.com>
Content-Length: 0
```

An example response to the OPTIONS request is:

```
SIP/2.0 200 OK
Via: SIP/2.0/UDP pc10.newton.com;branch=z9hG4bK1ea
 ;received=10.0.0.1
Max-Forwards: 70
To: Bob <sip:bob@gauss.com>;tag=3843
From: Alice <sip:alice@newton.com>;tag=14124
Call-ID: a8a931aa@pc10.newton.com
CSeq: 1000 OPTIONS
Contact: <sip:bob@host2.gauss.com>
Allow: INVITE, ACK, CANCEL, OPTIONS, BYE
Accept: application/sdp
Supported: 100rel
Content-Type: application/sdp
Content-Length: 146
```

```
v=0
o=bob 3812844327 3812844327 IN IP4 host2.gauss.com
s=-
t=0 0
c=IN IP4 host2.gauss.com
m=audio 0 RTP/AVP 0 8
a=rtpmap:0 PCMU/8000
a=rtpmap:8 PCMA/8000
```

The Allow header field lists the SIP methods that Bob's user agent accepts. The Accept header field indicates that Bob's user agent accepts SDP. The Supported header field indicates that the option tag 100rel is supported. The media capabilities are advertised in the SDP. Note that the m-line port number is set to 0 to prevent media streams inadvertently being set up. Bob's user agent indicates that it supports two audio codecs, namely G.711 PCM μ-law and A-law.

The OPTIONS request can be issued either within a SIP dialog or outside of a SIP dialog. When an OPTIONS request is issued within an existing SIP dialog, the response is always 200 OK and has no effect on the dialog itself. When an OPTIONS request is issued outside of a dialog, the response code returned is the same as would have occurred if an INVITE request was made instead. For example, a

`200 OK` would be returned if the UAS is ready to accept a call, a `486 Busy Here` would be returned if the UAS is busy, etc. This allows the UAC also to use the `OPTIONS` request to determine the basic call state of the UAS before making an actual call attempt via `INVITE`.

4.8 Security

In this section, we take a brief look at security mechanisms employed in SIP signalling. Security issues related to media and control sessions are covered in Chapter 6 and Chapter 7 as part of the general discussions there. While full encryption of messages certainly provides the best means to ensure confidentiality and integrity of signalling, SIP requests and responses cannot be simply encrypted end-to-end in their entirety because message fields such as the Request-URI, `Route` and `Via` need to be visible to proxies to facilitate correct routing of SIP requests and responses. Proxy servers must therefore be trusted to some degree by SIP user agents. To this end, SIP makes use of low-level transport and network security mechanisms on a hop-to-hop basis. This mechanism also allows endpoints to verify the identity of proxy servers to whom they send requests. In addition, SIP provides an authentication mechanism based on HTTP digest authentication to enable SIP entities to identify one another securely. Finally, SIP borrows a technique called S/MIME from the Simple Mail Transport Protocol (SMTP) to encrypt MIME bodies and provide end-to-end confidentiality and integrity. We look at each of these three techniques in the following subsections.

4.8.1 Transport and network layer security

Transport or network layer security encrypts signalling traffic, guaranteeing message confidentiality and integrity. In addition, certificates can be used in the establishment of lower-layer security, and these certificates can also be used to provide a means of authentication.

TLS provides for transport layer security over connection-oriented protocols – SIP uses TLS over TCP. TLS is most suited to providing hop-to-hop security between hosts with no pre-existing trust association.

Network layer security can be implemented at the operating system level or through a security gateway that provides confidentiality and integrity for all traffic it receives from a particular interface (as with Virtual Private Network, VPN, architectures). IP Security (IPSec) is a set of protocols developed by the IETF that collectively provides for a secure replacement of traditional IP and facilitates network layer security. Network layer security is often used in architectures where sets of hosts or administrative domains have an existing trust relationship with each other.

4.8.2 Authentication

SIP borrows HTTP's digest challenge–response authentication mechanism to provide authentication between a user agent and a registrar, proxy server, redirect server, or another user agent. Any time a SIP server or another user agent receives a request, it may challenge the issuer of the request to provide assurance of identity. Once the originator has been identified, the recipient of the request can ascertain whether or not the user is authorised to make the request. Digest access authentication verifies that both parties to a communication know a shared secret (a password) but avoids sending the password in the clear (to avoid interception and discovery of the password).

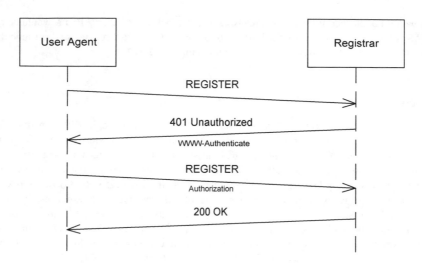

Figure 4.8 Registration with authentication.

When a request is made to a registrar, redirect server or another user agent that wishes to authenticate the originator of the request, a `401 Unauthorized` response is returned including the `WWW-Authenticate` header field to challenge the identity of the issuer of the request. The `WWW-Authenticate` header includes, among other things, the name of the security realm and a so-called nonce value. The realm string identifies the protection zone. The nonce value is used to create a response to the challenge by creating a checksum (also called the digest) built from the nonce value, the username, the password, and other parameters. This way, the password is never passed in the clear. The SIP request is reissued with an `Authorization` header field containing the response to the challenge. The reissued request uses the same `Call-ID` but increments the `CSeq` value. Proxy servers use the same mechanism to challenge issuers of requests but instead return a `407 Proxy Authentication Required` message in place of the `401 Unauthorized` response and with a `Proxy-Authorization` header field in place of the `WWW-Authenticate` header field. Figure 4.8 illustrates an example of a registrar using authentication.

4.8.3 S/MIME

S/MIME allows user agents to encrypt message bodies without otherwise affecting the SIP headers. This allows network intermediaries such as proxy servers to view certain header fields in order to route messages correctly while, at the same time, allowing the message body to be encrypted. S/MIME [48] provides end-to-end confidentiality and integrity for message bodies in addition to mutual authentication by using S/MIME certificates. S/MIME bodies are identifiable by the `multipart/signed` and `application/pkcs7-mime` content types.

4.9 Summary

In this chapter, we introduced SIP and SDP, which are key specifications leveraged by MRCP to establish and manage its media and control sessions with media resource servers. The introduction to SIP in this chapter focused on key concepts and those features relevant to MRCP. Interested readers who wish to learn more about SIP can find many good books on the subject [28, 29] or alternatively may wish to consult directly the RFCs referenced in this chapter (RFC 3261 [2] is a recommended starting point). Armed with sufficient knowledge of SIP, we are now ready to move to the next chapter, which focuses on the specifics of SIP applied within the MRCP framework.

5

Session initiation in MRCP

Building upon the last chapter, which introduced the Session Initiation Protocol (SIP), this chapter discusses the specific details of how SIP is used within the MRCP framework for session management. Several examples of real-world cases are given. This chapter also discusses mechanisms available to the MRCP client for locating media resource servers.

5.1 Introduction

The generalised MRCP architecture is illustrated in Figure 3.1. SIP is used to establish media and control sessions between the MRCP client and media resource server. The standard three-way `INVITE-200 OK-ACK` handshake is used to establish media and control sessions and the standard `BYE-200 OK` exchange is used to terminate the sessions. The SDP offer/answer model (see Chapter 4) is used to negotiate characteristics of the media and control sessions and is the subject of Sections 5.2 and 5.3.

5.2 Initiating the media session

The media session consists of one or more media streams and is established in the same way as per the standard IP telephony mechanisms discussed in Chapter 4. Examples of flows for both a co-located and a distributed media source/sink are given in Section 5.4.

5.3 Initiating the control session

The control session consists of one or more logical control channels (each logical control channel is distinguished by its unique `Channel-Identifier`). MRCP leverages the 'TCP-Based Media Transport in the Session Description Protocol' described in RFC 4145 [31] to establish a TCP

---

control channel to the media resource server. TCP provides a reliable conduit for MRCP protocol messages. TLS may be used for secure control sessions, in which case the related specification [32] is used. Below is an SDP offer excerpt consisting of typical media level information for a control channel:

```
c=IN IP4 10.0.0.1
m=application 9 TCP/MRCPv2
a=setup:active
a=connection:new
a=resource:speechsynth
a=cmid:1
```

This SDP is part of an offer from the MRCP client to the media resource server, e.g. contained in the message body of the SIP INVITE message. In this example, the MRCP wishes to allocate a speech synthesiser resource on the media resource server. The control session is identified with one or more m-lines, with the media type sub-field set to application (recall that the media session uses a sub-field type of audio). The IP address of the MRCP client is contained in the preceding c-line. Each m-line represents a single control channel with a one-to-one mapping with a media resource. The transport sub-field is TCP/MRCPv2 indicating the MRCPv2 protocol running over TCP transport (for TLS, the transport sub-field is TCP/TLS/MRCPv2). The port number from the MRCP client is always either 9 or 0. The port number 9 is known as the 'discard' port and is simply a placeholder used by the MRCP client (since the port that will be used by the client is not yet known nor particularly relevant to the media resource server). The port number 0 has special significance: it indicates that the channel is to be disabled. An MRCP client can issue an SDP offer with port number 0 to disable an MRCP control channel (a port number of 0 is also used to disable media streams as discussed in Chapter 4). Several a-line attributes are associated with the control channel's m-line. The setup attribute indicates which of the end points should initiate the TCP connection. The MRCP client always initiates the connection and thus the SDP offer employs a value of active – the corresponding answer from the media resource server will use a value of passive. The connection attribute specifies whether a new TCP connection is to be established (new) or whether an existing one is to be used (existing). MRCP allows multiple control channels to share the same connection since messages are labelled with a unique channel identifier. The MRCP client indicates the type of media resource it wishes to allocate and connect to via the resource attribute (media resource types are defined in Chapter 3). Finally, the cmid attribute relates this control channel to a media stream. The cmid attribute value must match a mid attribute value belonging to a media stream (i.e. associated with an m-line with a media type sub-field of audio). Note that it is possible for multiple control channels to use the same cmid attribute value, thus indicating that more than one control channel applies to a given media stream. This would occur if a speech synthesiser and speech recogniser were allocated in the same SIP dialog and used a single sendrecv media stream, for example.

An excerpt from the corresponding SDP answer from the media resource server, e.g. obtained from the SIP 200 OK message to response to the INVITE, is illustrated below:

```
c=IN IP4 10.0.0.22
m=application 43251 TCP/MRCPv2
a=setup:passive
a=connection:new
a=channel:43b9ae17@speechsynth
a=cmid:1
```

The `m=application` line indicates the server port number (43251) to which the MRCP client must initiate a connection. The corresponding IP address is specified in the preceding c-line (10.0.0.22). The `setup` attribute indicates `passive` – this is always the case for the answer from a media resource server. The `connection` attribute value of `new` indicates that the media resource server wishes a new connection to be used. If the MRCP client specifies a value of `new`, the media resource server must always respond with a value of `new`. If, however, the MRCP client specifies a value of `existing`, the media resource server may respond with `existing` if it wishes the MRCP client to reuse an existing connection, or alternatively with a value of `new` if it wishes the MRCP client to create a new connection. The `channel` attribute value specifies the unique control channel identifier consisting of a string prefix that identifies the MRCP session, a '@' character, and the media resource type. The string prefix is the same for all resources allocated within the same SIP dialog. The channel identifier appears in the `Channel-Identifier` header field in all MRCP messages, thus enabling the media resource server to determine which resource and media stream an MRCP message applies to. The `cmid` and corresponding `mid` attributes remain unchanged in the answer.

5.4 Session initiation examples

This section provides some common examples of session initiation in MRCP.

5.4.1 Single media resource

This example considers the case where an MRCP client allocates a speech synthesiser resource. For simplicity, we assume the MRCP client and media source/sink are co-located. Figure 5.1 illustrates the sequence of SIP messages used to establish the media and control sessions.

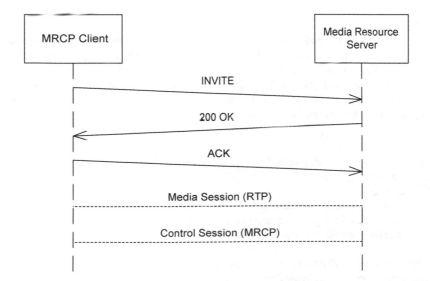

Figure 5.1   Media and control session initiation.

The INVITE message is as follows:

```
INVITE sip:mrcpv2@example.com SIP/2.0
Via: SIP/2.0/UDP host1.example.com;branch=z9hG4bKabc
Max-Forwards: 70
To: <sip:mrcpv2@example.com>
From: <sip:client@example.com>;tag=12425
Call-ID: 43fb8aec@host1.example.com
CSeq: 1 INVITE
Contact: <sip:client@host1.example.com>
Content-Type: application/sdp
Content-Length: 264


v=0
o=client 2890844527 2890844527 IN IP4
host1.example.com
s=-
c=IN IP4 host1.example.com
t=0 0
m=audio 5324 RTP/AVP 0
a=rtpmap:0 PCMU/8000
a=recvonly
a=mid:1
m=application 9 TCP/MRCPv2
a=setup:active
a=connection:new
a=resource:speechsynth
a=cmid:1
```

Looking at the SDP message body, there are two m-lines. The first, concerned with the media session, is used to establish a media stream in the direction of the MRCP client (a=recvonly) and is identified with a mid attribute value of 1. The second m-line is concerned with the control session. A speechsynth resource is requested. The MRCP client has elected to use a new connection for this control channel (a=connection:new). The control channel is related to the media stream via the cmid attribute. The 200 OK response message is illustrated below:

```
SIP/2.0 200 OK
Via: SIP/2.0/UDP host1.example.com;branch=z9hG4bKabc
 ;received=192.168.1.11
Max-Forwards: 70
To: <sip:mrcpv2@example.com>;tag=98452
From: <sip:client@example.com>;tag=12425
Call-ID: 43fb8aec@host1.example.com
CSeq: 1 INVITE
Contact: <sip:mrcpv2@host100.example.com>
Content-Type: application/sdp
Content-Length: 274
```

```
v=0
o=server 31412312 31412312 IN IP4 host100.example.com
s=-
c=IN IP4 host100.example.com
t=0 0
m=audio 4620 RTP/AVP 0
a=rtpmap:0 PCMU/8000
a=sendonly
a=mid:1
m=application 9001 TCP/MRCPv2
a=setup:passive
a=connection:new
a=channel:153af6@speechsynth
a=cmid:1
```

The media resource server confirms that it will be sending audio for the media session. The second
m-line indicates the server port for the MRCP control channel for the speech synthesiser resource.
The `channel` attribute specifies the unique channel identifier corresponding to the allocated media
resources. This channel identifier will appear in MRCP messages to identify the media resource target
of those messages. Finally, the MRCP client acknowledges the final response to the INVITE with an
ACK message:

```
ACK sip:mrcpv2@host100.example.com SIP/2.0
Via: SIP/2.0/UDP host1.example.com;branch=z9hG4bK214
Max-Forwards: 70
To: <sip:mrcpv2@example.com>;tag=98452
From: <sip:client@example.com>;tag=12425
Call-ID: 43fb8aec@host1.example.com
CSeq: 1 ACK
Contact: <sip:mrcpv2@host1.example.com>
Content-Length: 0
```

5.4.2 Adding and removing media resources

This example builds on the previous one to illustrate how a media resource may be added to a session
temporarily and subsequently removed through the use of re-INVITE messages to change the session
characteristics. Recall that the re-INVITE message enables the session characteristics to be changed
for an existing SIP dialog. The MRCP client wishes to allocate a `recorder` resource to record a
message. Figure 5.2 summarises the message flow:

The MRCP client issues a re-INVITE message:

```
INVITE sip:mrcpv2@host100.example.com SIP/2.0
Via: SIP/2.0/UDP host1.example.com;branch=z9hG4bK452
Max-Forwards: 70
To: <sip:mrcpv2@example.com>;tag=98452
From: <sip:client@example.com>;tag=12425
```

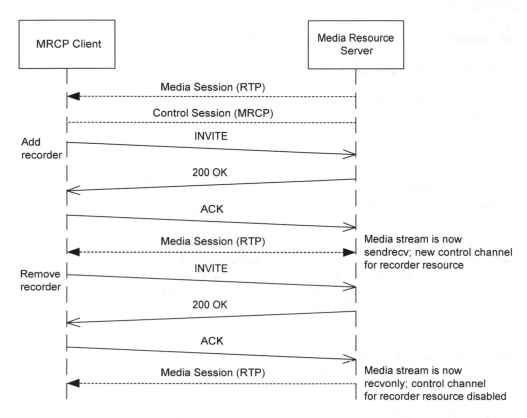

Figure 5.2 Using re-INVITE to allocate and de-allocate a resource (the direction of the media flow is illustrated to highlight mid-session changes).

```
Call-ID: 43fb8aec@host1.example.com
CSeq: 2 INVITE
Contact: <sip:client@host1.example.com>
Content-Type: application/sdp
Content-Length: 367

v=0
o=client  2890844527  2890844528  IN  IP4
host1.example.com
s=-
c=IN IP4 host1.example.com
t=0 0
m=audio 5324 RTP/AVP 0
a=rtpmap:0 PCMU/8000
a=sendrecv
a=mid:1
m=application 9 TCP/MRCPv2
a=setup:active
```

```
a=connection:existing
a=resource:speechsynth
a=cmid:1
m=application 9 TCP/MRCPv2
a=setup:active
a=connection:existing
a=resource:recorder
a=cmid:1
```

The re-INVITE message looks almost identical to an initial INVITE – the giveaway sign that this is a re-INVITE is the presence of tag parameters on both the To and From header fields indicating that this message is part of an existing SIP dialog. Note that the offer/answer model requires that the SDP fully describes the characteristics of the desired session and not just the change to the session. The version number (third sub-field) on the o-line is incremented by one to indicate a change in the session. The media session requires a small update to the media session: the MRCP client uses the sendrecv attribute in place of a recvonly attribute. When the MRCP client previously established a media stream for the speechsynth resource, it only required it to be recvonly. Now, with the addition of a recorder resource, audio must be sent to the media resource server and hence the MRCP client specifies the bi-directional sendrecv attribute. Next, the m-line for the speechsynth control channel is repeated. Note that the connection attribute specifies using the existing connection (the MRCP client wishes to make no change to the existing control channel for the speech synthesiser resource it previously allocated). The third m-line is new and specifies a control channel for a new recorder resource. The MRCP client requests using an existing connection for this resource. The recorder resource will share the same media stream being used by the speechsynth resource as is evident by the fact that both control channels share the same cmid value.

The media resource server accepts the updated session characteristics, and responds with a 200 OK message:

```
SIP/2.0 200 OK
Via: SIP/2.0/UDP host1.example.com;branch=z9hG4bK452
 ;received=192.168.1.1
To: <sip:mrcpv2@example.com>;tag=98452
From: <sip:client@example.com>;tag=12425
Call-ID: 43fb8aec@host1.example.com
CSeq: 2 INVITE
Contact: <sip:client@host1.example.com>
Content-Type: application/sdp
Content-Length: 387

v=0
o=client 31412312 31412313 IN IP4 host100.example.com
s=-
c=IN IP4 host100.example.com
t=0 0
m=audio 4620 RTP/AVP 0
a=rtpmap:0 PCMU/8000
a=sendrecv
a=mid:1
```

```
m=application 9001 TCP/MRCPv2
a=setup:passive
a=connection:existing
a=channel:153af6@speechsynth
a=cmid:1
m=application 9001 TCP/MRCPv2
a=setup:passive
a=connection:existing
a=channel: 153af6@recorder
a=cmid:1
```

The media resource server responds with the same channel identifier for the speechsynth resource, since no change has been made to this media resource allocation. The newly allocated recorder resource is assigned a unique channel identifier. The MRCP client responds with an ACK message:

```
ACK sip:mrcpv2@host100.example.com SIP/2.0
Via: SIP/2.0/UDP host1.example.com;branch=z9hG4bK554
Max-Forwards: 70
To: <sip:mrcpv2@example.com>;tag=98452
From: <sip:client@example.com>;tag=12425
Call-ID: 43fb8aec@host1.example.com
CSeq: 2 ACK
Contact: <sip:mrcpv2@host1.example.com>
Content-Length: 0
```

After recording the message, the MRCP client wishes to deallocate the recorder resource. It does so by issuing another re-INVITE:

```
INVITE sip:mrcpv2@host100.example.com SIP/2.0
Via: SIP/2.0/UDP host1.example.com;branch=z9hG4bK763
Max-Forwards: 70
To: <sip:mrcpv2@example.com>;tag=98452
From: <sip:client@example.com>;tag=12425
Call-ID: 43fb8aec@host1.example.com
CSeq: 3 INVITE
Contact: <sip:client@host1.example.com>
Content-Type: application/sdp
Content-Length: 367

v=0
o=client 2890844527 2890844529 IN IP4
host1.example.com
s=-
c=IN IP4 host1.example.com
t=0 0
m=audio 5324 RTP/AVP 0
a=rtpmap:0 PCMU/8000
```

```
a=recvonly
a=mid:1
m=application 9 TCP/MRCPv2
a=setup:active
a=connection:existing
a=resource:speechsynth
a=cmid:1
m=application 0 TCP/MRCPv2
a=setup:active
a=connection:existing
a=resource:recorder
a=cmid:1
```

The recorder resource is deallocated simply by specifying a port number of 0 on the appropriate
m-line. The media session is also reverted back to recvonly, since the MRCP client will now only
be receiving audio for this session. The media resource server accepts the session change and responds
with a 200 OK message:

```
SIP/2.0 200 OK
Via: SIP/2.0/UDP host1.example.com;branch=z9hG4bK763
 ;received=192.168.1.1
To: <sip:mrcpv2@example.com>;tag=98452
From: <sip:client@example.com>;tag=12425
Call-ID: 43fb8aec@host1.example.com
CSeq: 3 INVITE
Contact: <sip:client@host1.example.com>
Content-Type: application/sdp
Content-Length: 384

v=0
o=client 31412312 31412314 IN IP4 host100.example.com
s=-
c=IN IP4 host100.example.com
t=0 0
m=audio 4620 RTP/AVP 0
a=rtpmap:0 PCMU/8000
a=sendonly
a=mid:1
m=application 9001 TCP/MRCPv2
a=setup:passive
a=connection:existing
a=channel:153af6@speechsynth
a=cmid:1
m=application 0 TCP/MRCPv2
a=setup:passive
a=connection:existing
a=channel: 153af6@recorder
a=cmid:1
```

The media resource server confirms the deallocation of the recorder resource by also specifying a port number of zero on its m-line. It also confirms the change in direction of the media session, which is `sendonly` from its perspective. Finally, the MRCP client responds with an ACK message:

```
ACK sip:mrcpv2@host100.example.com SIP/2.0
Via: SIP/2.0/UDP host1.example.com;branch=z9hG4bK432
Max-Forwards: 70
To: <sip:mrcpv2@example.com>;tag=98452
From: <sip:client@example.com>;tag=12425
Call-ID: 43fb8aec@host1.example.com
CSeq: 3 ACK
Contact: <sip:mrcpv2@host1.example.com>
Content-Length: 0
```

At the end of the call, the MRCP client can simply issue a BYE message to teardown the remaining media and control sessions (not shown).

5.4.3 Distributed media source/sink

This example looks at the case where the media source / sink is distributed from the MRCP client. Figure 5.3 illustrates a typical flow.

The media source/sink might be a SIP phone, a media gateway, etc. The MRCP client behaves as a back-to-back user agent (B2BUA). In other words, it presents one SIP user agent to the media source/sink and another SIP user agent to the media resource server. The media session flows directly between the media source/sink and media resource server. The control session is established between

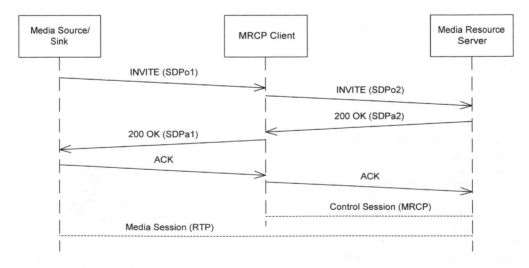

Figure 5.3 Session initiation with a distributed media source/sink.

the MRCP client and media resource server. The SIP messages themselves are standard and are not repeated here. Rather, the key to establish such a configuration of media session and control session is the MRCP client's modification of the SDP messages passed across its two user agents.

The SDP offer from the media source/sink (SDPo1) is illustrated below:

```
v=0
o=Bob 31245351 31245352 IN IP4 pc02.newton.com
s=-
c=IN IP4 pc02.newton.com
t=0 0
m=audio 5632 RTP/AVP 8
a=rtpmap:8 PCMA/8000
a=sendrecv
```

The MRCP client wishes to accept the INVITE message and bring a speech synthesiser resource and speech recogniser resource into the session. It generates a new SDP offer (SDPo2) using the media characteristics from SDPo1:

```
v=0
o=Client 43523532 43523532 IN IP4 host01.example.com
s=-
t=0 0
m=audio 5632 RTP/AVP 8
c=IP IP4 pc02.newton.com
a=rtpmap:8 PCMA/8000
a=sendrecv
a=mid:1
m=application 9 TCP/MRCPv2
c=IN IP4 host01.example.com
a=setup:active
a=connection:existing
a=resource:speechsynth
a=cmid:1
m=application 9 TCP/MRCPv2
c=IN IP4 host01.example.com
a=setup:active
a=connection:existing
a=resource:speechrecog
a=cmid:1
```

Note that each m-line has an associated c-line. The c-line associated with the audio m-line specifies the address of the media source/sink while the c-line associated with the MRCP control channels specifies the address of the MRCP client. The port number on the audio m-line is the same port as offered by the media source/sink. In this offer, the MRCP client wishes to use a speechsynth and speechrecog resource via an existing TCP connection. Both resources share the bi-directional media stream directly to the media source/sink.

The media resource server responds with a 200 OK message with the following SDP in the message body (SDPa2):

```
v=0
o=Server 15454326 15454326 IN IP4 host99.example.com
s=-
t=0 0
m=audio 87242 RTP/AVP 8
c=IP IP4 host99.example.com
a=rtpmap:8 PCMA/8000
a=sendrecv
a=mid:1
m=application 56723 TCP/MRCPv2
c=IN IP4 host99.example.com
a=setup:passive
a=connection:existing
a=channel:42a6f3e9@speechsynth
a=cmid:1
m=application 56723 TCP/MRCPv2
c=IN IP4 host99.example.com
a=setup:passive
a=connection:existing
a=channel: 42a6f3e9@speechrecog
a=cmid:1
```

The first m-line specifies the port number for the media source/sink to send audio to. The two subsequent m-lines relate to the control channels for the speechsynth and speechrecog resources respectively and specify the server port the client had previously connected to. The channel attributes specify the channel identifiers for the newly allocated resources.

The MRCP client then responds with a 200 OK to the media source/sink with SDPa1 that is based on aspects of SDPa2:

```
v=0
o=Client 24356331 24356331 IN IP4 host01.example.com
s=-
c=IP IP4 host99.example.com
t=0 0
m=audio 87242 RTP/AVP 8
a=rtpmap:8 PCMA/8000
a=sendrecv
```

Note that the c-line specifies the address of the media resource server. As far as the media source/sink is concerned, it has established a regular IP telephony call with the MRCP client where the SIP flow and media flow have different target IP addresses. The media source/sink responds with an ACK message and the MRCP client generates a separate ACK to the media resource server. There is now a media session directly between the media source/sink and media resource server and a control session between the MRCP client and media resource server.

5.5 Locating media resource servers

In general, there will be more than one media resource server available to the MRCP client for reliability and scalability. It is also conceivable that different media resource servers will have differing capabilities. In this section, we look at standard SIP mechanisms that can be used to determine the capabilities of the media resource server and to distribute the load from MRCP clients over multiple media resources servers.

5.5.1 Requesting server capabilities

The MRCP client may issue an OPTIONS request to the media resource server to query its capabilities (see also Chapter 4). This could be useful, for example, when the MRCP client has access to URIs to several media resource servers and wishes to establish their capabilities to influence later routing decisions. The SDP in the OPTIONS response indicates the media resource types supported in addition to the usual description of media capabilities such as audio codecs. An example of an OPTIONS request from an MRCP client is:

```
OPTIONS sip:mrcpv2@example.com SIP/2.0
Via: SIP/2.0/UDP host1.example.com;branch=z9hG4bKab3
Max-Forwards: 70
To: <sip:mrcpv2@example.com>
From: <sip:client@example.com>;tag=21342
Call-ID: 12fa3421@host1.example.com
CSeq: 21342 OPTIONS
Contact: <sip:client@host1.example.com>
Accept: application/sdp
Content-Length: 0
```

In response, one might see a message similar to:

```
SIP/2.0 200 OK
Via: SIP/2.0/UDP host1.example.com;branch=z9hG4bKab3
Max-Forwards: 70
To: <sip:mrcpv2@example.com>;tag=32421
From: <sip:client@example.com>;tag=21342
Call-ID: 12fa3421@host1.example.com
CSeq: 21342 OPTIONS
Contact: <sip:client@host99.example.com>
Allow: INVITE, BYE, OPTIONS, ACK, CANCEL
Content-Type: application/sdp
Content-Length: 288

v=0
o=Server 289123140 289123140 IN IP4 host99.example.com
s=-
t=0 0
m=application 0 TCP/MRCPv2
a=resource:speechsynth
```

```
a=resource:speechrecog
a=resource:recorder
m=audio 0 RTP/AVP 0 8 96
a=rtpmap:0 PCMU/8000
a=rtpmap:8 PCMA/8000
a=rtpmap:96 telephone-event/8000
a=fmtp:96 0-15
```

This response indicates that the media resource server supports three media resource types: the speechsynth, speechrecog, and recorder resources. The media resource server supports audio in PCM μ-law and A-law formats, and supports RFC 2833 telephone events (see Chapter 6 for more information on the media session).

5.5.2 Media resource brokers

A Media Resource Broker (MRB) can be inserted between an MRCP client and a pool of media resource servers to provide routing functions. An *inline* MRB is based on the SIP proxy role – see Figure 5.4. At its simplest, the MRB can query the location service database for the Request-URI specified in the INVITE request (e.g. sip:mrcp@production.example.com) and retrieve a set of Contact addresses, one for each media resource server (the media resource servers would have each previously REGISTERed the address-of-record sip:mrcp@production.example.com along with their individual Contact address binding). Given a set of Contact addresses, the MRB could use a selection strategy such as round-robin or random-robin to distribute the load over the set of available media resource servers. In the event of a failure (e.g. a non-*2xx* SIP response), the MRB could choose another Contact address to provide transparent failover.

As an alternative to the inline MRB, a redirect server can be used as the basis of a *query* MRB – see Figure 5.5. In this case, the redirect server responds with a 302 Moved Temporarily response

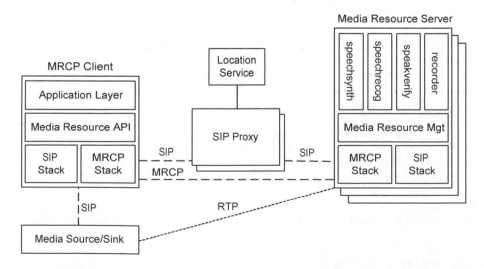

Figure 5.4 Using a SIP proxy as an inline MRB in MRCP architectures.

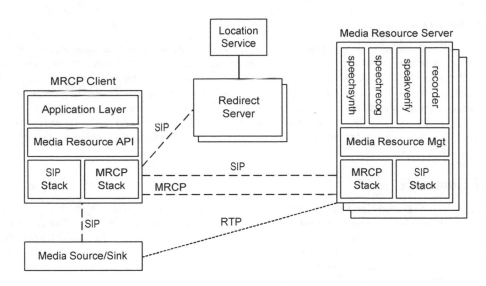

Figure 5.5   Using a SIP redirect server as a query MRB in MRCP architectures.

indicating the URI of the optimal media resource server to contact in the Content header. Again, a load balancing strategy such as round-robin or random-robin could be used to select the URI.

So far we have tacitly assumed that the pool of media resource servers registered to the same address-of-record provides the same set of capabilities. This may not necessarily be the case in reality. For example, one media resource server may incorporate a speech synthesiser and speech recogniser resource while another may include only a speech recogniser. To support heterogenous pools of media resource servers, we require a way for the MRCP client to indicate its preferences in influencing the routing, e.g. the MRCP client might indicate that the media resource server must incorporate a recorder resource. Similarly, we also need a mechanism to allow the media resource servers to advertise their capabilities when they register with the location service. In principle, it would be possible for the MRB to issue OPTION requests to different URIs in the location service database to learn of different capabilities. On receipt of an INVITE from an MRCP client, the MRB could inspect the SDP offer to determine an appropriate match of server in terms of capabilities. This mechanism is not ideal since the SDP body may not be accessible to the MRB (e.g. it could be encrypted using S/MIME).

Another solution to capability-based routing is to use SIP extensions designed precisely for this purpose. RFC 3841 [46], 'Caller Preferences for the Session Initiation Protocol (SIP)', provides a mechanism to allow a SIP user agent client (such as the one in the MRCP client) to express preferences for the capabilities of the SIP user agent server (such as the media resource server) that should be reached as a result of sending a SIP INVITE request. RFC 3840 [45], 'Indicating User Agent Capabilities in the Session Initiation Protocol (SIP)', enables the SIP user agent server (i.e. media resource server) to convey its feature set to the registrar of the domain. The basic idea is to append URI parameters called feature tags that express capabilities to the Content URI during registration. The MRCP client can indicate preferences by including an Accept-Contact header field specifying the desired features in the INVITE request. The MRB can subsequently consult the location service to selecting the URI of an appropriate media resource server, based on the preferences indicated by the MRCP client and the feature set registered by the media resource server. A set of feature tags for MRCP has been proposed [47].

5.6 Security

One of the benefits MRCP enjoys from using SIP for session establishment and management is SIP's provision of security mechanisms. All the security threats that apply to SIP generally also apply to MRCP. For example, an attacker impersonating a media resource server with a recorder resource could be used to obtain sensitive information from a user; denial of service attacks could render a media resource server defunct; a proxy server maliciously modifying SDP message bodies could direct audio to a wiretap device, etc. MRCP specifically requires that MRCP clients and media resource servers support SIP's digest authentication mechanism and SIPs transport layer security mechanism using TLS.

5.7 Summary

This chapter, in conjunction with the previous one, introduced the mechanisms by which the media sessions and control sessions within MRCP are established. SIP, together with the SDP offer/answer model, provides flexible session establishment and management features within the MRCP framework. We also described how SIP extensions for indicating user agent capabilities and caller preferences can be exploited to provide intelligent routing of MRCP client requests to appropriate media resource servers.

With a firm understanding of how the sessions are established in MRCP, we are now ready to focus on the low-level details of the media and control sessions themselves. Chapter 6 discusses the media session and Chapter 7 covers the control session.

6

The media session

In this chapter we take a closer look at the media session established between the media source/sink and the media resource server. The principles here apply just as readily to regular IP telephony as they do to MRCP. We focus first on the media encoding formats and subsequently on how the media is transported in real-time between endpoints.

6.1 Media encoding

The goal of media coding is to represent audio or video in as few bits as possible while, at the same time, maintaining a reasonable level of quality. In the context of telecommunication systems, speech coding forms an important area of study within the general umbrella of media coding. There are many speech encoding standards in use in both traditional PSTN and newer IP networks today, and several share similar underlying algorithms. In the following subsections, we focus on those encoding schemes most relevant to MRCP and summarise other important codecs that are commonly found in IP networks where MRCP systems can be found.

6.1.1 Pulse code modulation

Without doubt, the ITU-T Recommendation G.711 [38] is the most common encoding used in conjunction with MRCP media resources. There are basically three reasons for this. The first is that G.711 has been the standard for digitising voice signals by telephone companies since the 1960s. The second is that G.711 is a high bit rate codec and therefore yields a good quality audio signal, thus aiding speech recognition and verification rates and perceived synthesised audio quality. Thirdly, since G.711 is not a complex codec, it presents little burden to transcode to or from linear audio at the media resource, thereby facilitating the use of standard computer hardware to process speech signals in real-time without recourse to specialised costly DSPs.

G.711 is a Pulse Code Modulation (PCM) scheme for digitising audio signals. Audio signals are sampled at a rate of 8 kHz (equivalent to once every 0.125 ms) and each sample is represented as a binary code. Figure 6.1 illustrates a signal sampled at a regular interval. According to the sampling theorem, signals must be sampled at a rate greater than twice the highest frequency component to

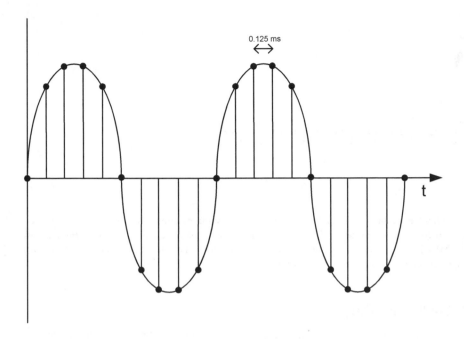

Figure 6.1 Sampling an analogue signal.

be properly reproduced. Thus, audio frequencies up to 4 kHz may be represented at a sampling rate of 8 kHz – sufficient to represent human speech with satisfactory quality. Contrast this with a high fidelity audio CD that employs sampled data at 44.1 kHz allowing frequencies as high as 22 kHz to be reproduced.

In addition to the sampling rate, the quantisation affects the quality of the signal. Quantisation refers to the approximation of the amplitude of a signal at a sampling instant by its closest available digital value. G.711 employs 8-bit resolution to represent each value at a sampling instant, corresponding to just 2^8 or 256 quantisation levels. The discrepancy between the actual value of the analogue signal at a sampling instant and its nearest, quantised, value results in distortion entering the digital representation of the signal. In fact, for uniform quantisation, the smaller the signal, the larger the quantisation noise contribution is relative to the signal. The PCM scheme employed by G.711 does not distribute the 256 levels linearly across the range of amplitudes but rather uses logarithmic quantisation (also known as companding). Low amplitudes are quantised with a high resolution (more quantisation levels) and high amplitudes are quantised with a low resolution (fewer quantisation levels). This choice is motivated by the observation that speech signals are statistically more likely to be near a low signal level than a high signal level. Logarithmic quantisation has the effect of evenly distributing the signal amplitude to quantisation noise ratio across the input range. Figure 6.2 illustrates how quantisation noise occurs and shows an example of non-uniform quantisation (dotted lines indicate quantisation levels).

G.711 specifies two encoding schemes known as μ-law and A-law, each using slightly different logarithmic functions. The μ-law encoding is used in the US and Japan while A-law is used in Europe and the rest of the world. When a voice signal traverses international boundaries, it is the responsibility of the country using μ-law to provide conversion to A-law. The μ-law encoding maps a 14-bit linear PCM sample into an 8-bit sample. The A-law encoding maps a 13-bit linear sample into an 8-bit

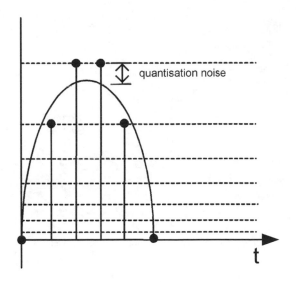

Figure 6.2   Quantisation noise and non-uniform quantisation.

sample. The encoding algorithms themselves don't actually apply a logarithmic function directly but rather approximate one by dividing the input range into so-called segments. Uniform quantisation is applied within a segment. Segments close to zero amplitude have small quantisation intervals and segments away from zero amplitude have large quantisation intervals with the quantisation interval doubling in each segment away from zero. A-law provides a larger segment around the zero amplitude level than does μ-law.

Generating 8-bits of data at a sampling rate of 8 kHz results in a bit rate of 64 kbit/s. A reduction in the bit rate can be obtained using Differential PCM (DPCM). This encoding looks at the difference between the value of the signal at successive sampling instants and only transmits the difference. Since the variance of differences is smaller than the variance of actual samples, they can be accurately quantised with fewer bits. At the decoder the quantised difference signal is added to the predicted signal to give the reconstructed speech signal. The ITU-T G.726 encoding harnesses DPCM in addition to an adaptive quantisation scheme known as Adaptive DPCM (ADPCM) to produce similar quality audio to G.711 but at half the bit rate (32 kbit/s).

6.1.2 Linear predictive coding

Linear predictive coding (LPC) is one of the most fundamental techniques used in speech coding and forms the basis of almost all modern low bit-rate codecs. In this section, we review the basics of LPC before introducing some important codecs. The core idea of linear prediction is that a given speech sample may be expressed as a linear combination of previous samples. Said differently, there is redundancy in the speech signal and this redundancy can be exploited to produce a compact representation of the signal. Analysed in the frequency domain, LPC can be used to construct a simplified model of speech production loosely based on anatomical details of the speech production system.

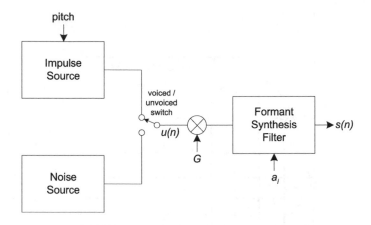

Figure 6.3   Acoustic filter model of speech synthesis.

As we saw in Chapter 2 (especially Figures 2.1 and 2.2), speech sounds are produced when air is forced from the lungs through the vocal cords, and passes through the vocal and nasal tracts. Speech production can be thought of in terms of acoustic filtering: the vocal tract (and nasal tract) behaves as a time-varying filter excited by the air forced into it through the vocal cords. The vocal tract shapes the spectrum and is resonant at certain frequencies called *formants*. Figure 6.3 illustrates a simplified model of speech synthesis. The vocal tract is modelled as a simple time-varying filter defined by a number of parameters a_i. This filter is also known as the formant synthesis filter since it mimics the effect that vocal tract has on shaping the spectrum. The excitation is either a periodic impulse train (corresponding to voiced excitation) or noise (corresponding to unvoiced excitation). The gain is specified by G.

Assuming a simple, all-pole structure for the formant synthesis filter, its output, $s(n)$, the synthesised speech signal can be given by:

$$s(n) = a_1 s(n-1) + a_2 s(n-2) + \ldots + a_m s(n-M) + G u(n)$$

We can estimate sample $\hat{s}(n)$ from a linear combination of M past samples:

$$\hat{s}(n) = a_1 s(n-1) + a_2 s(n-2) + \ldots + a_m s(n-M)$$

The prediction error, $e(n)$ is the difference

$$e(n) = s(n) - \hat{s}(n) = s(n) - \sum_{i=1}^{M} a_i s(n-i)$$

Minimising the mean square prediction error results in the optimum values for a_i (called the LPC parameters). Different algorithms exist for performing this step efficiently, for example the well-known Levinson–Durbin algorithm. Using the z-transform, we can restate this equation as:

$$S(z) = \frac{E(z)}{A(z)}$$

where $1/A(z)$ is the transfer function of the vocal tract filter and $A(z)$ is given by:

$$A(z) = 1 - \sum_{i=1}^{M} a_i z^{-i}$$

The parameter M is the called the order of the filter and typically takes on a value between 8 and 12. The problem of LPC is in estimating the parameters of the formant synthesis filter such that the spectrum of the output signal optimally matches the spectrum of the original signal given the chosen error signal $Gu(n)$. Speech signals are non-stationary by nature and must be analysed in short quasi-stationary periods of about 20–30 ms called a *frame*. Each frame is represented by a different set of LPC parameters.

The model presented in Figure 6.3 suffers from two major shortcomings. The first results from the fact that it is an oversimplification to say that speech sounds are simply either voiced or unvoiced. Secondly, the model does not capture phase information present in the speech signal because it simply attempts to match the frequency spectrum magnitude of the original signal. Modern codecs use a technique called Code Excited Linear Prediction (CELP[1]) to circumvent these shortcomings. Rather than using a two-state voiced / unvoiced excitation model, CELP codecs employ an excitation sequence given by an entry in a large vector quantised codebook in addition to a gain term to specify its power. Typically the codebook is represented by 10-bits to yield a size of 1024 entries and the gain is presented by a 5-bit term. The excitation sequences resemble white Gaussian noise. CELP codecs require a pitch synthesis filter to allow reproduction of voiced features in the speech signal. The pitch synthesis filter is placed in cascade with the formant synthesis filter. The structure of the pitch synthesis filter is similar to that of the formant synthesis filter presented except simpler. For the pitch synthesis filter, a sample value can be predicted from a scaled version of a single sample T units (the pitch period) in the past:

$$\hat{s}(n) = bs(n - T)$$

Conceptually, the parameters of the pitch filter are estimated using the residual signal obtained from subtracting the reconstructed signal using the formant synthesis filter from the original speech signal. This way, the pitch synthesis filter seeks to model the long-term behaviour of the speech signal while the formant synthesis filter captures the short-term characteristics.

To assist identification of the optimum excitation vector, and to encode some partial phase information, CELP codecs use a technique called *analysis-by-synthesis* during the encoding phase. Essentially, the synthesis part of the codec is included in the encoder and each excitation vector in the codebook is tested to find the one that results in the closest match between the reconstructed or synthesised speech and the original waveform. In practice, the error minimisation step uses a special weighting to emphasise parts of the spectrum where the energy is low since the human auditory system is much more sensitive to noise in parts of the spectrum where the energy of the speech signal is low.

Figure 6.4 illustrates the general encoding scheme. The input speech is divided into 20 ms or 30 ms frames. The LPC parameters for the formant synthesis filter are obtained for the frame. The frame is further divided into four sub-frames of length 5 ms or 7.5 ms. The LPC parameters for the pitch synthesis filter are obtained for each sub-frame and the excitation codebook entry also selected on a per-sub-frame basis (the length of each excitation sequence matches the length of the sub-frames). The index of the excitation codebook entry, gain value, long-term LPC parameters, and formant LPC parameters are encoded, packed, and transmitted as the CELP bit-stream.

---

[1] Pronounced as either 'kelp' or 'selp'

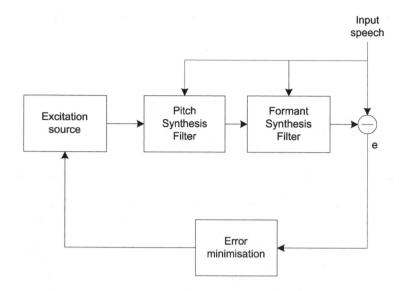

Figure 6.4   CELP encoding scheme.

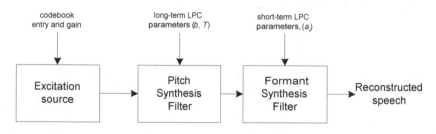

Figure 6.5   CELP decoding scheme.

Figure 6.5 illustrates the decoding scheme which uses the feed-forward path of the encoder to reconstruct the speech waveform.

The CELP codec has been successfully used in several standard codecs, producing equivalent quality to 64 kbits/s PCM at just 16 kbits/s and providing intelligible quality at bit-rates as low as 2.4 kbits/s. The following subsections briefly review some modern standard codecs used in both circuit-switched and packet-switched networks.

6.1.2.1 G.723.1

G.723.1 is an ITU-T recommendation [36] described as 'dual rate speech coder for multimedia communications transmitting at 5.3 and 6.3 kbit/s'. G.723.1 is part of the overall H.324 family of standards for multimedia communications using video phones over the PSTN. Third generation mobile networks also use a variant of these standards called H.324M that are specifically adapted for wireless networks. The algorithm has a silence compression algorithm in Annex A to G.723.1, a floating point specification in Annex B to G.723.1, and a scalable channel coding scheme for wireless applications in G.723.1 Annex C. With G.723.1, it is possible to switch between the two bit rates on a frame boundary – the lower bit rate yields good quality and the higher bit rate naturally results in better quality speech. G.723.1 encodes speech or other audio signals in frames using linear predictive analysis-by-synthesis coding. The excitation signal for the high rate coder is Multipulse Maximum Likelihood Quantisation (MP-MLQ) and for the low rate coder is Algebraic-Code-Excited Linear-Prediction (ACELP). Audio is sampled at a rate of 8 kHz and encoded in 30 ms frames, with an additional delay of 7.5 ms due to look-ahead. A G.723.1 frame can be one of three sizes: 24 octets (6.3 kbit/s), 20 octets (5.3 kbit/s), or 4 octets. The 4-octet frames are called SID frames (Silence Insertion Descriptor) and are used to specify comfort noise parameters. The Real-Time Protocol (RTP) payload format for G.723.1 is standardised in RFC 3551 [17] (see Section 6.2.1 for a discussion on RTP).

6.1.2.2 G.729

G.729 is an ITU-T Recommendation [37] described as 'coding of speech at 8 kbit/s using conjugate-structure algebraic-code-excited linear-prediction (ACELP)'. A reduced-complexity version of the G.729 algorithm is specified in Annex A to G.729 (known as G.729A) resulting in an approximate 50 % complexity reduction for a slight decrease in quality. Annex B to G.729 (known as G.729B) specifies a voice activity detection (VAD) and comfort noise generator (CNG) algorithm. Annexes D and E provide additional bit-rates – G.729D operates at a 6.4 kbit/s rate for handling momentary reductions in channel capacity and G.729E operates at a 11.8 kbit/s rate for improved performance over a wide range of input signals such as music. The G.729 coder is based on the CELP coding model. The coder uses a sampling frequency of 8 kHz and operates on speech frames of 10 ms duration. A G.729 or G.729 Annex A frame contains ten octets, while the G.729 Annex B comfort noise frame occupies two octets. The RTP payload format for G.729 is standardised in RFC 3551 [17].

6.1.2.3 AMR

The Adaptive Multi-Rate (AMR) codec was originally developed and standardised by the European Telecommunications Standards Institute (ETSI) for GSM cellular systems. The 3GPP has adopted AMR as a mandatory codec for third generation (3G) cellular systems – AMR is specified in 3G TS 26.090 [40]. The AMR codec is also suitable for packet networks and is notably a required codec for terminals of the packet-based 3GPP IP Multimedia Subsystem (IMS). There are two AMR codecs: the narrowband version (referred to simply as AMR or AMR-NB) and the wideband version (referred to as AMR-WB). AMR is a multi-mode codec that supports 8 narrow band speech encoding modes with bit-rates of 12.20, 10.20, 7.95, 7.40, 6.70, 5.90, 5.15 and 4.75 kbit/s. Three of the AMR bit-rates are adopted as standards elsewhere: the 6.7 kbit/s mode is adopted as PDC-EFR, the 7.4 kbit/s mode as IS-641 codec in TDMA, and the 12.2 kbit/s mode as GSM-EFR specified in ETSI GSM 06.60. The AMR-WB codec [41] was developed by the 3GPP for use in GSM and 3G cellular systems. AMR-WB supports nine wideband speech encoding modes ranging from 6.60 to 23.85 kbit/s. The AMR codec is based on the CELP coding model. For AMR, a sampling frequency of 8 kHz is employed while AMR-WB uses a sampling frequency of 16 kHz. Both AMR and AMR-WB perform speech encoding on 20 ms frames. The RTP payload format for AMR and AMR-WB is standardised in RFC 3267 [42].

6.2 Media transport

Media transport refers to the mechanism by which encoded media is transported across a network. RTP provides the core protocol for achieving this task in IP networks.

6.2.1 Real-time protocol (RTP)

RTP is a lightweight protocol that provides end-to-end network transport functions for real-time data such as audio, video, and simulation data. RTP is specified in RFC 3550 [11]. RTP usually depends on other protocols such as SIP, H.323, or RTSP together with SDP to provide rendezvous functionality – i.e. to negotiate all the parameters necessary for data exchange over RTP to commence. For example, a SIP telephone call between two participants uses RTP to transport the audio data between the participants in real-time. Each participant will send audio data to the other in small 20 ms or 30 ms chunks. Each chunk of audio data is encoded in some manner (for example, using PCM encoding). An RTP packet is constructed with an RTP header followed by the audio data payload. The RTP header and payload are in turn contained in a User Datagram Protocol (UDP) packet, which is sent across the network. The RTP header includes critical information such as an identifier for the payload called the payload type, a timestamp for the audio to indicate its position in a continuous stream, a sequence number so that the receiver can reconstruct the correct ordering used by the source, and a synchronisation source to uniquely identify the stream. The RTP specification does not specify the format of the payload data but rather delegates this to separate payload format specification documents. The definition of the payload type numbers is also delegated to separate profile specification documents that map the payload type numbers to payload formats. RFC 3551 [17] defines a set of default mappings.

The term 'RTP session' is used to refer to an association among a set of participants communicating over RTP. RTP may be used to transport data to a single destination, called unicast, or to multiple destinations, called multicast. Unicast distribution is used for point-to-point applications and multicast is appropriate for applications that need to broadcast data to multiple participants, e.g. a loosely coupled audio conference or a live video broadcast of a sporting event over the Internet. RTP also defines two intermediary systems known as 'mixers' and 'translators'. A mixer receives streams of RTP data packets from one or more sources, possibly alters the formats, combines the streams in some manner, and forwards the new stream. A conference bridge may use a mixer to combine the audio streams of multiple unicast participants, for example. A translator is an entity that forwards RTP packets it receives after possibly modifying the data in some way. Examples of translators include media firewalls and transcoding devices.

RTP works in conjunction with the RTP Control Protocol (RTCP) whose main function it is to provide feedback on the quality of the data distribution by sending statistics on data received at the receiver to the sender of the data. The sender can use this data for different purposes e.g. augmenting the encoding characteristics such as the bit rate. While RTP and RTCP are independent of the underlying transport and network layers, they usually run on top of UDP. RTCP operates on a separate port to RTP. By convention, RTP uses an even port number and RTCP uses the next higher odd port number. UDP does not provide reliability and ordering guarantees – transmitted packets may never arrive at their destination nor are they guaranteed to arrive in the order that they were sent. UDP does, however, ensure that those packets that do arrive and are made available to the application layer are free from data corruption (the transport layer achieves this through the use of checksums). Since UDP does not provide reliability and ordering guarantees, it is faster and more efficient than protocols such as Transmission Control Protocol (TCP) and hence more suitable for transporting time-sensitive RTP.

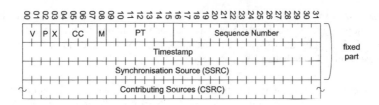

Figure 6.6 Format of the RTP header.

Indeed, for real-time applications, it is preferable to mask transmission errors and just simply get on with processing the next chunk of data rather than requesting a re-transmit.

The RTP header is comprised of a fixed part and a list of contributing sources if the packet originated from a mixer. The payload data follows the header. Figure 6.6 illustrates the format of the RTP header. The most significant bit is numbered 0 and the least significant bit is numbered 31 on the right. The bytes are transmitted in the following order: bits 0–7 first, then bits 8–15, then 16–23, then 24–31, then 32–39, etc. This is called 'big endian' byte ordering (and is equivalent to 'network byte order' used in IP).

All RTP headers include the first 12 bytes. The variable length CSRC field is present only if the packet originated from a mixer. The fields are defined as follows:

Version (*V*): (2 bits). This specifies the version of the protocol and is set to 2.

Padding (*P*): (1 bit). If this bit is set, there are one or more padding octets at the end of the payload that must be ignored. The last octet indicates the number of padding bits including itself.

Extension (*X*): (1 bit). This bit is used to indicate that the header has been extended with a protocol-specific header extension.

CSRC Count (*CC*): (4 bits). This indicates the number of Contributing Source (CSRC) Identifiers that follow the fixed header.

Marker (*M*): (1 bit). The definition of this bit is profile specific but is usually used to designate a significant event, e.g. the start of a talk-spurt.

Payload Type (*PT*): (7 bits). This field identifies the format of the payload. Payload types may be defined statically for a profile or may be dynamically assigned (see Section 4.5.2).

Sequence Number: (16 bits). The sequence number increases by one for each RTP data packet sent and is used by the receiver to detect packet loss. The receiver can use the sequence number to reorder packets that arrive out of sequence before passing the payload data on to other parts of the application.

Timestamp: (32 bits). The timestamp reflects the sampling instant of the first octet in the RTP payload. The timestamp refers specifically to the sampling clock and not to the system clock at the time the packet was sent. The timestamp clock frequency depends on the definition of the profile but the clock always increases monotonically and linearly. For fixed-rate codecs, the timestamp usually increments by one for each sampling period. For example, if audio is read in blocks of 240 samples from an input device and each block is encoded at a fixed rate, the timestamp would increment by 240 for each block encoded regardless of whether a packet was actually sent. The initial value of the timestamp is random. Note that although the timestamp clock is monotonic, the timestamp of consecutively transmitted packets may not be monotonic as is the case for MPEG interpolated video frames for example. This also explains why the timestamp field can not perform the function of the sequence number field.

SSRC: (32 bits). The SSRC field identifies the source of the stream and is chosen at random. Packets with the same SSRC belong to the same timestamp and sequence number space.

CSRC: (list of 0 to 15 sub-fields, each of 32 bits). This field is used by mixers. Each CSRC identifies a contributing source (an SSRC identifier) to the payload contained in this packet. The number of CSRC fields is indicated by the CC field. For example, for audio packets, the CSRC identifies all the sources that were mixed together to create a packet.

Consider a concrete example of transporting G.711 audio data across a network between two IP phones. Audio data is sampled at a rate of 8000 Hz. Each sample is represented with 8-bits (a byte) – resulting in 8000 bytes that must be transmitted to the receiver every second. Data may be conveniently divided into 20-ms frames consisting of 160 bytes per frame and transmitted in the payload of a single RTP packet. The timestamp on each packet will therefore increase by 160 and each packet must be sent every 20 ms. Network latencies may introduce jitter resulting in a variance in the inter-packet arrival time. In addition, some packets may arrive out-of-sequence or not at all. Receivers usually employ a jitter buffer before rendering the audio to the phone's loudspeaker. The jitter buffer introduces a small delay during which time the receiver can reorder any packets which arrive out of sequence (by inspecting the sequence number in the RTP header) and account for small variances in packet arrival time. The receiver uses the sequence number to reconstruct the audio stream in the correct order. In the case that no packet arrives, different strategies may be adopted. For G.711, some receivers will simply insert the last received packet in place of the missing one in an attempt to mask the loss of data. Other RTP payload formats include specific strategies for accommodating packet loss.

6.2.1.1 RTP control protocol (RTCP)

The primary function of RTCP is to provide feedback on the quality of the data distribution. RTCP also serves to identify uniquely each RTP source and may be used to perform synchronisation of multiple RTP sources. There are several RTCP packet types:

SR (Sender Report): Sent from participants who are active senders of data. Contains both sender information and reception reports.

RR (Receiver Report): Sent from participants who are not active senders of data. Similar format as the SR packet but without the sender information.

SDES: Contains the SSRC and associated source description items such as the CNAME. Although RTP streams are identified by an SSRC, the SSRC may change if the sender detects a collision (i.e. another participant is using the same SSRC in the same RTP session). The canonical name or CNAME is used to identify each participant unambiguously.

BYE: Indicates end of participation in an RTP session. A participant sends this packet when it wishes to leave an RTP session to inform the other participants of the event.

APP: Application-specific functions.

Each RTCP packet commences with a fixed part followed by packet-specific items. RTCP packets include a length field and hence may be concatenated without any intervening delimiters to form compound RTCP packets. For example, an SR and SDES packet may be placed in the same UDP packet and transmitted across the network. RTCP packets are sent compounded at a rate of no greater than every 5 seconds. As the number of participants increases in an RTP session, the rate is decreased linearly (if this were not the case, control traffic would grow linearly with the number of participants). RTCP uses an algorithm to ensure that the control packets occupy just 5 % of the overall bandwidth for the RTP session. In addition, the algorithm introduces a random variation to avoid synchronisation of transmission times.

The SR packet contains sender information which includes the following information:

NTP timestamp: (64 bits). This field indicates the absolute wallclock time of when the packet was sent measured in seconds relative to 0 h UTC on 1 January 1900 using the timestamp format of the Network Time Protocol (NTP) [35]. The higher 32 bits contain the integer part of the time and the lower 32 bits contain the fractional part.

RTP timestamp: (32 bits). Indicates the value of the RTP timestamp or sampling clock at the instant identified by the wallclock time.

Sender's packet count: (32 bits). The number of RTP packets sent since the start of transmission up until the time of the SR packet.

Sender's octet count: (32 bits). The number of octets sent in the RTP payload data up until the time of the SR packet.

The relationship between the wallclock time and the RTP timestamp is invaluable to the receiver for synchronisation of multiple streams such as audio and video. It allows the receiver to relate the RTP timestamp of one media stream with that of another media stream through the common currency of the wallclock time (which is assumed to be the same for all senders). By knowing the sampling clock frequency for the stream and the RTP timestamp at a given wallclock time (from the SR), one can calculate the wallclock time for when the data in a received RTP packet was originally sampled. This allows the receiver to 'line up' separate streams for synchronous rendering.

Both the SR and RR packets contain reception reports that convey statistics on the reception of RTP packets from a single synchronisation source. The reception report includes the following information:

Source identifier: (32 bits). This field identifies the SSRC of the source for which the reception report relates to.

Fraction lost: (32 bits). Indicates the fraction of RTP packets lost since the last SR or RR packet. This fraction is defined as the number of packets lost divided by the number of packets expected. The number of packets expected is obtained from the highest sequence number received in an RTP data packet less the initial sequence number received.

Cumulative number of packets lost: (24 bits). The total number of RTP data packets from the source that have been lost since the beginning of reception. This is defined as the number of packets expected less the number of packets actually received.

Extended highest sequence number received: (32 bits). The low 16 bits contain the highest sequence number received in an RTP data packet from the source. The high 16 bits contain the number of sequence number cycles (i.e. how many times the sequence number has wrapped around its 16-bit range).

Interarrival jitter: (32 bits). An estimate of the statistical variance of the RTP data packet interarrival time. The interarrival jitter is defined to be the smoothed absolute value of the difference in packet spacing at the receiver compared with the sender for a pair of packets.

Last SR timestamp: (32 bits). The middle 32-bits in the NTP timestamp of the last SR packet received. The roundtrip time can be calculated by subtracting this value from the arrival time of the reception report.

Delay since last SR: (32 bits). The delay in receiving the last SR packet from the source and sending this reception report. The roundtrip propagation delay can be calculated from subtracting this field from the roundtrip time (defined in the preceding paragraph).

Receiver reports provide valuable feedback to senders. A sender can modify the characteristics of its transmission based on the feedback. For example, a sender could decrease the transmission rate if it is informed about network congestion, or increase the amount of redundancy in the data if it is informed about packet loss (assuming the particular RTP payload profile supports this feature). Receivers may use sender reports to determine if network problems are local or global. Network management software may analyse RTCP packets on the network to report statistics pertaining to data transmission on the network.

6.2.2 DTMF

Dual-Tone Multifrequency (DTMF) is used for signalling during call setup to indicate the dialled number and also mid-call for interacting with IVR systems. On the PSTN, DTMF is carried within the voice band (called in-band) and may also be carried out-of-band in certain signalling systems – for example, an ISDN terminal may supply digits in Q.931 INFORMATION messages. DTMF uses eight different frequencies sent in pairs to represent 16 different symbols. The frequencies are chosen to prevent naturally occurring harmonics in the voice band being erroneously misconstrued as DTMF events. Figure 6.7 illustrates the keypad frequencies.

On packet networks, DTMF may be carried in the voice band or as events carried using a special RTP payload format. A separate RTP payload format for DTMF events is desirable since low bit-rate codecs are not guaranteed to reproduce the DTMF tones sufficiently faithfully to be accurately recognised by a DTMF tone detector. RFC 2833 [44] specifies an RTP payload format for carrying DTMF digits in addition to other line and trunk signals, and general multifrequency tones. It is common for media gateways to include DTMF tone detectors, clamp the inband DTMF tones, and produce RFC 2833 DTMF events inbound to the IP network. MRCP speech recogniser resources supporting DTMF are required to support detection of DTMF events through RFC 2833 and may optionally support recognising DTMF tones in the audio itself.

The payload format for RFC 2833 named events such as DTMF tones is illustrated in Figure 6.8.

	High-group frequencies			
	1209 Hz	1336 Hz	1477 Hz	1633 Hz
697 Hz	1	2	3	A
770 Hz	4	5	6	B
852 Hz	7	8	9	C
941 Hz	*	0	#	D

Figure 6.7   DTMF keypad frequencies.

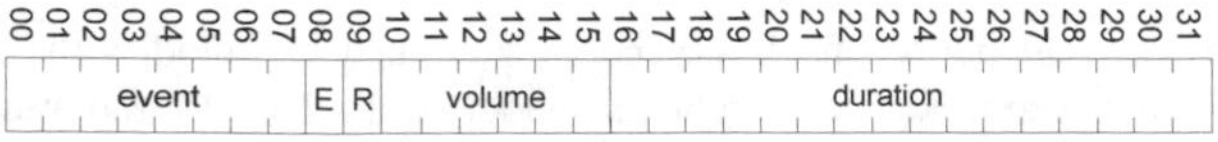

event	E	R	volume	duration

Figure 6.8   RFC 2833 payload format for named events.

The fields are defined as follows:

event: (8 bits). This field specifies the named event. Values 0–9 inclusive represent DTMF tones 0 to 9. The '*' tone is represented by value 10 and the '#' tone with value 11. The symbols A–D are represented with values 12–15.

E: (1 bit). This bit is set in the packet that demarcates the end of the event.

R: (1 bit). This bit is reserved for future use and is set to 0.

volume: (6 bits). This field specifies the power level of the DTMF tone expressed in dBm0 after dropping the sign. Power levels range from 0 to −63 dBm0. Larger values of this field indicate lower volumes.

duration: (16 bits). This field specifies the cumulative duration of the event, measured in timestamp units and expressed as an unsigned integer in network byte order. The start of the event is indicated by the RTP timestamp and has so far lasted the value of this field. This mechanism allows DTMF events to be sent incrementally without the receiver having to wait until completion of the event.

DTMF events are typically transmitted over several packets at the same rate as the audio codec (although they may alternatively be transmitted at a recommended period of 50 ms). The event packets use the same sequence number and timestamp base as the audio channel. An audio source will start transmitting event packets as soon as it recognises an event. For example, an IP phone will start transmitting event packets as soon as the user presses a DTMF key and will continue to send update packets while the key is pressed. The RTP header timestamp in the event packets indicates the position in the audio stream where the DTMF key was first pressed and all subsequent update event packets employ the same timestamp. The RTP header sequence number increases by 1 for each packet as usual. The first event packet will have the marker bit set in the RTP header to indicate the start of the event. The duration of the event thus far will be reflected in the duration field in each event packet. The update event packets will have increased durations as the key is held down. For example, if a G.711 codec is used with 20 ms frames (corresponding to 160 samples or timestamp units per RTP packet), the duration of event packets will increase by 160 in each new event packet. When the DTMF key is released, the E bit will be set on the next event packet. If there are no new DTMF key presses, the last packet (with the E bit set and final duration) is retransmitted three times. This ensures that the duration of the event can be recognised even in the face of some packet loss.

SIP endpoints indicate support of RFC 2833 events in their SDP offer (see also Section 4.5.2) using the `telephone-event` MIME type. The event values supported are indicated with the `fmtp` attribute. For example, a SIP endpoint supporting PCM μ-law and all 16 DTMF events will produce an offer similar to:

```
m=audio 32245 RTP/AVP 0 96
a=rtpmap:0 PCMU/8000
a=rtpmap:96 telephone-event/8000
a=fmtp:96 0-15
```

Note that the payload type number for RFC 2833 events is dynamic (the example chose the value 96).

6.3 Security

Certain applications might request sensitive information from the user and/or supply sensitive information to the user over the media stream. MRCP requires that media resource servers implement the Secure Real-time Transport Protocol (SRTP) [49]. SRTP may be used in place of RTP when

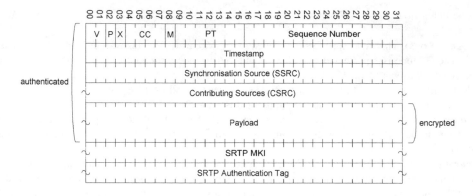

Figure 6.9 Format of an SRTP packet.

confidentiality and integrity protection of the media stream must be assured. Note that SRTP will also encrypt DTMF events transported over RTP using the mechanism in RFC 2833. Secure RTCP (SRTCP) provides the same security services to RTCP as SRTP does to RTP. As an alternative to SRTP, some MRCP implementations may instead depend on network layer security such as IPSec (discussed in Chapter 4).

SRTP is defined as a profile of RTP and is indicated on SDP m-lines with a transport protocol of RTP/SAVP in place of RTP/AVP. Figure 6.9 illustrates the format of an SRTP packet. SRTP encrypts the payload for confidentiality and authenticates the RTP header and payload for integrity. SRTP's default encryption algorithm sums a random stream of data derived from the sequence number of the packet (called the keystream) with the payload data to produce the cyphertext. The stream is decrypted at the receiver by subtracting the keystream. The XOR function is used for addition and subtraction of the keystream. Thus the size of the payload does not change. The definition of how the keystream is generated depends on the cipher and its mode of operation (the default cipher for SRTP is the Advanced Encryption Standard, AES). The RTP header and payload is authenticated by running a one-way function on them using a cryptographic key to produce a hash. The sender writes the hash into the SRTP authentication tag, which is between 4 and 10 bytes long. This allows the receiver to ensure that the sender of the message possessed a secret key and that no party lacking that cryptographic key modified the message enroute. The MKI portion of the SRTP packet stands for Master Key Identifier and is optional (the MKI may be used by key management for the purpose of re-keying). Re-keying is not usually used for IP telephony since IP telephony employs relatively short-lived sessions.

To commence using SRTP, the receiver needs decryption and authentication keys along with other information such as the encryption cipher, the authentication transform and related parameters. The keys use for decryption and authentication (called 'session keys') are derived from a 'master key'. The SDP crypto attribute described in *Session Description Protocol Security Descriptions for Media Streams* [50] may be used to associate a master key and associated parameters with a media stream in each direction as part of the regular SDP offer/answer model. An example is illustrated below:

```
m=audio 3241 RTP/SAVP 0
a=rtpmap:0 PCMU/8000
a=crypto:1 AES_CM_128_HMAC_SHA1_32
    inline:NzB4d1BINUAvLEw6UzF3WSJ+PSdFcGdUJShpX1Zj|2^20|1:32
```

The RTP/SAVP transport protocol identifier on the m-line indicates that the media stream is to use SRTP. The crypto attribute is used to signal the cryptographic parameters associated with this media stream. The first number (1) is a tag value to identify a particular crypto attribute. It is used with the offer/answer model to determine which of several offered crypto attributes were chosen by the answerer. The second field describes the encryption and authentication algorithms. The AES_CM_128_HMAC_SHA1_32 value indicates AES-CM encryption with a 128-bit key and HMAC-SHA1 message authentication with a 32-bit tag for both RTP and RTCP packets. The final sub-field contains keying information. The inline token signals that the actual keying information is provided here. The key comprises a 16-byte master key concatenated with a 14-byte so-called salting key, which are both base-64 encoded. The receiver uses this information to derive the encryption key and authentication key to decode the payload data. Note that since the key information is provided in plaintext in the SDP, it is critical that the SDP body itself be encrypted, e.g. through the use of S/MIME (see Chapter 4).

6.4 Summary

This chapter focuses on the media session – the conduit for speech data between the media source/sink and the media resource server. The two fundamental concepts related to the media session are that of media encoding and media transport. There is a variety of different media encoding schemes used to carry speech data on IP networks ranging from simple, high bit-rate, codecs such as PCM encoding to complex, low bit-rate, encodings such as AMR. Since low-bit rate codecs incur a high computational overhead and many media resource servers are implemented in software running on general-purpose hardware, PCM encoding tends to be the preferred encoding. Interested readers can find more information on speech coding elsewhere [39]. For media transport, RTP over UDP is the standard approach, providing real-time transport services over IP networks. RTP is a lightweight protocol that favours efficiency over redundancy for delivering real-time data. Several RTP payload formats for different encoding schemes are specified in RFC 3551 [17] and many codecs have dedicated RFCs defining the corresponding payload format.

Having covered the media session, the next chapter focuses on the control session between the MRCP client and the media resource server. The control session allows the MRCP client to issue instructions to the media resource server either to generate audio data over the media session (e.g. speech synthesiser resource) or to operate on the audio data carried over the media session (e.g. a speech recogniser, recorder or speaker verification resource).

7

The control session

Chapter 5 introduced the mechanisms employed by MRCP to set up and manage a control channel running over TCP between the MRCP client and media resource server. In this chapter we focus on the MRCP protocol that runs over this control channel. We begin by introducing the different message types, their structure and the message exchange patterns. We then look at the mechanism used for defining and retrieving parameters for a session. Finally, we discuss the generic headers that are used across different media resource types.

7.1 Message structure

There are three types of MRCP message: (i) request; (ii) response, and (iii) event.

MRCP clients originate requests toward the media resource server. The media resource server issues a response message to the request. Subsequently, one or more events may be used to report updates related to the original request back to the MRCP client. All messages commence with a start line (there is a different start line for each of the three message types), one or more header fields, an empty line, and an optional message body. Each line is terminated by a carriage return and line feed (CRLF). The header fields use the format `header-name: header-value CRLF`. The message body contains message specific data (e.g. grammars may be supplied to a speech recogniser, or text to synthesise may be sent to a speech synthesiser). The message body is a string of octets whose length is specified in the `Content-Length` header field. The message body type is identified by the `Content-Type` header.

Requests received by a media resource server are assigned an associated request state. A request may be in either the `PENDING`, `IN-PROGRESS` or `COMPLETE` state. The `PENDING` state indicates that the request has been placed in a queue, will be processed in first-in–first-out order, and further updates in the form of events are expected. The `IN-PROGRESS` state means that the request is currently being processed, is not yet complete, and further updates in the form of events are expected. The `COMPLETE` state indicates that the request was processed to completion and that there will be no further messages (i.e. events) from the media resource server to the MRCP client related to that request. Figure 7.1 illustrates the message exchange pattern for the case where the request is processed to completion immediately.

Speech Processing for IP Networks   Dave Burke
© 2007 John Wiley & Sons, Ltd

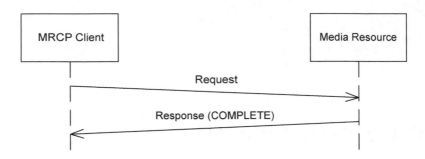

Figure 7.1   Message exchange pattern where the request is immediately processed to completion.

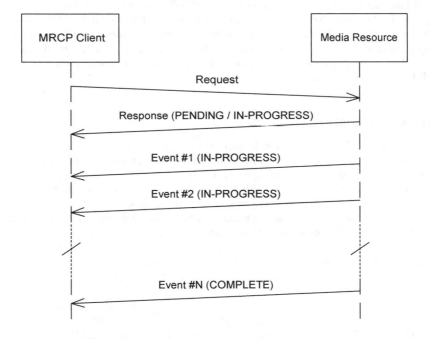

Figure 7.2   Message exchange pattern with update events.

Figure 7.2 shows the message exchange pattern for the case where the response indicates that the request is either in the PENDING state (i.e. waiting to be processed) or the IN-PROGRESS state (i.e. is currently being processed).

One or more updates in the form of events are issued after the response. The intermediate events indicate a request state of IN-PROGRESS. The final event is identified with a request state of COMPLETE, which signals to the MRCP client that no more events will be received for this request.

7.1.1 Request message

Request messages (sometimes referred to as 'request methods') are issued by the MRCP client to the media resource server to request that it perform a specific action. The start line of the request message is formatted according to:

```
MRCP/2.0 message-length method-name request-id
```

The first token identifies the protocol version. This is the same for all message types. The `message-length` specifies the length of the message in bytes including the start line, header fields, and message body (if any). The `message-length` is always the second token for all message types to make framing and parsing simpler. The `method-name` identifies the specific request from the MRCP client, i.e. the verb. Different media resource types support different methods, for example `SPEAK` is specific to the speech synthesiser resource and `RECOGNIZE` is specific to the speech recogniser resource. There are also two methods that are general to all resources (`SET-PARAMS` and `GET-PARAMS` and are described later). The `request-id` is a unique identifier for the request that is created by the MRCP client. The `request-id` is a 32-bit integer that is initialised to a small value in each session and increases monotonically during the session for each new request. An example request message is:

```
MRCP/2.0 267 SPEAK 10000
Channel-Identifier: 43b9ae17@speechsynth
Content-Type: application/ssml+xml
Content-Length: 150

<?xml version="1.0" encoding="UTF-8"?>
<speak version="1.0"
       xmlns="http://www.w3.org/2001/10/synthesis">
   Hello world!
</speak>
```

The `Channel-Identifier` header is mandatory on all request messages; its value is the same as the value returned in the SDP `channel` attribute (see Chapter 5). This message includes a body whose type is identified by the `Content-Type` header field as SSML. The length of the body, in bytes, is indicated in the `Content-Length` header field.

7.1.2 Response message

The response message is issued by the media resource server in response to a request. The start line of the response message is formatted according to:

```
MRCP/2.0 message-length request-id status-code request-state
```

The first token identifies the protocol version. The second token is the message length as before. The third token, `request-id`, identifies the request to which this response corresponds to. The `status-code` is a three-digit code representing success or failure of the request. Success codes are in the range of 200 – 299. Client errors are in the range 400 – 499 and relate to problems such as bad syntax in the request and problems that prevent the request from being fulfilled at a particular server.

Table 7.1 2xx success codes

Code	Description
200	Success
201	Success with some optional headers ignored

Table 7.2 4xx client failure codes

Code	Description
401	Method not allowed
402	Method not valid in this state
403	Unsupported header
404	Illegal value for header (i.e. the value has a syntax error)
405	Resource not allocated for this session or does not exist
406	Mandatory header missing
407	Method or operation failed
408	Unrecognised or unsupported message entity
409	Unsupported header value (i.e. the value syntax is correct but exceeds the resource's capabilities)
410	Non-monotonic or out of order sequence number in request
411–420	Reserved

Table 7.3 5xx server failure codes

Code	Description
501	Server internal error
502	Protocol version not supported
503	Proxy timeout
504	Message too large

Server errors are in the range 500 – 599 and indicate that the server failed to fulfil an apparently valid request. The status codes are summarised in Tables 7.1 to 7.3.

The request-state indicates the current state of the request. A request may be in the PENDING, IN-PROGRESS or COMPLETE state as described in Section 7.1 above. An example of a response message is:

```
MRCP/2.0 79 10000 200 IN-PROGRESS
Channel-Identifier: 43b9ae17@speechsynth
```

This response might be generated in response to a SPEAK request with request-id 10000.

7.1.3 Event message

Events are used by the media resource server to communicate a change in state of a media resource or the occurrence of some other event to the MRCP client. Events are only issued if the response to the request indicated a request state of PENDING or IN-PROGRESS. The start line of the event message is formatted according to:

MRCP/2.0 message-length event-name request-id request-state

The first and second tokens identify the protocol version and message length similar to the request and response messages. The event-name identifies the specific event. Different media resource types support different events, for example the SPEAK-COMPLETE event is specific to the speech synthesiser resource and the RECOGNITION-COMPLETE event is specific to the speech recogniser resource. The fourth token, request-id, identifies the request to which this event corresponds – events are never sent unsolicited and are only sent as a result of a request. The request-state indicates the current state of the request similar to the response message. For events, the request-state is either IN-PROGRESS, in which case further events will be sent, or COMPLETE, in which case no further events will be sent for the corresponding request-id. An example of an event message is:

MRCP/2.0 109 START-OF-INPUT 10000 IN-PROGRESS
Channel-Identifier: 43b9ae17@speechrecog
Input-Type: dtmf

This event might be generated by a speech recogniser resource for request-id 10000 when the speech recogniser first detects DTMF input.

7.1.4 Message bodies

We have already seen how an optional body may be added to any MRCP message. The message body is simply a sequence of bytes following the header fields (the header fields and message body are separated by a new line indicated by a CRLF). The number of bytes in the message body is specified in the Content-Length header. The receiver of the message decodes the message by inspecting the Content-Type header and applying the appropriate decoder. This approach is very similar to the one taken by the HTTP and SIP protocols to encapsulate message-specific data. Table 7.4 indicates some common content types used with MRCP messages.

Table 7.4 Common content types used in MRCP

Content type	Description
text/plain	Plain text for synthesising.
application/ssml+xml	SSML document. See Chapter 8.
application/srgs+xml	SRGS document (XML form). See Chapter 9.
application/srgs	SRGS document (ABNF form).
application/nlsml+xml	NLSML results. See Chapter 10.
application/pls+xml	PLS document. See Chapter 11.
multipart/mixed	Multipart message body. See Section 7.1.4.1.
text/uri-list	List of URIs. See Section 7.1.4.2.
text/grammar-ref-list	List of grammar URIs allowing each URI to specify a weight. See Section 7.1.4.2.

7.1.4.1 Multipart message body

In certain circumstances there is a need for multiple part messages where one or more different sets of data are combined into a single message body. In such cases, a multipart message is used. The multipart message body is identified with a `Content-Type` value of `multipart/mixed`. A typical multipart `Content-Type` header field might look like this:

```
Content-Type: multipart/mixed; boundary=a4V3bnq129s8
```

This header field indicates that the message body consists of several parts delimited by an encapsulation boundary of:

```
--a4V3bnq129s8
```

Note the additional prefix of two hyphens. The encapsulation boundary must start at the beginning of a new line, i.e. immediately following a `CRLF` and must terminate with a `CRLF`. The boundary string is chosen such that it is does not clash with the message body. Each body part may optionally include its own header fields although only headers commencing with `Content-` are recognised; MRCP-specific headers are not allowed. An additional `CRLF` separates these headers from the body data. If the body part contains no header fields, it is considered to be of type `text/plain`. The last body part is terminated with an encapsulation boundary with an additional two hyphens as a suffix to indicate no more body parts, for example:

```
--a4V3bnq129s8--
```

An example of a multipart message body is illustrated below:

```
MRCP/2.0 521 SPEAK 20000
Channel-Identifier: 43b9ae17@speechsynth
Content-Type: multipart/mixed; boundary=a4V3bnq129s8
Content-Length: 374

--a4V3bnq129s8
Content-Type: text/uri-list
Content-Length: 68

http://www.example.com/file1.ssml
http://www.example.com/file2.ssml

--a4V3bnq129s8
Content-Type: application/ssml+xml
Content-Length: 140

<?xml version="1.0" encoding="UTF-8"?>
<speak version="1.0"
       xmlns="http://www.w3.org/2001/10/synthesis">
    Hello world!
</speak>
--a4V3bnq129s8--
```

In this example, the MRCP client is sending a list of SSML URIs to synthesise in addition to a body part containing inline SSML to synthesise.

7.1.4.2 URI list

MRCP messages can use the `text/uri-list` content type for specifying one or more URIs in the message body. This is useful for supplying a list of external SSML documents in a `SPEAK` request or a list of external lexicon documents in a `DEFINE-LEXICON` request. Example:

```
Content-Type: text/uri-list
Content-Length: 67

http://www.example.com/names.pls
http://www.example.com/places.pls
```

A close cousin of the `text/uri-list` content type is the `text/grammar-ref-list` content type. This type is used to allow grammar URIs to be specified with different associated weights (see Chapter 9 for an explanation of grammar weights). For example:

```
Content-Type: text/grammar-ref-list
Content-Length: 99

<http://www.example.com/menu.grxml>;weight="2.0"
<http://www.example.com/names.grxml>;weight="1.5"
```

7.2 Generic methods

Some parameters do not change value during a session or else change value infrequently. For this purpose, MRCP provides two generic request methods for defining and retrieving 'sticky' parameters for a session called `SET-PARAMS` and `GET-PARAMS`. This allows the MRCP client to configure general parameters such as the volume of synthesised speech, recognition timer settings, etc. Values placed in subsequent requests override the session values defined by `SET-PARAMS` for that request only.

The `SET-PARAMS` message uses header fields to specify parameter values. The media resource server responds with a status code of 200 if the request is successful. A response with status code 403 indicates that the server did not recognise one or more parameters. A response with status code 404 means that the header value was invalid. For response messages with status code 403 or 404, the offending headers are included in the response. Figure 7.3 illustrates the `SET-PARAMS` flow.

An example of a `SET-PARAMS` request to set the language for a speech synthesiser is illustrated below:

```
MRCP/2.0 99 SET-PARAMS 12309
Channel-Identifier: 23fa32fg1@speechsynth
Speech-Language: fr-FR
```

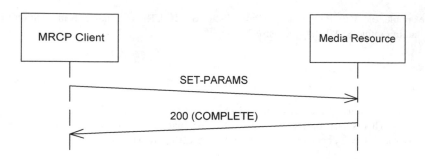

Figure 7.3 SET-PARAMS example.

A successful response is:

```
MRCP/2.0 77 12309 200 COMPLETE
Channel-Identifier: 23fa32fg1@speechsynth
```

The GET-PARAMS request message is used to retrieve parameters for a session. The parameters for which to retrieve values are specified in header fields with the value omitted. A successful response with status code 200 will include the same header fields with their values defined for the session filled in. If one or more header fields are not recognised, a response with status code 403 is returned, including the offending header(s). If no headers are set (except the mandatory Channel-Identifier), the media resource server will return the entire collection of settable parameters for a session. Note that this operation can be quite inefficient and it is not recommended that it be performed frequently. Figure 7.4 illustrates the GET-PARAMS flow.

An example of a GET-PARAMS request to get the volume for a speech synthesiser for the session is illustrated below:

```
MRCP/2.0 94 GET-PARAMS 12310
Channel-Identifier: 23fa32fg1@speechsynth
Speech-Language:
```

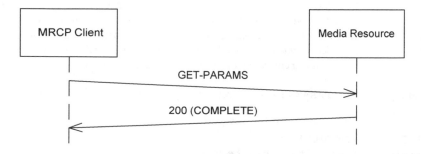

Figure 7.4 GET-PARAMS example.

A successful response is:

```
MRCP/2.0 102 12310 200 COMPLETE
Channel-Identifier: 23fa32fg1@speechsynth
Speech-Language: fr-FR
```

7.3 Generic headers

MRCP defines two categories of headers: those that apply to a specific media resource type (called resource-specific headers), and generic headers that apply more generally. Note carefully that the generic methods, GET-PARAMS and SET-PARAMS, are not limited in any way to generic headers (described in the next section) – in fact these methods are most often useful for setting resource-specific parameters. Table 7.5 lists the generic headers and the messages in which they are allowed to appear in.

As described earlier, each header field consists of a name followed by a colon and the value, and is terminated with a CRLF. The header name is case insensitive. White space occurring before the first character and after the last character of the value is ignored. It is convention to place a single space after the colon. The following subsections describe each generic header.

7.3.1 Channel-Identifier

The Channel-Identifier header field is mandatory in all MRCP requests, responses, and events. The Channel-Identifier uniquely identifies the channel to which the message applies. Since there is a one-to-one mapping of control channel to media resource, the Channel-Identifier also identifies an allocated media resource for the session. The media resource server allocates its

Table 7.5 Generic headers

Header name	SET-PARAMS / GET-PARAMS
Active-Request-Id-List	No
Content-Base	No
Content-Encoding	No
Content-ID	No
Content-Length	No
Content-Location	No
Content-Type	No
Proxy-Sync-Id	No
Channel-Identifier	Yes
Accept	Yes
Accept-Charset	No
Fetch-Timeout	Yes
Cache-Control	Yes
Set-Cookie / Set-Cookie2	Yes
Vendor-Specific	Yes
Logging-Tag	Yes

value when the MRCP client establishes the control session (it is contained in the channel attribute in the SDP answer from the media resource server – see Chapter 5). The Channel-Identifier value is comprised of two parts delimited by an '@' character. A string prefix identifying the MRCP session is placed to the left of the '@' character, and the media resource type is placed to the right (see Table 3.1 for a list of resource type names). The string prefix is the same for all resources allocated within the same SIP dialog. The Channel-Identifier enables connection sharing by virtue of the fact that all messages are unambiguously identified and hence can be run over the same connection (the media resource server is responsible for ensuring that each active control channel is assigned a unique identifier).

Example:

```
Channel-Identifier: 466453af837@speechrecog
```

7.3.2 Active-Request-Id-List

The Active-Request-Id-List may appear in requests and responses (but not events). In a request, this header field's value indicates a list of request-ids to which the request applies. This is useful when there are multiple requests in the PENDING state and the MRCP client wants this request to apply specifically to one or more of these requests. In a response, this header field's value indicates the list of request-ids (comma delimited) that the corresponding request affected.

The Active-Request-Id-List header field most commonly appears in the response to a STOP request to a resource when one or more requests have been stopped. In this case, no further events are sent (e.g. SPEAK-COMPLETE in the case of a SPEAK request or RECOGNITION-COMPLETE in the case of a RECOGNIZE request). The Active-Request-Id-List header field may optionally appear in the STOP request for resources that queue requests (such as the speech synthesiser resource) to allow specific requests to be targeted. The Active-Request-Id-List header field also appears in the responses to the BARGE-IN-OCCURRED, CONTROL, PAUSE and RESUME requests for the speech synthesiser resource to indicate the requests affected by the method.

For example, consider a SPEAK request with request-id 00001 that is currently IN-PROGRESS, and a second SPEAK request with request-id 00002 in the PENDING state. If the MRCP client issues a STOP request and omits the Active-Request-Id-List header field, the request will apply to all PENDING and IN-PROGRESS requests. The response will include an Active-Request-Id-List header field to indicate those requests affected by the STOP request. No SPEAK-COMPLETE events will be generated for these requests. Alternatively, the MRCP client could specify exactly which requests it wanted to stop by including the Active-Request-Id-List header field in the STOP request.

Example:

```
Active-Request-Id-List: 00001,00002
```

7.3.3 Content-Base

The Content-Base URI may be used to specify the base URI to which relative URIs should be resolved. For example, a relative URI contained in an SRGS grammar may be interpreted relative to the absolute URI of the Content-Base used in the request method specifying that grammar. Note that individual parts in a multipart message may contain their own `Content-Base` URI, in which case this header field value takes precedence over that of the request.

    Example:

```
Content-Base: http://www.example.com/myapp/
```

7.3.4 *Content-Encoding*

The `Content-Encoding` header field is used as a modifier to the media type specified in the `Content-Type`. When present, its value indicates additional content encoding that has been applied to the message body and, hence, what decoding operations must be carried out to obtain the type specified in the `Content-Type`. If multiple encodings have been performed, then the encodings are listed (comma-delimited) in the order in which they were applied. This header field's semantics are inherited from HTTP [12].

    Example:

```
Content-Encoding: gzip
```

7.3.5 *Content-ID*

The `Content-ID` header field value is used as an identifier for the content in the message body or, in the case of a multipart message, to identify an individual body part. The presence of this header in a request indicates that the media resource server must store, for the duration of the session, the content for later retrieval through MRCP's `session:` URI scheme. This mechanism is typically used in conjunction with the `DEFINE-GRAMMAR` and `DEFINE-LEXICON` methods to define and prepare potentially large inline content (i.e. content in the message body) for use later in the session. The `Content-ID` value essentially specifies a 'hook' to allow the content to be referenced later with a `session:` URI derived from the `Content-ID` value. Stored content can be overwritten by reusing the same `Content-ID` with different inline content. Specifying a `Content-ID` while omitting the inline content indicates to the media resource that it must delete the previously defined content. The `Content-ID` header field can also be specified in `RECOGNIZE` requests whose message bodies contain inline grammars. This allows later `RECOGNIZE` requests in the same session to specify `session:` URIs instead of resending potentially large data on each request. See Chapter 12 for more information on `DEFINE-LEXICON` and Chapter 13 for more information on `DEFINE-GRAMMAR` and `RECOGNIZE`.

    The MRCP specification constrains the syntax of the `session:` URI scheme (and therefore the `Content-ID` that it is referencing) to follow a format of:

```
local-part@domain
```

where both `local-part` and `domain` may contain words separated by a dot. For example, when a grammar is specified inline in the message body of a `RECOGNIZE` request, a `Content-ID` must be supplied, e.g.

```
Content-ID: menu@example.com
```

This indicates that the media resource server must store the grammar in the message body. The MRCP client can subsequently reuse the grammar in another `RECOGNIZE` by referring it to via the `session:` URI scheme. For example:

```
Content-Type: text/uri-list
Content-Length: 24

session:menu@example.com
```

The `session:` URI scheme is also used in the NLSML results in a `RECOGNITION-COMPLETE` event to identify an inline grammar that was matched.

7.3.6 *Content-Length*

This header field specifies the length of the message body in bytes. It is required to be present on all messages that contain a message body. If the header field is omitted (i.e. when no message body is present), a default value of 0 is assumed. This header field's semantics are inherited from HTTP [12].
Example:

```
Content-Length: 128
```

7.3.7 *Content-Location*

The `Content-Location` may be used to state the location of a resource contained in the message body. The media resource server can use this information to optimise certain operations, e.g. caching of the content. For example, if a grammar was provided inline in a `RECOGNIZE` request, the `Content-Location` might be used to specify the URI used by the MRCP client to retrieve that grammar. If the `Content-Location` is a relative URI, then it is interpreted relative to the `Content-Base` URI. This header field's semantics are inherited from HTTP [12].
Example:

```
Content-Location: http://example.com/menu_grammar.grxml
```

7.3.8 *Content-Type*

This header field is used to specify the IANA registered MIME type of the message body. MRCP supports a restricted set of MIME types for representing content such as speech synthesis markup, speech grammars, and recognition results. The `multipart/mixed` content type is used to allow an MRCP message body to contain multiple content types in the same message (see Section 7.1.4 above). This header field's semantics are inherited from HTTP [12].
Example:

```
Content-Type: application/ssml+xml
```

7.3.9 *Proxy-Sync-Id*

The `Proxy-Sync-Id` header field is used to provide coordination of barge-in detection across a resource that detects the start of speech or DTMF input (i.e. a speech synthesiser, recorder or speaker verification resource) and a speech synthesiser. The MRCP client receives a `START-OF-INPUT` event from the resource detecting input when that resource first encounters speech or DTMF input in the media stream. The MRCP client subsequently sends a `BARGE-IN-OCCURRED` request to the speech synthesiser resource which will then decide whether to stop playback of audio (this depends on whether the current `SPEAK` request has the `Kill-On-Barge-In` parameter set to `true`). In some implementations, the speech synthesiser resource might be tightly coupled with the resource detecting input (they might be hosted on the same media resource server from the same vendor, for example). A faster barge-in response may be possible by having the speech synthesiser resource and the resource detecting input coordinate internally. In this case, the media resource server needs a way to know that the `BARGE-IN-OCCURRED` request sent by the MRCP client relates to a barge-in that the media resource server has already handled. This is achieved by having the media resource server add an identifier to the `START-OF-INPUT` event in the form of the `Proxy-Sync-Id` header field. The MRCP client relays the same `Proxy-Sync-Id` header field in the subsequent `BARGE-IN-OCCURRED` request, thus allowing the media resource to correlate this request with a particular barge-in event.

    Example:

```
Proxy-Sync-Id: 1241256
```

7.3.10 *Accept*

The `Accept` header field is used to specify certain media types that are acceptable in the response. Accept headers can be used to indicate that the request is specifically limited to a small set of desired types. If the `Accept` header is omitted, the media resource server will usually assume a default value that is specific to the resource type that is being controlled. The default value can be set through `SET-PARAMS` for a resource for the duration of the session. The current default value defined for the session can be retrieved via `GET-PARAMS`. This header field's semantics are inherited from HTTP [12].

    Example:

```
Accept: application/sdp
```

7.3.11 *Accept-Charset*

The `Accept-Charset` header field is used in a request to specify the acceptable character set(s) for message bodies returned in a response or the events associated with this request. For example, this header field can be used to set the character set of the NLSML returned in the message body of the `RECOGNITON-COMPLETE` event. This header field's semantics are inherited from HTTP [12].

    Example:

```
Accept-Charset: iso-8859-1
```

7.3.12 *Fetch-Timeout*

The `Fetch-Timeout` header field allows the MRCP client to configure the timeout (in milliseconds) used by the media resource for accessing the content specified by a URI. The `Fetch-Timeout` header can be set for the session or used on a per-request basis (on the `SPEAK` request for the speech synthesiser resource and the `DEFINE-GRAMMAR` and `RECOGNIZE` requests for the speech recogniser resource). The default value if omitted is implementation specific.
   Example:

```
Fetch-Timeout: 5000
```

7.3.13 *Cache-Control*

Media resource servers usually implement content caching for HTTP resources using the caching rules defined in RFC 2616 [12]. This allows content authors to add headers to HTTP responses to indicate for how long the contained document can be considered fresh, and permits performance improvements by obviating the need for the media resource server to re-fetch the document unnecessarily and instead use a locally cached version (see Appendix C for more information). In the context of MRCP, these documents are SSML documents, audio files, lexicon documents, SRGS grammars, and voiceprints.
   The `Cache-Control` header field in MRCP is inherited from the `Cache-Control` header field used in HTTP requests (see Appendix C, located on the Web at http://www.daveburke.org/speechprocessing/). Specifically, this header allows the MRCP client to override default caching behaviour used in the media resource server by putting a limit on the maximum age for usable documents in the cache, extending the use of stale documents in the cache by a number of seconds, or requesting that documents usable in the cache are going to be fresh for a minimum amount of time. The header field value is written in the form of `<cache directive>=<delta seconds>` where `<cache-directive>` is one of `max-age`, `max-stale` or `min-fresh`, and `<delta-seconds>` is the number of seconds. More than one `<cache-directive>` can be specified using a comma-delimited list. The meaning of the three directives are given below:

- `max-age`: Indicates that the MRCP client can tolerate the media resource server using content whose age is no greater than the specified time in seconds. Unless a `max-stale` directive is also included, the media resource server must not use stale content. This defaults to 'not defined'.
- `max-stale`: Indicates that the MRCP client is willing to allow a server to utilise cached data that has exceeded its expiration time by `max-stale` number of seconds. This usually defaults to 0.
- `min-fresh`: Indicates that the MRCP client is willing to allow the media resource server to accept a HTTP server response with cached data whose expiration is no less than its current age plus the specified time in seconds. This defaults to 'not defined'.

   The `Cache-Control` header field can be used in `SET-PARAMS` to set a default for the session (the session default can be retrieved via `GET-PARAMS`) or it can be used on a per-request basis.
   Example:

```
Cache-Control: max-age=86400
```

7.3.14 *Set-Cookie / Set-Cookie2*

Cookies are used to maintain state over multiple HTTP requests (see Appendix C for more information, located on the Web at http://www.daveburke.org/speechprocessing/). Within the MRCP framework,

HTTP requests can originate from several places. For example, a speech synthesiser resource may make HTTP requests for SSML documents, lexicon documents or audio files. A speech recogniser resource may make HTTP requests for grammar files and lexicon documents. In both these cases, the media resource server is behaving as a HTTP client. The MRCP client may also take on the role of an HTTP client and make HTTP requests, for example if it incorporates a VoiceXML interpreter. With no coordination from these different HTTP clients, the HTTP server would not be able to use stateful sessions encompassing the multiplicity of HTTP clients comprising the MRCP session, since any cookie set in one HTTP client would not apply to the other HTTP clients. The MRCP protocol provides a way for the MRCP client to update the cookie store on the media resource server with new cookies and also to return new cookies in MRCP responses and events to the MRCP client. In this way, the MRCP client can effectively manage a single logical cookie store. The cookie store on the media resource server only persists for the duration of the session and is associated with the particular media resource.

MRCP simple inherits the `Set-Cookie` and `Set-Cookie2` headers from HTTP. These headers can be placed in any request that subsequently results in the media resource performing an HTTP access. When a server receives new cookie information from an HTTP server, the media resource server returns the new cookie information in the final response or event as appropriate to allow the MRCP client to update its own cookie store. The `Set-Cookie` and `Set-Cookie2` header fields can be used with the `SET-PARAMS` request to update the cookie store on a server. The `GET-PARAMS` request can be used to return the entire cookie store of `Set-Cookie` or `Set-Cookie2` type cookies.

Example:

```
Set-Cookie2:   sessionid="1234"   Version="1";
Path="/test"
```

7.3.15 Vendor-Specific

The `Vendor-Specific` header field may be used to set or retrieve parameters that are implementation specific. `Vendor-Specific` parameters can be sent in any method to manage implementation-specific parameters exposed by the media resource. The header field value consists of `<name>=<value>` pairs delimited by a semicolon. The `<name>` format follows the reverse Internet domain name convention. When retrieving parameters, the `Vendor-Specific` header field contains a semicolon-delimited list of `<name>`s.

Example:

```
Vendor-Specific: com.mycompany.fastmode=true
 ;com.mycompany.lowpower=false
```

7.3.16 Logging-Tag

The `Logging-Tag` header field is used in `SET-PARAMS` and `GET-PARAMS` requests only. This header field specifies a tag to associate with log messages produced for a session. Media resource servers may use this header to annotate logs related to a particular session (e.g. to specify the caller ID of the user). This feature allows administrators to extract logs particular to a single call for analysis and debugging purposes.

Example:

```
Logging-Tag: 1234821
```

7.4 Security

The control session consists of one or more connections to one or more media resource servers. Depending on the application, it can be quite common for sensitive information to be passed over a control channel. For example, an application that requires a password to be spoken or entered via DTMF will use a speech recogniser resource to collect the information. The input collected from the user will be returned to the MRCP client over the control channel represented by NLSML data in a RECOGNITION-COMPLETE event message. Another application might synthesise private information such as an account number to be used in future transactions. In this case, the MRCP client could send the account number as part of an SSML body of a SPEAK request.

MRCP specifies two alternative mechanisms to protect the confidentiality and integrity of data passed over the control channel. The first mechanism uses a transport layer approach called Transport Layer Security (TLS). TLS is an evolution of the popular Secure Sockets Layer (SSL) traditionally used to power e-commerce security on the Web. TLS runs beneath the MRCP protocol but above the TCP transport layer and offers bidirectional authentication and communication privacy through the use of cryptography. An MRCP client can establish a secure control channel by specifying a transport sub-field of TCP/TLS/MRCPv2 on the m-line corresponding to the control channel (see Chapter 5 for more information).

An alternative mechanism for securing the control channel involves the use of network layer security in the form of IPSec. IPSec is a collection of network-layer protocol tools that serves as a secure replacement of the IP layer. IPSec is usually deployed between a set of hosts of administrative domains, either at the operating system level or on a security gateway that provides confidentiality and integrity for all traffic received from a particular interface (as in Virtual Private Network, VPN, architectures). An advantage of IPSec is that is can be deployed independently of the MRCP components since security is supplied at a low level.

7.5 Summary

In this chapter, we reviewed the MRCP protocol, which is an integral part of the overall MRCP framework. The three types of message were introduced, namely the request, response and event messages, in addition to the message exchange patterns. We discussed how message bodies with heterogeneous parts are encapsulated in multipart structures. We presented the two methods for setting and getting session parameters – 'sticky' variables that maintain their value for the life of the session. Finally, we talked about how MRCP differentiates headers into those that are media resource type-specific and those that apply more generally, and we summarised those generic headers.

This chapter concludes Part II. Next, we move on to Part III, 'Data Representation Formats'. This series of chapters looks at the XML representation data formats used in MRCP message bodies to convey information to and from media resources.

Part III

Data Representation Formats

8

Speech Synthesis Markup Language (SSML)

The MRCP speech synthesiser resource supports both plain text and marked-up text as input to be spoken. This chapter introduces the standard markup format that speech synthesiser resources are required to support, namely the W3C's Speech Synthesis Markup Language (SSML) [8]. SSML also comes up in the context of VoiceXML (see Chapter 16) as an author-friendly language for providing control over the speech synthesis process. Readers new to XML might find it useful to review Appendix B at this point.

8.1 Introduction

SSML is a standard, XML-based, markup annotation for instructing speech synthesisers how to convert written language input into spoken language output. SSML is primarily intended to help application authors improve the quality of synthesised text by providing a standard way of controlling aspects of the speech output such as pronunciation, volume, pitch and rate. SSML can also express playback of prerecorded audio. A number of key design criteria were chosen in the standardisation efforts that led to the development of SSML:

- *Consistency*: Provide predictable control of voice output across different speech synthesisers.
- *Interoperability*: Work with other W3C specifications such as VoiceXML.
- *Generality*: Support a wide range of applications with varied speech content.
- *Internationalisation*: Enable speech output in a large number of languages.
- *Generation and readability*: Support both automatic generation and hand authoring. The latter implies human readability.
- *Implementable*: The specification is designed to be implementable with existing, generally available technology.

Different speech synthesiser implementations use different underlying speech processing techniques (see Chapter 2). Hence the synthesised result for a given input text will inevitably vary across different implementations. SSML does not attempt to provide specific control of speech synthesisers but rather just provides a way for the application author to make prosodic and other information available to the speech synthesiser that it would otherwise not be able to acquire on its own. In the end, it is up to the speech synthesiser to determine whether and in what way to use this information.

8.2 Document structure

SSML documents are identified by the media type `application/ssml+xml`. Table 8.1 summarises the elements and attributes defined in SSML.

The basic structure of an SSML document is illustrated below:

```
<?xml version="1.0" encoding="UTF-8"?>
<speak version="1.0"
       xmlns="http://www.w3.org/2001/10/synthesis"
       xmlns:xsi="http://www.w3.org/2001/XMLSchema-instance"
       xsi:schemaLocation="http://www.w3.org/2001/10/synthesis
               http://www.w3.org/TR/speech-synthesis/synthesis.xsd"
       xml:lang="en-GB">
    Hello world!
</speak>
```

All SSML documents include the root element `<speak>`. The `version` attribute indicates the version of SSML and is fixed at 1.0 for the specification described in [8]. The default namespace for the SSML `<speak>` element and its children is indicated by the `xmlns` attribute and is defined as `http://www.w3.org/2001/10/synthesis`.

The `xmlns:xsi` attribute associates the namespace prefix of `xsi` to the namespace name `http://www.w3.org/2001/XMLSchema-instance`. The namespace prefix is defined since it is needed for the attribut, `xsi:schemaLocation`. The `xsi:schemaLocation` attribute indicates the location of the schema to validate the SSML document against. The first attribute value contains the SSML namespace followed by a hint to the location of the schema document. In the rest of the examples in this chapter, we will omit the `xmlns:xsi` and `xsi:schemaLocation` attributes for clarity. In doing so, this does not invalidate the markup although the SSML specification does recommend that both attributes be present. In reality, high-density systems typically may not validate against a schema for each and every document that arrives due to the significant processing overhead that is required for validation.

The `xml:lang` attribute indicates the language for the document and optionally also indicates a country or other variation. The format for the `xml:lang` value follows the language tag syntax described in RFC 3066 [55]. The syntax includes a two- or three-letter language identifier followed by a hyphen, and a two-letter country identifier (although other information can be registered here also). The language tag syntax is case insensitive although by convention the language subtag tends to be written in lower case and the country identifier in upper case characters. Table 8.2 illustrates examples of language identifiers.

It is possible to override the language within the document by including the `xml:lang` attribute on other elements as we will see. If the `xml:lang` attribute is omitted, the speech synthesis uses a default language. The contents of the `<speak>` element contains the text to synthesis and optionally further markup elements.

The `<p>` element and `<s>` element can be used explicitly to demarcate paragraphs and sentences. If they are omitted, the synthesiser will attempt to determine the structure using language-specific knowledge of the format of the plain text (e.g. the presence of a full stop, '.', to mark the end of the sentence). The `xml:lang` attribute may be specified on the `<p>` and `<s>` elements to indicate the language of the contents.

Table 8.1   Elements and attributes defined in SSML

Elements	Attributes	Description

`<speak>`	`version`
	`xmlns`
	`xml:lang`
	`xmlns:xsi`
	`xsi:schemaLocation`
	`xml:base`
		Root element for SSML documents.

`<lexicon>`	`uri` `type`	References an external pronunciation lexicon document.
`<p>`	`xml:lang`	Explicitly demarcates a paragraph.
`<s>`	`xml:lang`	Explicitly demarcates a sentence.
`<audio>`	`src`	Inserts a recorded audio file.
`<desc>`	`xml:lang`	Container element for a description of the audio source.
`<phoneme>`	`ph` `alphabet`	Provides a phonemic/phonetic pronunciation for the contained text.
`<sub>`	`alias`	Provides acronym / abbreviation expansions.

`<say-as>`	`interpret-as`
	`format`
	`detail`
		Used to indicate information on the type of text construct contained within the element.

`<break>`	`time` `strength`	Controls the pausing or other prosodic boundaries between words.
`<emphasis>`	`level`	Requests that the contained text be spoken with emphasis.

`<voice>`	`xml:lang`
	`gender`
	`age`
	`variant`
	`name`
		Requests a change to the speaking voice.

`<prosody>`	`pitch`
	`contour`
	`range`
	`rate`
	`duration`
	`volume`
		Provides control of the pitch, speaking rate and volume of the speech output.

`<mark>`	`name`	Places a marker into the text/tag sequence.

`<meta>`	`name`
	`http-equiv`
	`content`
		Contains metadata for the document.

`<metadata>`	—	Contains metadata for the document.

Table 8.2 Language identifier
examples

Identifer	Meaning
en-GB	English, Great Britain
en-US	English, United States
en-IE	English, Ireland
fr-FR	French, France
fr-CA	French, Canada
de-DE	German, Germany
de-CH	German, Switzerland
it-IT	Italian, Italy
da-DK	Danish, Denmark
nl-NL	Dutch, Netherlands
sv-SE	Swedish, Sweden
es-ES	Spanish, Spain
pt-BR	Portuguese, Brazil

An example SSML document using these elements is illustrated below:

```
<?xml version="1.0" encoding="UTF-8"?>
<speak version="1.0"
       xmlns="http://www.w3.org/2001/10/synthesis"
       xml:lang="en-GB">
   <p>
      <s>This is the first sentence in a paragraph</s>
      <s>This is a second sentence in a paragraph</s>
   </p>
</speak>
```

Most speech synthesisers internally provide extensive high quality lexicons with pronunciation information for many words or phrases. The <lexicon> element can be used to indicate an additional external pronunciation lexicon document identified by a URI for use in conjunction with the SSML document. A lexicon maps the orthography or written form of words and phrases to an associated pronunciation expressed in some sort of phonetic alphabet. The W3C has created a standard pronunciation lexicon format (see Chapter 11 for more information). One or more <lexicon> elements may be present as immediate children of the <speak> element. The uri attribute indicates the location of the external lexicon document and the optional type attribute specifies the type of the document (application/pls+xml is the standard media type for the W3C's Pronunciation Lexicon Specification [9]). For example:

```
<?xml version="1.0" encoding="UTF-8"?>
<speak version="1.0"
       xmlns="http://www.w3.org/2001/10/synthesis"
       xml:lang="en-GB">
   <lexicon uri="http://www.example.com/lexicon.pls"
               type="application/pls+xml"/>
      The centre is equidistant from the corners.
</speak>
```

The xml:base attribute may be specified on the <speak> element to resolve URIs used within <lexicon> and <audio> elements against. For example, if the uri attribute on <lexicon> specifies names.pls and the xml:base attribute on <speak> specifies http://www.example.com/, then the URI http://www.example.com/names.pls is used to identify the external lexicon.

8.3 Recorded audio

A simple yet powerful capability in SSML is its ability to play recorded audio referenced by a URI using the <audio> element. This allows the insertion of recorded audio within synthesised speech output or even creation of SSML documents that just play prerecorded audio. Table 8.3 lists the audio formats that need to be supported by SSML-compliant implementations. Other formats may be supported. The audio format can be identified[1] by its media type e.g. as supplied in the Content-Type header in a HTTP response.

Although SSML makes no explicit requirements in this area, common URI formats supported are http: and https: for fetching of files via the Hyper Text Transfer Protocol (HTTP) (see Appendix C, located on the Web at http://www.daveburke.org/speechprocessing/), file: for accessing local files, and rtsp: for streaming files using the Real-Time Streaming Protocol (RTSP) [54]. Unfortunately, SSML does not include capabilities to control fetching such as timeouts and cache control settings. VoiceXML (see Chapter 16), for example, extends SSML by adding the fetchtimeout, fetchhint, maxage and maxstale attributes to <audio>. The fetchtimeout attribute specifies the interval to wait for the content to be returned before giving up. The fetchhint attribute defines when the content should be retrieved from the server: a value of prefetch indicates that the content may be fetched when the page is loaded and a value of safe indicates that the content need only be fetched when actually needed. The maxage attribute originates from HTTP [12] and indicates that the SSML document is willing to use content whose age is no greater than the specified time in seconds. The maxstale attribute also originates from HTTP and indicates that the document is willing to use content that has exceeded its expiration time by the specified number of seconds. Without these extensions, the author is left to use the defaults of the speech synthesiser. A future version of SSML may address these shortcoming by adding similar attributes.

Table 8.3 SSML audio formats

Audio Format	Media type
Raw (headerless) 8 kHz 8-bit mono mu-law (PCM) single channel (G.711).	audio/basic
Raw (headerless) 8 kHz 8 bit mono A-law (PCM) single channel (G.711).	audio/x-alaw-basic
WAV (RIFF header) 8 kHz 8-bit mono mu-law (PCM) single channel.	audio/x-wav
WAV (RIFF header) 8 kHz 8-bit mono A-law (PCM) single channel.	audio/x-wav

---

[1] There are cases when the speech synthesiser has to determine the audio type itself, e.g. if the Content-Type is omitted in a HTTP response or the audio URI is a local file. In these cases, some speech synthesisers are able to open the file and analyse it, e.g. look for an identifying header.

An example of an SSML document containing synthesised and pre-recorded audio is illustrated below:

```
<?xml version="1.0" encoding="UTF-8"?>
<speak version="1.0"
       xmlns="http://www.w3.org/2001/10/synthesis"
       xml:lang="en-GB">
    Please say your name after the beep.
    <audio src="http://www.example.com/beep.wav"/>
</speak>
```

SSML also provides a fallback capability that allows the application author to specify alternate content to be spoken if an audio resource is not available. In the following example, the `company-intro.wav` file is unavailable (e.g. the Web server is malfunctioning) and as a result the text within the audio is spoken instead:

```
<?xml version="1.0" encoding="UTF-8"?>
<speak version="1.0"
       xmlns="http://www.w3.org/2001/10/synthesis"
       xml:lang="en-GB">
  <audio src="http://www.example.com/company-intro.wav">
      Welcome to AJAX, the asynchronous company.
  </audio>
</speak>
```

SSML includes an element called `<desc>`, which can be placed as a child of `<audio>` and whose textual content is a description of the audio source (e.g. 'screeching brakes'). A speech synthesiser, operating in a text-only output mode (as part of an accessible solution for the hearing impaired, for example), can use this element to provide descriptive text in place of recorded audio. The `<desc>` element also has an `xml:lang` attribute that may be used to specify the language of the descriptive text.

8.4 Pronunciation

In this section, we look at SSML mechanisms for controlling pronunciation.

8.4.1 Phonemic/phonetic content

The `<phoneme>` element allows an application author to insert content to be spoken in terms of a phonemic or phonetic alphabet. A phonemic alphabet consists of phonemes – language-dependent speech units that represent all the sounds needed to distinguish one word from another. A phonetic alphabet, on the other hand, consists of phones – language-independent speech units that characterise the manner (puff of air, click, vocalized, etc.) and place (front, middle, back, etc.) of articulation within the human vocal tract. Speech synthesisers implementing SSML usually support the International Phonetic Alphabet (IPA), which is mostly expressible in Unicode (IPA is discussed in Section 2.1.1). Many speech synthesisers also support their own proprietary alphabet formats. The `<phoneme>` element is useful for rendering words that a speech synthesiser is not able to render appropriately, for example some proper nouns or words from a different language.

An example using the `<phoneme>` element is illustrated below:

```
<?xml version="1.0" encoding="UTF-8"?>
<speak version="1.0"
       xmlns="http://www.w3.org/2001/10/synthesis"
       xml:lang="en-US">
    <phoneme alphabet="ipa"
             ph="t&#x259;mei&#x325;&#x27E;ou&#x325;">
        tomato
    </phoneme>
</speak>
```

The `alphabet` attribute on `<phoneme>` identifies the phonemic/phonetic alphabet in use – in this case IPA. Vendor-specific alphabets must commence with 'x-'. The ph attribute contains the string of phonetic characters to synthesise. The text content of the `<phoneme>` element is optional and is provided only to make the content more human readable – the speech synthesiser ignores it. For the ph attribute value, this example uses the XML entity escaped versions of the IPA characters encoded in UTF-8 (this is common practice since many editors cannot correctly cut and past Unicode characters).

8.4.2 Substitution

The purpose of the `<sub>` element is to allow an SSML document to contain both a written form and a spoken form – i.e. to make the document more human readable. The written form is contained as text within the `<sub>` element; the spoken form used by the speech synthesis is inserted into the alias attribute on `<sub>`. In the following example, the speech synthesiser speaks 'Media Resource Control Protocol':

```
<?xml version="1.0" encoding="UTF-8"?>
<speak version="1.0"
       xmlns="http://www.w3.org/2001/10/synthesis"
       xml:lang="en-US">
    <sub alias="Media Resource Control Protocol">
        MRCP
    </sub>
</speak>
```

Another common use of the `<sub>` element is to make sure an acronym is properly spoken. The following example ensures that the constituent letters of the acronym IPA are spoken:

```
<?xml version="1.0" encoding="UTF-8"?>
<speak version="1.0"
       xmlns="http://www.w3.org/2001/10/synthesis"
       xml:lang="en-US">
    <sub alias="I P A">IPA</sub>
</speak>
```

8.4.3 Interpreting text

The `<say-as>` element is used to indicate information about the type of text construct contained within the element and to help specify the level of detail for rendering the contained text. Interpreting the contained text in different ways will typically result in a different pronunciation of the content (although a speech synthesiser is still required to pronounce the contained text in a manner consistent with how such content is normally produced for the language).

The `<say-as>` element has three attributes: `interpret-as`, `format` and `detail`. The `format`, and `detail` attributes are optional. The `interpret-as` attribute indicates the content type of the contained text construct, e.g. `date` to indicate a date, or `telephone` to indicate a telephone number. The optional `format` attribute provides further hints on the precise formatting of the contained text, e.g. a value of `dmy` could be used to indicate that a date should be spoken in the format of date, then month, then year. The optional `detail` attribute indicates the level of detail to be spoken although it is not defined for many `interpret-as` types.

The values for the `<say-as>` attributes are not specified in the SSML specification itself but rather in a separate W3C Note [53]. Values for `interpret-as`, `format` and `detail` are simply ignored if they are not recognised by the speech synthesiser. Below are some common examples of `<say-as>`:

```
<say-as interpret-as="date" format="mdy">2/3/2006</say-as>
<!-- Interpreted as 3rd of February 2006 -->

<say-as interpret-as="time" format="hms24">01:59:59</say-as>
<!-- Interpreted as 1 second before 2 o'clock in the morning -->

<say-as interpret-as="ordinal">12</say-as>
<!-- Spoken in English as "twelfth" -->

<say-as interpret-as="cardinal">9</say-as>
<!-- Spoken in English as "nine" -->

<say-as interpret-as="characters" format="characters">WAY</say-as>
<!-- Spoken as letters W, A, Y and not the word "WAY" -->
```

8.5 Prosody

In this section, we look at the mechanisms SSML provides to control prosody – patterns of stress and intonation in the spoken audio.

8.5.1 Prosodic boundaries

The `<break>` element is used to insert explicit pauses or prosodic boundaries between words. When this element is not present, the speech synthesiser determines the break based on linguistic content. The `<break>` element has two attributes: `time` and `strength`. The `time` attribute indicates the duration to pause in seconds or milliseconds. The `strength` attribute indicates the strength of the prosodic break in the speech output. Allowed values are `none`, `x-weak`, `weak`, `medium` (the default value), `strong`, or `x-strong`. The value `none` indicates that no prosodic break boundary should

be outputted and is used to prevent a prosodic break in places where the speech synthesiser would
have otherwise inserted one. The other values indicate increasing break strength between words. The
stronger values are typically accompanied by pauses. The time and strength attributes are optional:
if both are omitted, a break will be produced by the speech synthesiser with a prosodic strength greater
than that which the processor would otherwise have used if no <break> element was supplied. A
simple example is:

```
<?xml version="1.0" encoding="UTF-8"?>
<speak version="1.0"
      xmlns="http://www.w3.org/2001/10/synthesis"
      xml:lang="en-GB">
   Take a deep breath and count to three <break time="3s"/>
   Now we are ready to proceed.
</speak>
```

8.5.2 Emphasis

The <emphasis> element is used to indicate that the contained text should be spoken with prominence
or stress. Emphasis may be realised with a combination of pitch changes, timing changes, volume
and other acoustic differences. The <emphasis> element has a single optional attribute called
level that indicates the strength of the emphasis to apply. Allowable values for this attribute are:
strong, moderate (the default value), none, and reduced. The none level is used to prevent
the speech synthesiser from emphasising words that it might otherwise emphasise. The reduced level
is effectively the opposite of emphasising a word. For example:

```
<?xml version="1.0" encoding="UTF-8"?>
<speak version="1.0"
      xmlns="http://www.w3.org/2001/10/synthesis"
      xml:lang="en-GB">
   That is the <emphasis level="strong">right</emphasis> answer!
</speak>
```

8.5.3 Speaking voice

The <voice> element is used to request a change in speaking voice. The <voice> element has five
optional attributes that can be used to indicate various aspects of the requested voice: xml:lang,
gender, age, name, and variant. The xml-lang attribute is used to indicate the language of
the content. The gender attribute indicates the preferred gender of the voice and may take a value of
male, female or neutral. The age specifies the preferred age in years since birth of the voice.
The name attribute can be used to indicate an implementation-specific voice name. The value may be
a space-separated list of names ordered from top preference down. Finally, the variant attribute can
be used to provide a positive integer indicating a preferred variant of the other voice characteristics to
speak the contained text (e.g. the third female voice for language en-GB).

When a speech synthesiser encounters a <voice> element and a single voice is available that
exactly matches the attributes specified, then that voice is used. Otherwise, the following steps are
used to find a voice:

(i) If the `xml:lang` attribute is specified and a voice is available for that language, it must be used. If there are multiple such voices available, the speech synthesis typically chooses the voice that best matches the specified values for `name`, `variant`, `gender` and `age`.

(ii) If there is no voice that matches the requested `xml:lang`, then the speech synthesiser may choose one that is closely related, for example a variation on the dialect for a language.

(iii) Otherwise an error may be returned. In the context of MRCP, this means a `Completion-Cause` of `005 language-unsupported` may be returned in the `SPEAK-COMPLETE` event (see Chapter 12).

Since changing voice is quite a drastic operation (different voices may have different defaults for pitch, volume, rate, etc.), it is usually recommended only on sentence or paragraph boundaries. The following example illustrates use of the `<voice>` element:

```
<?xml version="1.0" encoding="UTF-8"?>
<speak version="1.0"
       xmlns="http://www.w3.org/2001/10/synthesis"
       xml:lang="en-GB">
   <voice gender="male">
       Welcome to online shopping for him
   </voice>
   <voice gender="female">
       and her!
   </voice>
</speak>
```

8.5.4 Prosodic control

The `<prosody>` element enables the application author to control the volume, speaking rate and pitch of the speech output of the contained text within the element. The `<prosody>` element has six optional attributes: `volume` for indicating the volume for the contained text, `pitch`, `range` and `contour` to specify pitch information, and `rate` and `duration` to control the speaking rate. The `<prosody>` attribute values should be interpreted as 'hints' to the speech synthesiser. If the `<prosody>` attribute value is outside the permissible range for the speech synthesiser, a best effort to continue processing is made by setting that attribute value as close to the desired value as possible. A speech synthesiser may also choose to ignore a `<prosody>` attribute value if it would otherwise result in degraded speech quality.

8.5.4.1 Volume

The value of the `volume` attribute can either be a number, a relative change or a label. Legal number values may range from 0.0 to 100.0 where higher values are louder and 0 is silent. Relative changes are specified with a '+' or '−' character followed by a number and a '%' character (e.g. '+10%') or just a number (e.g. '+15'). Legal labels are `silent`, `x-soft`, `soft`, `medium`, `loud`, `x-loud`, or `default`. Labels `silent` through `x-loud` represent a linearly increasing sequence of volume levels. For example

```
<?xml version="1.0" encoding="UTF-8"?>
<speak version="1.0"
```

```
        xmlns="http://www.w3.org/2001/10/synthesis"
        xml:lang="en-GB">
    Please consult our website for job vacancies.
    <prosody volume="soft">
        ACME is an equal-opportunities employer
    </prosody>
</speak>
```

8.5.4.2 Speaking rate

The speaking rate is controlled by the `rate` and `duration` attributes. The value of the `rate` attribute can either be a relative change or a label. When a number is used as a relative change, its value is interpreted as a multiplier, e.g. '+2' indicates a doubling of the rate. Legal labels are `x-slow`, `slow`, `medium`, `fast`, `x-fast`, or `default`. Labels `x-slow` through `x-fast` represent an increasing sequence of rates. The `default` rate is the normal speaking rate for the voice when reading text aloud. Alternatively, the `duration` attribute can be used to express the rate indirectly by specifying a value in milliseconds or seconds for the desired time to read the text contained in the `<prosody>` element, e.g. '750 ms' or '5 s'. If both `rate` and `duration` are specified, `duration` takes precedence. For example

```
<?xml version="1.0" encoding="UTF-8"?>
<speak version="1.0"
        xmlns="http://www.w3.org/2001/10/synthesis"
        xml:lang="en-IE">
    Please select an option from the following.
    <prosody rate="-30%">
        Press 1 for sales or press 2 for support.
    </prosody>
    Otherwise hold the only line and a customer
    service representative will be with you shortly.
</speak>
```

8.5.4.3 Pitch

Pitch is controlled by the `pitch`, `range`, and `contour` attributes on the `<prosody>` element. The `pitch` attribute sets the baseline pitch for the contained text of the `<prosody>` element. Increasing the value for `pitch` will increase the approximate pitch of the spoken output. The value of the `pitch` attribute can either be a number followed by 'Hz', a relative change, or a label. As before, a relative change is indicated by a '+' or '−' followed by a number or a percentage e.g. '+3' or '+80%'. Legal label values are: `x-low`, `low`, `medium`, `high`, `x-high`, or `default`. Labels `x-low` through `x-high` represent increasing pitch levels. For example:

```
<?xml version="1.0" encoding="UTF-8"?>
<speak version="1.0"
        xmlns="http://www.w3.org/2001/10/synthesis"
        xml:lang="en-IE">
    <prosody pitch="x-high">
        When it hurts, I speak in a high-pitched voice!
```

```
    </prosody>
</speak>
```

The `range` attribute enables the pitch variability for the contained text of `<prosody>` to be altered. The values for the `range` attribute are the same as for the `pitch` attribute. Increasing the value for `range` will increase the dynamic range of the output pitch. For example:

```
<?xml version="1.0" encoding="UTF-8"?>
<speak version="1.0"
       xmlns="http://www.w3.org/2001/10/synthesis"
       xml:lang="en-IE">
    <prosody range="+100%">
        Life is full of ups and downs.
    </prosody>
</speak>
```

Finally, the `contour` attribute provides more advanced control of pitch. The `contour` attribute enables the application author to state the target pitch at different time positions for the contained text of the `<prosody>` element. The format for the `contour` attribute value is a space-delimited list of (`time-position`, `pitch-target`) pairs. The format for the `time-position` is a percentage ranging from 0 % to 100 %. The format for the `pitch-target` is the same as for the `pitch` attribute on `<prosody>`. The `pitch-target` value is taken relative to the pitch just before the `<prosody>` element. For example:

```
<?xml version="1.0" encoding="UTF-8"?>
<speak version="1.0"
       xmlns="http://www.w3.org/2001/10/synthesis"
       xml:lang="en-IE">
    <prosody contour="(0%,+10Hz) (50%,+50Hz) (100%,-50Hz)">
        Life is full of ups and downs.
    </prosody>
</speak>
```

8.6 Markers

SSML provides the `<mark>` element to allow markers to be inserted into the markup. The `<mark>` element has one required attribute: `name`. The `name` attribute specifies the name of the mark. In the context of MRCP, when the rendered audio output of the speech synthesiser resource encounters a `<mark>` element, the SPEECH-MARKER event is sent to the MRCP client with the `name` of the mark inserted in the `Speech-Marker` header field (see Chapter 12 for more information). Markers can be used to determine how much of the output actually gets rendered to the user when barge-in is activated. Mark information is useful to decide if an advertisement or important notice was played, or to mark a place in the content to resume from later, etc. For example:

```
<?xml version="1.0" encoding="UTF-8"?>
<speak version="1.0"
       xmlns="http://www.w3.org/2001/10/synthesis"
       xml:lang="en-GB">
```

```
    <mark name="before-ad"/>
    <audio src="advertisement.wav"/>
    <mark name="after-ad"/>
    Please select from the following options.
</speak>
```

8.7 Metadata

SSML[2] provides two container elements for metadata: the `<meta>` element and the `<metadata>` element. The `<meta>` element is an older mechanism while the `<metadata>` element provides a newer, more general and powerful mechanism for specifying metadata. It is likely that a future version of SSML will only support the `<metadata>` approach.

The `<meta>` element has three attributes: `name`, `http-equiv` and `content`. One of `name` or `http-equiv` is required. The `content` attribute is always present. The only defined `name` value in SSML is `seeAlso` – the corresponding content attribute value specifies a resource (i.e. a URI) that provides additional information about the content (SSML does not specify what format that additional information is in, however). The `http-equiv` attribute, on the other hand, is used to specify HTTP response header information. This is intended to be useful in cases where the author is unable to configure the HTTP header fields on their Web server. Note that if an application author specifies HTTP header field information inside an SSML document, there is no guarantee (in fact it is very unlikely) that the Web server (or intermediate proxy servers) will pick up that information and set the HTTP header fields themselves. This approach of specifying protocol information in the data markup is a legacy from HTML and not recommended for use in practice. An example of an SSML document using the `<meta>` element is illustrated below:

```
<?xml version="1.0" encoding="UTF-8"?>
<speak version="1.0"
       xmlns="http://www.w3.org/2001/10/synthesis"
       xml:lang="en-GB">
    <meta http-equiv="Cache-Control" content="max-age=0"/>
    <meta name="seeAlso"
          content="http://www.example.com/metadata.xml"/>
    Welcome to online banking.
</speak>
```

The `<metadata>` element is a container element where information about the document may be placed. That information is represented in XML syntax by using a different namespace to that of the rest of the SSML document. The SSML specification recommends the XML syntax of the Resource Description Framework (RDF) [63] be used in conjunction with the general metadata properties defined in the Dublin Core Metadata Initiative [64]. RDF provides a standard way to represent metadata in the form of statements and relationships. The Dublin Core Metadata Initiative provides generally applicable metadata properties such as title, subject, creator, etc. An SSML document using RDF and Dublin Core Metadata is illustrated below:

---

[2] Several other W3C specifications use the same metadata approach, including SRGS (described in Chapter 9), PLS (described in Chapter 11), and VoiceXML (described in Chapter 16).

```xml
<?xml version="1.0" encoding="UTF-8"?>
<speak version="1.0"
       xmlns="http://www.w3.org/2001/10/synthesis"
       xml:lang="en-GB">
    <metadata>
    <rdf:RDF
        xmlns:rdf="http://www.w3.org/1999/02/22-rdf-syntax-ns#"
        xmlns:rdfs="http://www.w3.org/2000/01/rdf-schema#"
        xmlns:dc="http://purl.org/dc/elements/1.1/">
        <rdf:Description
            rdf:about="http://www.example.com/meta.xml"
            dc:Title="Online banking welcome"
            dc:Description="Provides a welcome message"
            dc:Publisher="D Burke"
            dc:Language="en-US"
            dc:Date="2006-02-02"
            dc:Rights="Copyright (c) D Burke"
            dc:Format="application/ssml+xml" >
            <dc:Creator>
                <rdf:Seq ID="CreatorsAlphabeticalBySurname">
                    <rdf:li>Dave Burke</rdf:li>
                </rdf:Seq>
            </dc:Creator>
        </rdf:Description>
    </rdf:RDF>
    </metadata>

        Welcome to online banking.
</speak>
```

8.8 Summary

This chapter covered the main features of SSML. SSML is a simple yet flexible markup language designed to enable application authors control aspects of spoken output from speech synthesisers in a standard way. MRCP-compliant speech synthesisers are required to support SSML. Speech application authors using VoiceXML also employ SSML to express spoken and recorded audio to playback to the user.

Chapter 9, next, discusses a markup language that is used for expressing the permissible set of words and phrases from which a speech recogniser can recognise.

9

Speech Recognition Grammar Specification (SRGS)

This chapter introduces the Speech Recognition Grammar Specification (SRGS) [6], a W3C standard that defines a syntax for representing grammars for use in speech recognition. SRGS enables applications developers to specify the words, and patterns of words, to be listened for by a speech recogniser in a standard way. SRGS also supports the ability to express patterns of DTMF key presses. There are actually two equivalent formats of SRGS: the XML form and the ABNF form. In this chapter, and indeed this book, we are concerned only with the XML form since it is the format most widely used and the format that is required to be supported by MRCP speech recognisers and VoiceXML platforms. We introduce the basic concepts of SRGS grammars and also include a section on semantic interpretation – the mechanism by which annotations can be used to associate meaning with raw utterances to facilitate natural language understanding.

9.1 Introduction

Speech grammars are used to apply linguistic constraints to the speech recognition task. Linguistic constraints lessen the uncertainty of the content of sentences and hence facilitate recognition. Formally, a grammar is a set of rules by which symbols may be properly combined in a language. The language is the set of all possible combinations of symbols. A terminal symbol or token is the primitive component of a grammar. In the context of speech grammars, a word delimited by white space is a token. An SRGS grammar which is recursive (i.e. contains self-referencing rules) has the expressive power of a context free grammar. An SRGS grammar that is not recursive has the expressive power of a finite state or regular grammar. With finite state grammars, the rules always map a non-terminal into either just a terminal, or a terminal and a different non-terminal. (A context free grammar is less restrictive and can map a non-terminal into the same non-terminal and hence be recursive). As we saw in Chapter 2, the Hidden Markov Model (HMM), which forms the core of modern speech recognisers, is itself a finite state automaton – it can produce language described by a regular grammar. This explains why HMMs and regular grammars go hand in hand, and why both feature so prominently in modern speech recognition.

Speech Processing for IP Networks   Dave Burke
© 2007 John Wiley & Sons, Ltd

Modern speech recognisers can handle grammars with small vocabularies right up to vocabularies that contain hundreds of thousands of words. Accuracy and speed of a recogniser does slowly degrade with increasing vocabulary size – the difficulty of the recognition task increases logarithmically with the size of the vocabulary. Another dimension of a grammar is its perplexity. Perplexity roughly refers to the average number of branches at any decision point, in other words, the average number of terminals or non-terminals that may follow a non-terminal. Clearly, increasing grammar perplexity will have a negative impact on recognition accuracy and speed as, potentially, each path has to be considered. One difficulty with using regular grammars for speech recognition that ought to be called out is that they are usually hand written and require the author to foresee all the different sentence patterns that users can come up with in spontaneous speech. As a result, developing high quality grammars usually requires specialist input, pilot testing and follow-up tuning after an application has been put into production.

Finally, note that SRGS grammars on their own do not make any attempt to assign a meaning to a recognised string of words. This function is provided by a separate specification called Semantic Interpretation for Speech Recognition [7] and is discussed in Section 9.8.

9.2 Document structure

SRGS documents employing the XML form[1] are identified by the media type `application/srgs+xml`. Table 9.1 summarises the elements and attributes defined in SRGS.

The basic structure of an SRGS grammar is illustrated below:

```
<?xml version="1.0" encoding="UTF-8"?>
<grammar version="1.0"
         xmlns="http://www.w3.org/2001/06/grammar"
         xmlns:xsi="http://www.w3.org/2001/XMLSchema-instance"
         xsi:schemaLocation="http://www.w3.org/2001/06/grammar
             http://www.w3.org/TR/speech-grammar/grammar.xsd"
         mode="voice"
         xml:lang="en-GB"
         root="example">
    <rule id="example">
        yes
    </rule>
</grammar>
```

All SRGS documents include the root element `<grammar>`. The `version` attribute indicates the version of SRGS and is fixed at 1.0 for the specification described in [6]. The default namespace for the SRGS `<grammar>` element and its children is indicated by the `xmlns` attribute and is defined as `http://www.w3.org/2001/06/grammar`. The `xmlns:xsi` and `xsi:schemaLocation` attributes are used to indicate the XML schema for the document identical to the mechanism used for SSML described in the previous chapter. In the rest of the examples in this chapter, we will omit the `xmlns:xsi` and `xsi:schemaLocation` attributes for clarity.

The `mode` attribute specifies the type of input that the grammar applies to and may be one of `voice` or `dtmf`. A value of `voice` indicates that the input is speech and is the default. DTMF grammars are discussed in Section 9.7. Note that grammars cannot be mixed – they either express speech or DTMF input but not both. It is possible, however, to have both speech grammars and DTMF grammars

---

[1] The ABNF form of SRGS grammars are identified by the media type `application/srgs`.

Table 9.1   Elements and attributes defined in SRGS

Elements	Attributes	Description

`<grammar>`	`version`
	`xmlns`
	`root`
	`mode`
	`tag-format`
	`xml:lang`
	`xml:xsi`
	`xsi:schemaLocation`
	`xml:base`
		Root element for SRGS documents.

`<lexicon>`	`uri` `type`	References an external pronunciation lexicon document.
`<rule>`	`id` `scope`	Container element for a grammar rule definition.
`<token>`	`xml:lang`	Container element for explicitly demarcating a token.
`<one-of>`	`xml:lang`	Container element for alternates.

`<item>`	`repeat`
	`repeat-prob`
	`weight`
	`xml:lang`
		Container used to demarcate individual alternate rule expansions and to assign repeat information.

`<ruleref>`	`uri`
	`type`
	`special`
		Reference to another rule.

`<tag>`	–	Container element for semantic interpretation information.
`<example>`	–	Container element for example phrases that match containing rule definition.

`<meta>`	`name`
	`http-equiv`
	`content`
		Contains metadata for the document. See Chapter 8.

`<metadata>`	–	Contains metadata for the document. See Chapter 8.

activated at the same time in the recogniser, thus allowing the user to input speech or DTMF (see Chapter 13 for more information).

The `xml:lang` attribute specifies the primary language identifier. It is required for all grammars where the `mode` is of type `voice` (and is ignored for DTMF grammars). The `xml:lang` may be specified on other elements within SRGS (namely `<one-of>`, `<token>`, and `<item>`) to override the primary language identifier for that scope. The format of the `xml:lang` attribute value was described in the previous chapter on SSML (see Section 8.2).

All grammars have one or more rules identified via the `<rule>` element. The `id` attribute on the `<rule>` element indicates the name of the rule and must be unique within the grammar (this is enforced by XML). In the example above, the trivial rule called `example` can only match the token yes. The root rule is the starting rule to use to match an input and is indicated by the `root` attribute on the `<grammar>` element. Rules are described further in the next section.

The `<lexicon>` element can be used to indicate an external pronunciation lexicon document identified by a URI to use in conjunction with the SRGS document. This mechanism is exactly the same as the one described for SSML in the previous chapter. Also similar to SSML, the `xml:base` attribute may be specified on the `<grammar>` element to resolve URIs used within the grammar document. Specifically, this attribute can be used to specify the base URI of relative URIs given on `<lexicon>` and `<ruleref>` elements.

9.3 Rules, tokens and sequences

Grammars comprise one or more rule definitions. Each rule definition is represented by a `<rule>` element and assigned a rule name with the `id` attribute on that element. The content of the `<rule>` element is called a rule expansion. A rule expansion typically matches a portion of the full utterance.

A simple rule expansion consists of a sequence of tokens. A single word delimited by white space is the most common example of a token. For example, the following grammar accepts the input 'I want a hot pepperoni pizza':

```
<?xml version="1.0" encoding="UTF-8"?>
<grammar version="1.0"
         xmlns="http://www.w3.org/2001/06/grammar"
         mode="voice"
         xml:lang="en-GB"
         root="pizza">
    <rule id="pizza">
        I want a hot pepperoni pizza
    </rule>
</grammar>
```

In the above example, the speech utterance must match exactly the sequence of tokens. It is not possible to match a subset of the sequence (e.g. just saying 'pepperoni pizza' will not match the rule). Furthermore, the temporal order of the input must match the order written in the grammar (e.g. saying 'I want a pepperoni pizza hot' will not match). The `<token>` element can be used explicitly to designate a token and is a way of allowing white space within the token. For example, the occurrence of the words `New Zealand` would be considered as two tokens while the occurrence of `<token>New Zealand</token>` would be considered as a single token. The most common usage of the `<token>` element is to indicate the language of the token by including the `xml:lang` on the element.

SRGS includes an element called `<example>` to enable the grammar author to include examples of phrases that match rule definitions along with the definition. Zero or more `<example>` elements may be included as a child of `<rule>` as illustrated below:

```
<?xml version="1.0" encoding="UTF-8"?>
<grammar version="1.0"
         xmlns="http://www.w3.org/2001/06/grammar"
         mode="voice"
         xml:lang="en-GB"
```

```
            root="pizza">
    <rule id="pizza">
        I want a hot pepperoni pizza
        <example>I want a hot pepperoni pizza</example>
    </rule>
</grammar>
```

The content of the `<example>` element improves human readability of the grammar and may also be used by automated tools for regression testing and for the generation of grammar documentation.

The grammar examples in this section are, of course, very restrictive. The real power of SRGS comes from its ability to express more advanced rule expansions as we will see in the following sections.

9.4 Alternatives

Alternatives allow the expression of one of a set of alternative rule expansions. Alternatives are expressed with the `<one-of>` and `<item>` elements. Each `<item>` child of `<one-of>` contains an alternative rule expansion. For example:

```
<?xml version="1.0" encoding="UTF-8"?>
<grammar version="1.0"
        xmlns="http://www.w3.org/2001/06/grammar"
        mode="voice"
        xml:lang="en-GB"
        root="pizza">
    <rule id="pizza">
        I want a hot
        <one-of>
            <item>pepperoni</item>
            <item>vegetarian</item>
            <item>cheese</item>
        </one-of>
        pizza
    </rule>
</grammar>
```

This grammar accepts more flexible input – the following utterances match: 'I want a hot pepperoni pizza', 'I want a hot vegetarian pizza', 'I want a hot cheese pizza'. Note that the content of the `<item>` element is itself a rule expansion and may therefore contain any legal rule expansion including sequences of tokens and more alternatives.

Advanced grammars can assign weights to different alternatives. A weight is a multiplying factor in the likelihood domain of a speech recognition search.[2] A value of 1.0 is equivalent to no weight. A value greater than 1.0 positively biases the alternative. A value less than 1.0 negatively biases the alternative. For example, we might augment the previous grammar to positively bias one of the alternatives that is known to have a significantly higher a priori occurrence:

---

[2] This definition is intentionally vague since the specifics of how a speech recogniser uses the weight value tends to be implementation dependent.

```
<?xml version="1.0" encoding="UTF-8"?>
<grammar version="1.0"
         xmlns="http://www.w3.org/2001/06/grammar"
         mode="voice"
         xml:lang="en-GB"
         root="pizza">
   <rule id="pizza">
       I want a hot
       <one-of>
          <item>pepperoni</item>
          <item>vegetarian</item>
          <item weight="2.0">cheese</item>
       </one-of>
       pizza
   </rule>
</grammar>
```

The xml:lang attribute may be used on the <one-of> or <item> elements to express the language for the contained alternative rule expansions.

9.5 Rule references

Rule expansions may also contain rule references. A rule reference is used to 'pull in' another rule and is indicated by the <ruleref> element. The primary purpose of a rule reference is to permit reuse of a rule expansion. Following on from our pizza example, we can factor out the pizza type into a separate rule:

```
<?xml version="1.0" encoding="UTF-8"?>
<grammar version="1.0"
         xmlns="http://www.w3.org/2001/06/grammar"
         mode="voice"
         xml:lang="en-GB"
         root="pizza">
   <rule id="pizza">
       I want a hot <ruleref uri="#type"/> pizza
   </rule>
   <rule id="type">
       <one-of>
           <item>pepperoni</item>
           <item>vegetarian</item>
           <item>cheese</item>
       </one-of>
   </rule>
</grammar>
```

The referenced rule is indicated by the uri attribute value. The '#' character is used to indicate a fragment identifier corresponding to a rule name indicated by the id attribute on the referenced

`<rule>`. In this example, the referenced rule (`type`) is local to the referencing rule (`pizza`), i.e. it is within the same grammar file. It is also possible to refer to a rule in an external grammar file by including a grammar URI, for example:

```
<ruleref uri="http://www.example.com/file.grxml#type"/>
```

If the fragment identifier is omitted from the URI, then the root rule of the grammar is used (recall that the root rule is the rule indicated by the `root` attribute value on the `<grammar>` element). It is considered good practice always to include the `root` attribute in grammars. In the previous example, we designated the pizza rule as the root rule. Referencing a remote grammar implicitly (without a grammar fragment) when the referenced grammar does not include the `root` attribute on its `<grammar>` element will cause an error in the speech recogniser.

Each defined grammar `<rule>` within a grammar has a defined scope. The scope of a grammar is indicated by the `scope` attribute on the `<rule>` element. The scope attribute value may take on a value of `private` or `public` (the default is `private`). A public-scoped rule may be explicitly referenced (via `<ruleref>`) in the rule definitions of other grammars. A private-scoped rule, on the other hand, may only be referenced within the grammar. The use of scope designation allows grammar authors to leverage principles of encapsulation and reuse. The exposed (public) rules of a grammar can be thought of as an API which is establishing a contract with other grammars. By hiding the internals of a grammar using private-scoped rules, it easier to create libraries of grammars that can be reused. These hidden rules can be revised and improved without affecting the referencing grammars. Finally, by making rules private, advanced grammar compilers can perform certain intra-grammar optimisations that might not be possible were those rules public.

A powerful feature of the `<ruleref>` element is its ability to 'pull in' other grammar formats beyond SRGS. The `uri` attribute on `<ruleref>` may refer to other grammar formats, for example a proprietary voice-enrolled or SLM grammar. The format of the referenced grammar type is identified through its returned media type (with HTTP URIs, this is indicated by the `Content-Type` header in the HTTP response). The `<ruleref>` element also allows the grammar author to indicate the `type` for the grammar in the event that the media type is not indicated (e.g. a misconfigured web server or a file reference).

9.5.1 Special rules

SRGS defines three special rules identified by the reserved rule names of NULL, VOID, and GARBAGE. A special rule is indicated via the `special` attribute on the `<ruleref>` element (only one of the `uri` attribute or `special` attribute may be present). The NULL rule is automatically matched without the user speaking any word. The NULL rule serves as an authoring convenience to force a part of a grammar always to be matched. Inserting a VOID rule in a sequence has the opposite effect and will cause that sequence never to be matched. The VOID rule serves as an authoring convenience to disable part of a grammar. The GARBAGE rule will match any speech up until the next token, rule reference, or end of spoken input. For example, consider the following grammar:

```
<?xml version="1.0" encoding="UTF-8"?>
<grammar version="1.0"
        xmlns="http://www.w3.org/2001/06/grammar"
        mode="voice"
        xml:lang="en-GB"
```

```
        root="pizza">
  <rule id="pizza">
     I want a hot <ruleref special="GARBAGE"/> pizza
  </rule>
</grammar>
```

Any speech occurring between the words hot and pizza will be matched. For example, the following will match the above grammar: 'I want a hot something or other pizza'.

9.6 Repeats

SRGS provides a construct allowing a rule expansion to be repeated. This greatly enhances the compactness of the grammar. For example, a grammar rule defining the digits from zero to nine could be referenced by a `<ruleref>` element and repeated a number of times to allow a variable length digit string to be matched. The `<item>` element and its `repeat` attribute are used to indicate the number of times the contained rule expansion may be repeated. The three variations for the permitted syntax for the `repeat` attribute value are summarised in Table 9.2.

The `repeat` attribute values allow the grammar author to express common repeats. A repeat of '0-1' indicates that the contained rule expansion is optional. A repeat of '0-' indicates that the contained rule expansion can occur zero, one, or many times and corresponds to the 'Kleene star'. A repeat of '1-' indicates that the contained rule expansion can occur one, or more times and corresponds to 'positive closure'. It should be pointed out that a speech recogniser will usually not accept input indefinitely, thus putting an implementation-specific upper bound on the number of repeats accepted. A variant on our previous examples allows the user to say the word 'very' a number of times.

```
<?xml version="1.0" encoding="UTF-8"?>
<grammar version="1.0"
        xmlns="http://www.w3.org/2001/06/grammar"
        mode="voice"
        xml:lang="en-GB"
        root="pizza">
  <rule id="pizza">
     I want a <item repeat="0-3">very</item> hot
     <ruleref uri="#type"/> pizza
  </rule>
  <rule id="type">
     <one-of>
        <item>pepperoni</item>
        <item>vegetarian</item>
        <item>cheese</item>
     </one-of>
  </rule>
</grammar>
```

Table 9.2 Syntax for the `repeat` attribute value

Syntax	Meaning
repeat="n"	The rule expansion contained inside the `<item>` element is repeated exactly n times.
repeat="m-n"	The rule expansion contained inside the `<item>` element is repeated between m and n times.
repeat="m-"	The rule expansion contained inside the `<item>` element is repeated m times or more.

A second attribute of the `<item>` element, `repeat-prob`, allows the grammar author to express the probability that the rule expansion is repeated. For example, a repeat of '0–1' with a repeat probability of 0.75 indicates that the probability that the rule expansion will be matched is 75 % and conversely, the chance that the rule expansion will not be matched is 25 %. The `repeat-prob` attribute only applies when the repeat attribute value is in the form of a range. Note that for infinitely recurring ranges (e.g. '1-'), the effective probability repetition decays exponentially with each repeat (because the joint probability of occurrence of N independent events equals the probability of each event multiplied). Note also that how a speech recogniser implements repeat probabilities is not specified in SRGS and hence may vary from one implementation to another. Usually, repeat probabilities are assigned values based on statistical analysis of real patterns of user input.

9.7 DTMF grammars

SRGS grammars were designed for use with speech recognisers in telecommunication settings. It should come as little surprise that SRGS grammars also support the ability to express user input in terms of DTMF digits. Indeed, a common strategy for speech applications is to 'fallback' to using DTMF in the event that the environment proves very noisy. DTMF is also useful for entering sensitive information such as passwords. As noted earlier, SRGS grammars are declared either as speech grammars or DTMF grammars via the `mode` attribute on the `<grammar>` root element.

DTMF grammars behave almost identically to their speech counterparts. The main different is that the tokens are limited to a finite set of 12 characters: the digits '0' to '9' inclusive, the '#' character, the '*' character and the rarely used characters 'A', 'B', 'C', 'D' (see Section 6.2.2 for more information). As with speech grammars, the tokens are delimited by white space. Unlike speech grammars, the language declaration does not apply – it is simply ignored. A simple DTMF grammar that accepts a four digit entry is illustrated below:

```
<?xml version="1.0" encoding="UTF-8"?>
<grammar version="1.0"
         xmlns="http://www.w3.org/2001/06/grammar"
         mode="dtmf"
         root="digits">
   <rule id="digits">
      <item repeat="4"><ruleref uri="#digit"/></item>
   </rule>
   <rule id="digit">
      <one-of>
```

```
            <item>0</item>
            <item>1</item>
            <item>2</item>
            <item>3</item>
            <item>4</item>
            <item>5</item>
            <item>6</item>
            <item>7</item>
            <item>8</item>
            <item>9</item>
        </one-of>
    </rule>
</grammar>
```

9.8 Semantic interpretation

Semantic interpretation refers to the mechanism of associating semantics or meaning to matched utterances. While the sequence of words comprising the matched utterance is certainly of use to the speech application, it is perhaps not the most useful information that can be returned in the recognition results. Many different utterances may be semantically equivalent and placing the burden on the application to detect the subtle similarities and differences across utterances would result in an awkward and tight coupling of the grammar to the application. Semantic interpretation allows the grammar author to annotate the grammar in such a way that the speech recogniser can also return a computer processable semantic result for the utterance. As an example, consider a grammar that accepts all the different ways a person may indicate the affirmative or negative (e.g. yea, OK, sure, that's fine, no, nope, no way, etc.). By returning a canonical positive or negative result to the speech application, the application may free itself from being concerned with the nuances of user input and, instead, just focus on the semantics of what the user said. This has the advantage that the permissible user input and its associated meaning is maintained in one place (i.e. inside the grammar) and the grammar's coverage can be extended independently of the application in many cases.

SRGS provides the `<tag>` element to allow the grammar author to annotate parts of the grammar with semantics. Inserting `<tag>` elements within a grammar does not modify the set of input utterances that can be matched by the grammar in any way. A `<tag>` element is selected when the preceding token or non-terminal (e.g. an alternate, repeat, etc.) is matched. The `tag-format` attribute on the `<grammar>` element indicates the content type for all `<tag>` elements within the grammar. SRGS does not define the actual contents of the `<tag>` elements; rather it leaves that to other specifications. The Semantic Interpretation for Speech Recognition (SISR) [7] specification provides a standard definition for semantic interpretation and defines the `tag-format` values of `semantics/1.0-literals` and `semantics/1.0`. The `semantics/1.0-literals` syntax defines the `<tag>` content as ECMAScript strings and thus facilitates a way of producing semantic results consisting of simple strings. The `semantics/1.0` syntax is more powerful and provides the expressive power of a full scripting language. With this syntax, the content of the `<tag>` elements are defined to be ECMAScript code [56], allowing intra-grammar computations to be carried out.

For both syntaxes, the final semantic result returned from the grammar is an ECMAScript data structure. In the case of the `semantics/1.0-literals` syntax, the result is restricted to be an ECMAScript string. Obviously, for the result to be transferred 'over-the-wire', some sort of encoding of that ECMAScript result is required. SISR defines an XML transformation of the ECMAScript result that can be carried within the Natural Language Semantics Markup Language (NLSML) and returned in the MRCP `RECOGNITION-COMPLETE` event from the speech recogniser. NLSML and the ECMAScript to XML serialisation mechanism are covered in the next chapter. In the following two subsections, we take a closer look at each of the two SISR syntaxes.

9.8.1 Semantic literals

Grammars employing semantic literals use a `tag-format` of `semantics/1.0-literals`. The contents of the `<tag>` elements are treated as strings (in the ECMAScript sense). Semantic literals are used for simple annotation of grammars and are ideal for normalising content. For example, we can normalise a yes/no grammar by using semantic literals similar to the following:

```
<?xml version="1.0" encoding="UTF-8"?>
<grammar version="1.0"
         xmlns="http://www.w3.org/2001/06/grammar"
         mode="voice"
         xml:lang="en-GB"
         root="yesno"
         tag-format="semantics/1.0-literals">
    <rule id="yesno">
        <one-of>
            <item>yes</item>
            <item>yea<tag>yes</tag></item>
            <item>aye<tag>yes</tag></item>
            <item>no</item>
            <item>nope<tag>no</tag></item>
            <item>nah<tag>no</tag></item>
        </one-of>
    </rule>
</grammar>
```

Regardless of which of a number of ways the user says yes or no, the semantic result returned is the ECMAScript string 'yes' or 'no'. Note that for the tokens `yes` and `no`, we chose to omit what would essentially be a redundant `<tag>` element. SISR uses 'default assignment' to assign the text utterance matching the rule when no `<tag>` element is selected. There is one caveat to default assignment to be careful of: if a rule contains a `<ruleref>` element, then the text associated with the last matched `<ruleref>` will be automatically used as the semantic result for that rule.

9.8.2 Semantic scripts

Grammars employing semantic scripts use a `tag-format` of `semantics/1.0`. The contents of the `<tag>` elements are treated as ECMAScript code. ECMAScript is a scripting language based on several originating technologies, the most well known being JavaScript and JScript – both languages were made popular by their use for client-side scripting in HTML web pages. ECMAScript is also used within VoiceXML. SISR specifically employs the ECMA-327 'Compact Profile' [56] version of standard ECMAScript (itself defined in ECMA-262 [57]). ECMA-327 Compact Profile is a subset of ECMA-262 designed to meet the requirements of resource-constrained environments. Features that require proportionately large amounts of memory or processing power are constrained. The rationale for using ECMA-327 Compact Profile instead of ECMA-262 in SISR is to avoid placing large processing or memory overhead on the server when large numbers of concurrent recognition activities are taking place.

Each rule within a grammar has a single *rule variable* called out. The rule variable is initialised to an empty object (equivalent to executing the ECMAScript statement:

```
var out = new Object();
```

Typically, the grammar author will assign properties to the rule variable or, instead, assign a scalar value to it such as an ECMAScript number or string. Often, scalar types are used in lower level rules and more structured objects in higher-level rules (particularly root rules). The values of rule variables from referenced rules are available to the referencing rule and hence values of referenced rules can be propagated upwards to the referencing rule. The semantic result returned for the grammar is the value of root variable of the root rule after all computations have been carried out. Consider a simple example:

```
<?xml version="1.0" encoding="UTF-8"?>
<grammar version="1.0"
         xmlns="http://www.w3.org/2001/06/grammar"
         mode="voice"
         xml:lang="en-GB"
         root="pizza"
         tag-format="semantics/1.0">
   <rule id="pizza">
       I want a
       <one-of>
           <item>large<tag>out.size="large";</tag></item>
           <item>small<tag>out.size="small";</tag></item>
       </one-of>
       <one-of>
           <item>pepperoni<tag>out.sort=1;</tag></item>
           <item>vegetarian<tag>out.sort=2;</tag></item>
           <item>cheese<tag>out.sort=3;</tag></item>
       </one-of>
       pizza
   </rule>
</grammar>
```

Given the utterance 'I want a large vegetarian pizza', the semantic result for the grammar is an ECMAScript object with two properties called size and sort of type string and number and taking values of 'large' and 2 respectively. We can represent such an ECMAScript object using the following 'Object Initialiser' notation:[3]

---

[3] Actually, the object can be instantiated directly using the following ECMAScript code: var out = {size: "large", sort: 2}; This is equivalent to executing the three statements: var out = new Object(); out.size="large"; out.sort=2;

```
{
   size: "large",
   sort: 2
}
```

The curly braces indicate the presence of an object. Properties are contained within the curly braces and written in the form of name: value. Commas delimit multiple properties. String values are surrounded in quotes. The speech application can operate on this data, the semantics of the user's utterance. For example, the application might place a customer order by creating a record in a database using this information. This is a typical example of how semantic scripts are used, i.e. to aggregate semantically related information into a single data structure to pass back to the speech application.

SRGS also allows the placement of <tag> elements directly as children of the <grammar> element. The contents of these elements are used to define global variables that can be accessed as read-only[4] elsewhere. Global <tag> elements are executed once and before any <tag> elements in the rules. The following grammar accepts the utterance of either 'one' or 'two' and returns a semantic result consisting of a number after performing a computation to add an offset to the number spoken by the user:

```
<?xml version="1.0" encoding="UTF-8"?>
<grammar version="1.0"
        xmlns="http://www.w3.org/2001/06/grammar"
        mode="voice"
        xml:lang="en-GB"
        root="example"
        tag-format="semantics/1.0">
  <tag>var offset=10;</tag>
  <rule id="example">
     <one-of>
        <item>one<tag>out=1+offset;</tag></item>
        <item>two<tag>out=2+offset;</tag></item>
     </one-of>
  </rule>
</grammar>
```

In addition to the reserved rule variable out, SISR defines two other variables (both of which are objects). The rules object is used to access the rule variable of a previously matched referenced rule. The meta object is used to access meta information pertaining to the rule. The rules object has properties named by the previously matched referenced rule name that evaluates to that rule's rule variable. For example, we can modify our earlier example slightly to make use of the rules object:

```
<?xml version="1.0" encoding="UTF-8"?>
<grammar version="1.0"
        xmlns="http://www.w3.org/2001/06/grammar"
        mode="voice"
        xml:lang="en-GB"
```

---

[4] Assignments to a read-only variable in ECMAScript don't generate errors and are simply ignored.

```
        root="pizza"
        tag-format="semantics/1.0">
  <rule id="pizza">
      I want a
      <ruleref uri="#sizerule"/>
      <tag>out.size=rules.sizerule;</tag>
      <ruleref uri="#sortrule"/>
      <tag>out.sort=rules.sortrule;</tag>
      pizza
  </rule>
  <rule id="sizerule">
      <one-of>
          <item>large<tag>out="large";</tag></item>
          <item>small<tag>out="small";</tag></item>
      </one-of>
  </rule>
  <rule id="sortrule">
      <one-of>
          <item>pepperoni<tag>out=1;</tag></item>
          <item>vegetarian<tag>out=2;</tag></item>
          <item>cheese<tag>out=3;</tag></item>
      </one-of>
  </rule>
</grammar>
```

This example demonstrates how information from lower level rules is propagated upwards. Note also that we used scalar types in the lower level rules and a structure type (i.e. object) in the root rule – this is a typical authoring style. SISR also provides a convenient syntax for accessing the most recently matched referenced rule by replacing the rule name property of the `rules` object with the function call `latest()`. For example, we could rewrite the root rule in the previous example equivalently as:

```
<rule id="pizza">
    I want a
    <ruleref uri="#sizerule"/>
    <tag>out.size=rules.latest();</tag>
    <ruleref uri="#sortrule"/>
    <tag>out.sort=rules.latest();</tag>
    pizza
</rule>
```

Note that properties of the `rules` object are only populated after the corresponding referenced rule is matched. For example, if we placed the first `<tag>` element before the first `<ruleref>`, both `rules.sizerule` and `rules.latest()` would evaluate to `undefined`.

The `meta` object contains meta information related to previously matched referenced rules or the current rule. The syntax is similar to that of the `rules` object – e.g. meta information for `sizerule` is stored under the `meta.sizerule` property, and meta information for the latest matched referenced rule is stored under `meta.latest()`. The meta information for the current rule is available under

Table 9.3 Sub-properties of the `meta` object

meta sub-property	Description
text	The portion of the raw utterance matched by the rule. For example, the text variable for the current rule is `meta.current().text`.
score	The score variable for the rule. Score values refer to the confidence of the match – higher values indicate a higher confidence. For example, the score variable for the previously matched referenced rule called `example` is `meta.example.score`.

`meta.current()`. Table 9.3 summarises the meta information sub-properties. The properties of the `meta` object are read-only.

9.9 Summary

In this chapter, we discussed a standard format for authoring grammars for large vocabulary speech recognisers. SRGS grammars comprise one or more rule expansions. A rule expansion is any combination of token, rule reference or tag where combinations are expressed as sequences, alternatives and repeats. SRGS grammars are also applicable for expressing permissible DTMF input sequences. SRGS grammars dovetail with SISR to allow grammar authors to assign meaning or semantics to rules, thus allowing speech applications to operate on the semantics of the user input rather than the many (often semantically equivalent) variations in user input.

In the next chapter we look at NLSML, the data representation format employed in the output of a speech recogniser. NLSML encapsulates the recognition results including the raw utterance matched by the grammar and the corresponding semantic interpretation serialised to XML.

10

Natural Language Semantics Markup Language (NLSML)

MRCP employs the Natural Language Semantics Markup Language (NLSML) as a data representation format for encapsulating results from speech recognition resources and speaker verification resources. NLSML was originally published as a W3C Working Draft [58] in 2000. The W3C subsequently discontinued work on NLSML without fully completing it (a new more general specification called Extensible MultiModal Annotation (EMMA) [59] was created as its replacement). The MRCP specification has essentially 'imported' the original NLSML specification and tailored it specifically to its needs. This chapter describes MRCP's version of NLSML.

10.1 Introduction

NLSML is a data interchange format used between speech recognition and speaker verification resources residing on the media resource server, and the MRCP client. In the context of speech recognition, NLSML represents information such as the user utterance[1], semantic interpretation of the utterance, and related information such as confidence and timestamp data (sometimes called 'side information'). MRCP extends NLSML to represent data related to the training of voice enrolled grammars. Voice enrolled grammars (also called speaker dependent grammars) are grammars created by the user's voice that can be used in future recognition operations. NLSML is also used in conjunction with the speaker verification resource to represent verification and identification results such as speaker information and verification scores.

10.2 Document structure

NLSML documents are identified by the media type `application/nlsml+xml`. The basic structure of an NLSML document is illustrated below:

---

[1] We use the term 'utterance' loosely to cover both speech input and DTMF input.

Speech Processing for IP Networks   Dave Burke
© 2007 John Wiley & Sons, Ltd

```
<?xml version="1.0" encoding="UTF-8"?>
<result xmlns="http://www.ietf.org/xml/ns/mrcpv2">
   ...
</result>
```

All NLSML documents include the root element `<result>`. The namespace for MRCP's version of NLSML is `http://www.ietf.org/xml/ns/mrcpv2` as indicated by the `xmlns` attribute. The rest of the document structure depends on whether or not NLSML is being used to return speech recognition results, voice enrolment results, or speaker verification/identification results. We discuss each of the three cases in the following sections.

10.3 Speech recognition results

Speech recognition results encapsulated in NLSML are contained in the body of a `RECOGNITION-COMPLETE` event or `GET-RESULT` response. Speech recognition in MRCP was introduced in Chapter 3 and is discussed in detail in Chapter 13. NLSML documents containing speech recognition results are identifiable by the presence of one or more `<interpretation>` elements as children of the `<result>` element. Table 10.1 summarises the elements used for representing speech recognition results.

For speech recognition results, the root element `<result>` includes a `grammar` attribute that indicates which grammar URI matched. This is necessary so that the MRCP client can determine which one of possibly many activate grammars was matched. The format of the grammar URI follows that used by SRGS's `<ruleref>` element to reference external grammar files (see Chapter 9). If this attribute is omitted on the `<result>` element, it must appear on each `<interpretation>` element.

Table 10.1 NLSML elements and attributes for representing speech recognition results

Elements	Attributes	Description
`<result>`	xmlns grammar	Root element for NLSML documents.
`<interpretation>`	grammar confidence	Encapsulates an input hypothesis.
`<instance>`	—	Contains the semantic interpretation of the utterance.
`<input>`	mode confidence timestamp-start timestamp-end	Text representation of a user's input.
`<noinput>`	—	Indicates there was no input – a timeout expired in the speech recogniser while waiting for input.
`<nomatch>`	—	Indicates that the speech recogniser was unable to match any input with a confidence above a certain threshold.

Each NLSML document contains one or more <interpretation> elements. When more than one <interpretation> element is present (called an 'N-best list'), the interpretations are ordered in decreasing confidence. The optional confidence attribute indicates the confidence of the interpretation and ranges from 0.0 to 1.0 – a value of 1.0 indicates maximum confidence. The grammar attribute may be specified on <interpretation> if it is necessary to specify a per-interpretation grammar match. This is useful when the same utterance can match different active grammars with different levels of confidence. If the grammar attribute on <result> is omitted, there must be a grammar attribute on each <interpretation> element.

The <interpretation> element includes an <input> element to describe the text representation of the user's input, and an <instance> element to contain the corresponding semantic interpretation (see Section 10.3.1). For the simple case, the <input> element contains the text representation of the user's input for successful recognition, or a <noinput> or <nomatch> element for an unsuccessful recognition. The optional mode attribute on <input> indicates the modality of the input – in the context of MRCP this is either speech or dtmf. The optional confidence attribute indicates the confidence of the recogniser in the range of 0.0 and 1.0. The optional timestamp-start and timestamp-end attributes indicate the time at which the input began and ended respectively, using the ISO 8601 format (for example, the timestamp 2006-01-01T0:00:00 represents midnight on 1st of January, 2006). It is possible to have multi-level <input> elements where, for example, multiple <input> elements may appear as children of a parent <input> element. Multi-level <input> elements can be used to associate timestamps and confidences to different portions of the utterance. For example:

```
<input>
    <input mode="speech" timestamp-start="2000-04-03T0:00:00">
        to recognise
    </input>
    <input mode="speech" timestamp-start="2000-04-03T0:00:01">
        speech
    </input>
</input>
```

When the speech recogniser does not detect input for a certain period of time, recognition completes with a completion cause value of 002 no-input-timeout returned in the RECOGNITION-COMPLETE event (see Chapter 13). If NLSML results are returned in the event's message body, the NLSML will contain a <noinput> element as a child of <input>. If the speech recogniser does detect input but is not able to match any of the grammars with a confidence above a certain threshold, the recognition completes with a completion cause value of 001 no-match returned in the RECOGNITION-COMPLETE event (see Chapter 13). If NLSML results are returned in the event's message body, the NLSML will contain a <nomatch> element as a child of <input>. The <nomatch> element may optionally contain text for the best of the rejected matches.

Below is an example of an NLSML document corresponding to a successful speech recognition:

```
<?xml version="1.0" encoding="UTF-8"?>
<result xmlns="http://www.ietf.org/xml/ns/mrcpv2"
        grammar="http://www.example.com/demo.grxml">
    <interpretation confidence="0.81">
        <input mode="speech">
```

```
            to recognise speech
        </input>
        <instance>
            to recognise speech
        </instance>
    </interpretation>
    <interpretation confidence="0.75">
        <input mode="speech">
            to wreck a nice beach
        </input>
        <instance>
            to wreck a nice beach
        </instance>
    </interpretation>
</result>
```

In this example, there is an *N*-best list consisting of two different interpretations. The grammar does not perform any semantic interpretation and hence the contents of the `<instance>` element mirrors that of the `<input>` element. Next, we consider an example where recognition is unsuccessful (no-input):

```
<?xml version="1.0" encoding="UTF-8"?>
<result xmlns="http://www.ietf.org/xml/ns/mrcpv2">
    <interpretation>
        <input mode="speech">
            <noinput/>
        </input>
        <instance/>
    </interpretation>
</result>
```

In this case, there is a single indicating that the recognition did not succeed due to a no-input condition. Note that the `<instance>` element is empty. The final example illustrates the case for `<nomatch>` (i.e. the confidence of the recognition was not sufficiently high to be accepted):

```
<?xml version="1.0" encoding="UTF-8"?>
<result xmlns="http://www.ietf.org/xml/ns/mrcpv2"
        grammar="http://www.example.com/number.grxml">
    <interpretation confidence="0.31">
        <input mode="speech">
            <nomatch>one</nomatch>
        </input>
        <instance/>
    </interpretation>
</result>
```

With this example, the recognition is rejected but the best (rejected) match is indicated in the contents of the `<nomatch>` element.

10.3.1 *Serialising semantic interpretation results*

In this section we consider recognition results from grammars that include semantic interpretation (see also Chapter 9).

10.3.1.1 Semantic literals

When the grammar's `tag-format` is set to `semantics/1.0-literals`, the contents of the `<tag>` elements are interpreted as ECMAScript strings (see Section 9.8). The semantic result is obtained from the root rule of the grammar after all the matched `<tag>` elements have been executed. NLSML documents place the matched grammar's semantic interpretation result in the `<instance>` element of the NLSML document. For example, consider the grammar:

```
<?xml version="1.0" encoding="UTF-8"?>
<grammar version="1.0"
         xmlns="http://www.w3.org/2001/06/grammar"
         mode="voice"
         xml:lang="en-IE"
         root="yesno"
         tag-format="semantics/1.0-literals">
   <rule id="yesno">
       <one-of>
           <item>yes<tag>affirmative</tag></item>
           <item>yea<tag>affirmative</tag></item>
           <item>no<tag>negative</tag></item>
           <item>nah<tag>negative</tag></item>
       </one-of>
   </rule>
</grammar>
```

Assuming the user articulated the word 'yea', one might expect an NLSML result similar to:

```
<?xml version="1.0" encoding="UTF-8"?>
<result xmlns="http://www.ietf.org/xml/ns/mrcpv2"
        grammar="http://www.example.com/ycsno.grxml">
   <interpretation confidence="0.91">
       <input mode="speech">
           yea
       </input>
       <instance>affirmative</instance>
   </interpretation>
</result>
```

The matched token(s) in the grammar, 'yea', is contained in the NLSML `<input>` element and the associated semantic literal obtained from the corresponding `<tag>` element is contained in the NLSML `<instance>` element. A speech application would perform an action based on the value of the `<instance>` element, thus freeing it from having to be aware of all the different ways a person might indicate the affirmative.

10.3.1.2 Semantic scripts – scalars

Recall from Section 9.8 that for grammars where the `tag-format` is set to `semantics/1.0`, the contents of the `<tag>` elements are considered to be ECMAScript code. The semantic interpretation result for the grammar is taken as the value of the rule variable (designated by the reserved name `out`) of the root rule after all other selected `<tag>`s are executed. Which `<tag>` elements and the order they are executed in is dictated by the particular path through the grammar taken when matching the grammar. When the root rule variable evaluates to a scalar type (i.e. string, number, boolean, null or undefined), then the semantic result is interpreted as its string equivalent and it is this value that is contained in NLSML's `<instance>` element. For the ECMAScript number type, the string equivalent is the cardinal value (if the number is less than 0, a negative sign prefixes it). For type boolean, the string is either 'true' or 'false'. For type null it is 'null', and for type undefined it is 'undefined'. For example, the following grammar will produce the same semantic result as the previous grammar example:

```
<?xml version="1.0" encoding="UTF-8"?>
<grammar version="1.0"
         xmlns="http://www.w3.org/2001/06/grammar"
         mode="voice"
         xml:lang="en-IE"
         root="yesno"
         tag-format="semantics/1.0">
   <rule id="yesno">
      <one-of>
         <item>yes<tag>out="affirmative";</tag></item>
         <item>yea<tag>out="affirmative";</tag></item>
         <item>no<tag>out="negative";</tag></item>
         <item>nah<tag>out="negative";</tag></item>
      </one-of>
   </rule>
</grammar>
```

10.3.1.3 Semantic scripts – objects

When the root rule's rule variable is not a scalar type but is instead an ECMAScript object, a more complex representation is required. The Semantic Interpretation for Speech Recognition (SISR) [7] defines a transformation from an ECMAScript object to XML fragment – it is this resultant XML fragment that may appear under NLSML's `<instance>` element. The transformation rules for an object are fairly straightforward:

(i) Each property of the rule variable becomes an XML element. The name of the element will be the same as the name of the property.
(ii) If the value of the property is a simple scalar type (string, number, boolean, null or undefined) then the character data content of the XML element will equate to the string value of this property.
(iii) If the property is of type object, then each child property of this object becomes a child element, and the contents of these child elements are in turn processed.

Consider the following grammar:

```
<?xml version="1.0" encoding="UTF-8"?>
<grammar version="1.0"
         xmlns="http://www.w3.org/2001/06/grammar"
         mode="voice"
         xml:lang="en-GB"
         root="travel"
         tag-format="semantics/1.0">
    <rule id="travel">
        <tag>out.travel=new Object();</tag>
        I want a fly from
        <ruleref uri="#city"/>
        <tag>out.travel.orig=rules.city;</tag>
        to
        <ruleref uri="#city"/>
        <tag>out.travel.dest=rules.city;</tag>
    </rule>
    <rule id="city">
        <one-of>
            <item>Dublin<tag>out="Dublin";</tag></item>
            <item>London<tag>out="London";</tag></item>
            <item>Paris<tag>out="Paris";</tag></item>
        </one-of>
    </rule>
</grammar>
```

For an input of 'I want to fly from Dublin to Paris', the resulting rule variable of the root rule evaluates to:

```
{
    travel: {
        orig: Dublin,
        dest: Paris
    }
}
```

Using the preceding rules, the NLSML result will be similar to the following:

```
<?xml version="1.0" encoding="UTF-8"?>
<result xmlns="http://www.ietf.org/xml/ns/mrcpv2"
        grammar="http://www.example.com/travel.grxml">
    <interpretation confidence="0.93">
        <input mode="speech">
            I want to fly from Dublin to Paris
        </input>
        <instance>
```

```
        <travel>
            <orig>Dublin</orig>
            <dest>Paris</dest>
        </travel>
    </instance>
  </interpretation>
</result>
```

10.3.1.4 Semantic scripts – arrays

It is possible for the rule variable of the root rule to be of type Array (e.g. x[0], x[1], etc) or indeed for one of its properties to be of type Array. In this case, a special set of rules are defined for the transformation into XML:

(i) Indexed elements of an Array object (e.g. a[0], a[1], etc.) become XML child elements with name `<item>`.
(ii) Each `<item>` element has an attribute named `index`, which is the index of the corresponding element in the array.
(iii) The XML element containing the `<item>` elements includes an attribute named `length`, whose value is given by the length property of the ECMAScript Array object.

Consider a grammar that takes a string of one or more digits from the user and builds up an Array:

```
<?xml version="1.0" encoding="UTF-8"?>
<grammar version="1.0"
         xmlns="http://www.w3.org/2001/06/grammar"
         mode="voice"
         xml:lang="en-GB"
         root="string"
         tag-format="semantics/1.0">
    <rule id="string">
        <tag>out.digitstring = new Array();</tag>
        <item repeat="1-">
            <ruleref uri="#digits"/>
            <tag>out.digitstring.push(rules.digits);</tag>
        </item>
    </rule>
    <rule id="digits">
        <one-of>
            <item>one<tag>out=1;</tag></item>
            <item>two<tag>out=2;</tag></item>
            <item>three<tag>out=3;</tag></item>
            <item>four<tag>out=4;</tag></item>
            <item>five<tag>out=5;</tag></item>
        </one-of>
    </rule>
</grammar>
```

For the utterance 'five four three two one', the resulting Array object for the rule variable of the root rule is:

```
{
   digitstring: [ 5, 4, 3, 2, 1 ]
}
```

The corresponding NLSML document would be similar to (using the transformation rules defined above):

```
<?xml version="1.0" encoding="UTF-8"?>
<result xmlns="http://www.ietf.org/xml/ns/mrcpv2"
        grammar="http://www.example.com/yesno.grxml">
   <interpretation confidence="0.93">
       <input mode="speech">
           five four three two one
       </input>
       <instance>
           <digitstring length="5">
               <item index="0">5</item>
               <item index="1">4</item>
               <item index="2">3</item>
               <item index="3">2</item>
               <item index="4">1</item>
           </digitstring>
       </instance>
   </interpretation>
</result>
```

10.4 Voice enrolment results

Voice enrolment is the mechanism by which a user can create a speaker dependent, personal grammar by adding phrases to the grammar using their own voice. This voice enrolled grammar can be subsequently used in normal speech recognition (it is simply referenced via a grammar URI). A classic use-case for voice enrolled grammars is a personal address book whereby the user can add new entries by simply speaking the name of the person several times so as to train the recogniser. The same user can then later recall a previously trained name and retrieve some associated information such as the corresponding telephone number.

Conceptually, a voice enrolled grammar can be thought of as a simple SRGS grammar consisting of alternates, for example:

```
<?xml version="1.0" encoding="UTF-8"?>
<grammar version="1.0"
         xmlns="http://www.w3.org/2001/06/grammar"
         mode="voice"
         xml:lang="en-GB"
         root="address"
```

```
            tag-format="semantics/1.0-literals">
    <rule id="address">
        <one-of>
            <item>Mary<tag>user01</tag></item>
            <item>Anne<tag>user02</tag></item>
            <item>John Henry<tag>user03</tag></item>
            . . .
        </one-of>
    </rule>
</grammar>
```

Each of the alternate phrases within the grammar (e.g. Mary, Anne, John Henry) is added via an enrolment session along with its corresponding semantic interpretation (the semantic interpretation format is platform specific but here we assume a semantic literal string). Each phrase is kept track of by a unique phrase ID supplied by the MRCP client. Within an enrolment session, the user speaks the phrase several times. Each time a phrase is captured, it is compared for similarity with the other samples, and a clash test is performed against other entries in the personal grammar to ensure that there are no similar or confusable entries. When a sufficient number of similar utterances have been captured, the phrase is trained and committed to the personal grammar. The personal grammar can be used later in normal recognition tasks by specifying the personal grammar URI – when the phrase matches, the semantic interpretation result specified during the enrolment session is returned using NLSML as described earlier in this chapter. Note that the SRGS example above is only conceptual – the implementation will use some kind of platform-specific format for the grammar and the phrases themselves will be stored using some kind of acoustic representation.

A voice enrolment session commences when the MRCP client issues a START-PHRASE-ENROLLMENT request to a speech recogniser resource and ends when the MRCP client issues an END-PHRASE-ENROLLMENT request. Each capture of the utterance is triggered via a RECOGNIZE request including the header field/value of Enroll-Utterance: true. The RECOGNITION-COMPLETE event corresponding to each RECOGNIZE request, will contain NLSML that encapsulates the results of the enrolment attempt. The enrolment results consist of information such as the number of good repetitions of the phrase, the number of repetitions still required to train the phrase sufficiently, the number of clashes with other phrases in the grammar, etc. Voice enrolment is described in detail in Chapter 13 – here we are just concerned with the format of the NLSML document containing voice enrolment results.

NLSML documents containing voice enrolment results are identifiable by the presence of an <enrollment-result> element as a child of the <result> element. Table 10.2 summarises the elements used for representing voice enrolment results.

Below is an example of an NLSML result returned in a RECOGNITION-COMPLETE event during a voice enrollment session:

```
<?xml version="1.0" encoding="UTF-8"?>
<result xmlns="http://www.ietf.org/xml/ns/mrcpv2"
        grammar="http://example.com/ve/personal-grammar-01">
    <enrollment-result>
        <num-good-repetitions>1</num-good-repetitions>
        <num-repetitions-still-needed>2</num-repetitions-still-needed>
        <consistency-status>consistent</consistency-status>
        <transcriptions>
            <item>Marie</item>
```

Table 10.2 NLSML elements for representing voice enrolment results

Elements	Attributes	Description
`<result>`	`xmlns` `grammar`	Root element for NLSML documents.
`<enrollment-result>`	—	Container element for voice enrolment results.
`<num-good-repetitions>`	—	Indicates the number of consistent utterances received so far.
`<num-repetitions-still-needed>`	—	Indicates the number of consistent utterances still needed before the phrase can be added to the personal grammar.
`<consistency-status>`	—	Indicates how consistent the utterance repetition was. Contains either `consistent`, `inconsistent`, `undecided`.
`<transcriptions>`	—	Contains the platform-specific transcription of the last utterance. Each transcription is contained in an `<item>` element.
`<num-clashes>`	—	Indicates the number of clashes with other phrases in the personal grammar or absent if there are no clashes.
`<clash-phrase-ids>`	—	Contains the phrase IDs of clashing phrases or absent if there are no clashes.
`<confusable-phrases>`	—	Contains a list of phrases (originally specified in the `Confusable-Phrases-URI` header in RECOGNIZE – see Chapter 13) that are confusable with the phrase being added to the personal grammar.
`<item>`	—	Used to specify demarcate separate entries under `<clash-phrase-ids>` and `<confusable-phrases>`.

```
        </transcriptions>
        <num-clashes>2</num-clashes>
        <clash-phrase-ids>
            <item>Mary</item>
            <item>Madge</item>
        </clash-phrase-ids>
    </enrollment-result>
</result>
```

The grammar URI for the personal grammar is specified in the `grammar` attribute of the `<result>` element. The NLSML result indicates that there has been one good repetition so far and two more are needed. However, there are two clashes indicated by the phrase IDs Mary and Madge.

10.5 Speaker verification results

MRCP uses NLSML to encapsulate speaker verification/identification results contained in the message body of VERIFICATION-COMPLETE events and responses to GET-INTERMEDIATE-RESULT requests. Speaker verification in MRCP was introduced in Chapter 3 and is discussed in detail in Chapter 15. The same NLSML document structure is used to describe results of training a voiceprint, verifying utterances against a voiceprint, and for performing identification of utterances against multiple voiceprints and is identifiable by the presence of a `<verification-result>` element.

Table 10.3 summarises the elements used for representing speaker verification/identification results.

Below is an example of verification results returned during the training of a voiceprint. The NLSML is contained in either a `VERIFICATION-COMPLETE` event message body or the response to a `GET-INTERMEDIATE-RESULTS` request (see Chapter 15):

```
<?xml version="1.0" encoding="UTF-8"?>
<result xmlns="http://www.ietf.org/xml/ns/mrcpv2">
    <verification-result>
        <voiceprint id="joebloggs.voiceprint">
            <incremental>
                <verification-score>0.91</verification-score>
                <device>cellular-phone</device>
                <gender>male</gender>
                <utterance-length>751</utterance-length>
            </incremental>
            <cumulative>
                <verification-score>0.93</verification-score>
                <device>cellular-phone</device>
                <gender>male</gender>
                <utterance-length>1522</utterance-length>
                <need-more-data>true</need-more-data>
            </cumulative>
        </voiceprint>
    </verification-result>
</result>
```

In this example, there is a single voiceprint identified by `joebloggs.voiceprint` (this voiceprint would have been specified in the `START-SESSION` request using the `Voiceprint-Identifier` header field). The information contained in the `<incremental>` element refers to the analysis of the previous utterance. Separate tags are used to indicate information including the device type, gender of the speaker, and the utterance length data as determined from the last utterance. The information contained in the `<cumulative>` element refers to the analysis of all utterances over potentially multiple `VERIFY` or `VERIFY-FROM-BUFFER` calls. For a training session, the `<verification-score>` indicates the likelihood that the same speaker spoke all the utterances. The contained information for device and gender is estimated over all utterances. The `<utterance-length>` element contains the total length of utterance data used in milliseconds. Finally, the `<need-more-data>` value of `true` indicates that more data is required to finish training the voiceprint, i.e. the MRCP client needs to solicit further data from the user and issue more `VERIFY` or `VERIFY-FROM-BUFFER` requests.

Next, we consider an example of verification results. We assume that a single voiceprint was specified in `Voiceprint-Identifier` header in the `START-SESSION` request so that we expect the NLSML results to contain a single `<voiceprint>` element.

Table 10.3 NLSML elements for representing speaker verification/identification results

Elements	Attributes	Description
`<result>`	`xmlns`	Root element for NLSML documents.
`<verification-result>`	—	Contains the speaker verification/identification results namely one or more `<voiceprint>` elements.
`<voiceprint>`	`id`	Encapsulates information on how the speech data matched a single utterance. For verification and training, there will be only one `<voiceprint>` element in the results. For multi-verification or identification, there will be as many `<voiceprint>` results as there were voiceprints specified in the `START-SESSION` request listed in decreasing order of their cumulative `<verification-score>`.
`<adapted>`	—	Indicates that the voiceprint has been adapted as a consequence of analysing the source utterance during verification/identification.
`<cumulative>`	—	This element contains cumulative scores for how well multiple utterances (analysed over multiple `VERIFY` or `VERIFY-FROM-BUFFER` requests) matched the voiceprint.
`<incremental>`	—	This optional element indicates how well the last utterance (analysed from the last `VERIFY` or `VERIFY-FROM-BUFFER` request) matched the voiceprint.
`<verification-score>`	—	For verification/identification results, the higher the score the more likely it is that the speaker is the same person as the one who spoke the voiceprint utterances. For voiceprint training results, the higher the score the more likely the speaker is to have spoken all of the analysed utterances. The value varies from −1.0 to 1.0.
`<decision>`	—	Indicates the verification/identification decision. Can be one of `accepted`, `rejected`, or `undecided`. The `<decision>` element is specified within either the `<incremental>` or `<cumulative>` element under the first `<voiceprint>` only.
`<utterance-length>`	—	Indicates the duration of the last utterance or cumulative utterances in milliseconds.
`<device>`	—	Indicates the device used by the caller. Can be one of `cellular-phone`, `electret-phone`, `carbon-button-phone`, or `unknown`.
`<gender>`	—	Indicates the apparent gender of the speaker.
`<need-more-data>`	—	A contained value of `true` indicates that more data is needed to complete training of the voiceprint or to declare a positive decision for verification / identification. If omitted, no more data is needed.

```
<?xml version="1.0" encoding="UTF-8"?>
<result xmlns="http://www.ietf.org/xml/ns/mrcpv2">
    <verification-result>
        <voiceprint id="joebloggs.voiceprint">
            <incremental>
                <verification-score>0.85</verification-score>
                <device>carbon-button-phone</device>
                <gender>male</gender>
                <utterance-length>841</utterance-length>
            </incremental>
            <cumulative>
                <verification-score>0.81</verification-score>
                <device>carbon-button-phone</device>
                <gender>male</gender>
                <utterance-length>1619</utterance-length>
                <decision>accepted</decision>
            </cumulative>
        </voiceprint>
    </verification-result>
</result>
```

In this example, both a `<verification-score>` is present both for the latest utterance (represented under the `<incremental>` element) and the cumulative analysis (represented under the `<cumulative>` element). The cumulative analysis indicates that the decision is `accepted`, indicating that the speaker's utterances sufficiently match those of the voiceprint (the decision boundary is specified via the `Min-Verification-Score` header field specified in the `START-SESSION` of `SET-PARAMS` requests – see Chapter 15 for more information).

In our final example, we consider the case for speaker identification where we want to identify the speaker from a set of speakers. Let's assume that 'Joe Bloggs' in our previous example belongs to a family of 'Mary Bloggs' and 'Ted Bloggs'. The `Voiceprint-Identifier` specified in the `START-SESSION` request would list all three voiceprints and consequently the NLSML results will include three `<voiceprint>` elements:

```
<?xml version="1.0" encoding="UTF-8"?>
<result xmlns="http://www.ietf.org/xml/ns/mrcpv2">
    <verification-result>
        <voiceprint id="marybloggs.voiceprint">
            <incremental>
                <verification-score>0.85</verification-score>
                <device>cellular-phone</device>
                <gender>female</gender>
                <utterance-length>842</utterance-length>
            </incremental>
            <cumulative>
                <verification-score>0.85</verification-score>
                <device>cellular-phone</device>
                <gender>female</gender>
                <utterance-length>842</utterance-length>
```

```
                    <decision>accepted</decision>
                </cumulative>
            </voiceprint>
            <voiceprint id="tedbloggs.voiceprint">
                <cumulative>
                    <verification-score>0.31</verification-score>
                </cumulative>
            </voiceprint>
            <voiceprint id="joebloggs.voiceprint">
                <cumulative>
                    <verification-score>0.29</verification-score>
                </cumulative>
            </voiceprint>
        </verification-result>
</result>
```

The `<voiceprint>` elements are ordered in decreasing cumulative `<verification-score>`. Redundant information is omitted from subsequent voiceprints. Of the three voiceprints, the utterance data only matches the voiceprint of 'Mary Bloggs', thus identifying her as the speaker.

10.6 Summary

In this chapter, we reviewed MRCP's version of NLSML. NLSML provides a data representation format for encapsulating results from speech recognition, results from voice enrolment (i.e. enrolling speak-dependent phrases in personal grammars), and results from speaker verification (either training a voiceprint or matching utterances against one or more trained voiceprints for the purposes of verification or identification of the user). In a future version of MRCP, the W3C's EMMA specification [59] will be used in place of NLSML.[2] In this case, extensions to the original EMMA will most likely be needed to encapsulate MRCP-specific data such as results from voice enrolment and speaker verification results.

In the next chapter, we take a look at another data representation format: the W3C Pronunciation Lexicon Specification (PLS) [9], which allows lexicons to be specified for use in conjunction with speech recognisers and speech synthesisers.

---

[2] The MRCP specification suggests that the MRCP client could indicate support for the EMMA format by, for example, specifying a SIP `Accept` header field with a value of `application/emma+xml`.

11

Pronunciation Lexicon Specification (PLS)

The Pronunciation Lexicon Specification (PLS) [9] is a W3C specification that defines a standard syntax to specify pronunciation lexicons for use with speech synthesisers and speech recognisers. A pronunciation lexicon consists of words or phrases and their corresponding pronunciation, usually expressed in some phonetic alphabet (see Section 2.2.1). MRCP employs the PLS either indirectly from Speech Synthesis Markup Language (SSML) or Speech Recognition Grammar Specification (SRGS) documents, which reference PLS documents via the `<lexicon>` element (see Chapters 8 and 9) or directly through MRCP's `DEFINE-LEXICON` request issued to a speech synthesis resource (see Chapters 12 for more information). This chapter looks at the PLS document structure and presents some common examples.

11.1 Introduction

Modern speech synthesisers and speech recognisers usually come equipped internally with high quality lexicons that provide extensive coverage for pronunciations within the various languages supported by the product. Of course, achieving exhaustive coverage for words in a language is not realistic when one considers, for example, proper nouns or words or phrases originating from other languages. When a modern speech synthesiser or speech recogniser encounters a word for which it does not have an entry in its internal lexicon, it will typically perform some kind of automatic analysis to determine a pronunciation based on certain lexical rules. While this mechanism may work well in certain cases, it is often less than ideal since the rules for deriving a pronunciation from the orthography (written form) are not exact and uniform. This is where PLS comes in. A PLS document can be used to add or even alter the pronunciation of words and phrases by associating the desired phonemic or phonetic transcription with the words or phrases in question.

PLS documents are usually application specific and may be targeted for use with just speech synthesisers, just speech recognisers, or both. The target of a PLS document is not determined by the document itself but rather by how that document is referred to. For example, an SSML document referencing a PLS document applies the lexicon for speech synthesis while an SRGS document referencing a PLS document applies the lexicon for speech recognition. Similarly, a PLS document specified in a DEFINE-LEXICON request to a speech synthesiser is used for subsequent speech synthesis requests.

It should be noted that the PLS document described in [9] is intended as a simple syntax for describing lexicons. Many complex features are not available, such as the ability to restrict pronunciations to be used for certain tenses or parts of speech. However, the PLS document syntax does provide a convenient and useful place for application authors to maintain application-specific pronunciations that can be used for both speech synthesis and speech recognition. Note that while SSML does allow phonemic or phonetic strings to be used within its documents (via the SSML <phoneme> element), SRGS has no such mechanism and must rely on PLS documents for phonemic/phonetic transcriptions of words and phrases.

11.2 Document structure

PLS documents are identified by the media type application/pls+xml. Table 11.1 summarises the elements and attributes defined in PLS.

The basic structure of a PLS document is illustrated below:

```
<?xml version="1.0" encoding="UTF-8"?>
<lexicon version="1.0"
        xmlns="http://www.w3.org/2005/01/pronunciation-lexicon"
        alphabet="ipa"
        xml:lang="en-US">
   ...
</lexicon>
```

Table 11.1 Elements and attributes defined in PLS

Elements	Attributes	Description
<lexicon>	version xmlns xml:lang alphabet xml:base	Root element for PLS documents.
<lexeme>	role xml:id	Container element for lexical entries.
<grapheme>	orthography	Contains orthographic information for a lexical entry.
<phoneme>	prefer alphabet	Contains pronunciation information for a lexical entry.
<example>	—	Contains example usage for the lexical entry.
<alias>	Prefer	Contains acronym / abbreviation expansions.
<meta>	name http-equiv content	Contains metadata for the document. See Chapter 8.
<metadata>	—	Contains metadata for the document. See Chapter 8.

All PLS documents include the root element <lexicon>. The version attribute indicates the version of PLS and is fixed at 1.0 for the specification described in [9]. The default namespace for the PLS <lexicon> element and its children is indicated by the xmlns attribute and is defined as http://www.w3.org/2005/01/pronunciation-lexicon.

The alphabet attribute is required on the <lexicon> element and specifies the alphabet used within the pronunciation lexicon. The value ipa indicates the IPA alphabet and is required to be supported (see Section 2.2.1 for a discussion on IPA). Vendor defined strings must be of the form x-organisation or x-organisation-alphabet, e.g. x-acme-phoneyabc. It is possible to override the default alphabet within individual lexical entries as we will see in the next section.

The xml:lang attribute is also required to be present and specifies the language for which the pronunciation lexicon is relevant. The format of the xml:lang attribute value was described in Chapter 8. Note that the xml:lang attribute specifies a single unique language for the pronunciation lexicon. This does not preclude the referencing document from being multilingual. For example, an SSML document may include several languages and can reference separate PLS documents for each language.

Similarly to SSML and SRGS, the xml:base attribute may be specified on the <lexicon> element to resolve relative URIs used within lexicon document against. The lexical entries are contained as children of the <lexicon> root element and their syntax is discussed next.

11.3 Lexical entries

The lexical entries are contained within separate <lexeme> elements. A <lexeme> contains one or more orthographies and one or more corresponding pronunciations. The orthography is contained within the <grapheme> element. An optional attribute orthography may be used on the <grapheme> element to define the script code used for writing the orthography (script codes are defined in ISO 15924; for example a value of Latn is used to indicate Latin, Cyrl for Cyrillic, Kana for Katakana). The corresponding pronunciation is contained in the <phoneme> element using the alphabet specified in the alphabet attribute on the root element <lexicon>. A <phoneme> element may optionally specify its own alphabet attribute to specify the phonetic alphabet used for the contents of the <phoneme> element. An example of a simple PLS document with two lexical entries is illustrated below:

```
<?xml version="1.0" encoding="UTF-8"?>
<lexicon version="1.0"
         xmlns="http://www.w3.org/2005/01/pronunciation-lexicon"
         alphabet="ipa"
         xml:lang="en-GB">
    <lexeme>
        <grapheme>ceramic</grapheme>
        <phoneme>sɪˈramɪk </phoneme>
    </lexeme>
    <lexeme>
        <grapheme>cavatina</grapheme>
        <phoneme>kavəˈtiːnə </phoneme>
    </lexeme>
</lexicon>
```

Note that some editors may not be able to display certain Unicode characters and as a result it is common for authors of lexicons to use the XML entity escape equivalents written in the form of `&#xN;` where N is the character code in hexadecimal for the encoding used in the XML document. For example, we can rewrite the phonetic symbols for the first lexical entry in the previous example using the following sequence:

```
<lexeme>
    <grapheme>ceramic</grapheme>
    <phoneme>s&#x026A;&#x2C8;r&#x061;m&#x026A;k</phoneme>
</lexeme>
```

As an authoring convenience, a lexical entry may include zero or more `<example>` elements whose contents are a sentence illustrating the use of the entry. The contents of `<example>` elements may be useful for regression testing tools and for documentation purposes. For example:

```
<?xml version="1.0" encoding="UTF-8"?>
<lexicon version="1.0"
        xmlns="http://www.w3.org/2005/01/pronunciation-lexicon"
        alphabet="ipa"
        xml:lang="en-GB">
    <lexeme>
        <grapheme>read</grapheme>
        <phoneme>rɛd</phoneme>
        <example>I have read many books</example>
    </lexeme>
</lexicon>
```

The `<lexeme>` element may also include an optional attribute called `role` to associate additional information with the lexeme. This attribute is primarily useful for disambiguating homographs.[1] For example, the `role` attribute could specify the part of speech (e.g. verb versus noun) to disambiguate lexical entries for the words 'refuse' (the verb) and 'refuse' (the noun). The optional `xml:id`, attribute provides an alternative mechanism for associating additional information with a lexeme. This attribute allows the author to assign a unique identifier to each element (the `xml:id` specification [62] requires that the identifier value is unique within the document). This identifier allows an external document to refer to lexemes and associate further information with them.

11.4 Abbreviations and acronyms

A lexical entry may contain one or more `<alias>` elements to indicate the pronunciation of an acronym or abbreviated term contained in the `<grapheme>` element. The `<alias>` element allows the expansion of the acronym or abbreviated term to be expressed in the written form. Thus, the `<alias>` element is similar to `<phoneme>` element in that it expresses a pronunciation but differs in that it uses the orthographic form to represent the pronunciation instead of a phonetic alphabet (if

---

[1] Homographs are words with the same spelling but different pronunciation

the author requires explicit control over the pronunciation, then the `<phoneme>` element should be used instead). For example:

```
<?xml version="1.0" encoding="UTF-8"?>
<lexicon version="1.0"
          xmlns="http://www.w3.org/2005/01/pronunciation-lexicon"
          alphabet="ipa"
          xml:lang="en-GB">
   <lexeme>
       <grapheme>MRCP</grapheme>
       <alias>Media Resource Control Protocol</alias>
   </lexeme>
</lexicon>
```

Note that recursion is supported when expanding `<alias>` elements. For example, an additional `<lexeme>` entry could be added to the above lexicon to specify the pronunciation of the grapheme 'Media'. This pronunciation would, in turn, be employed as part of the `<alias>` expansion of MRCP.

11.5 Multiple orthographies

There may be more than one `<grapheme>` element in the lexical entry to account for different written forms of a word or phrase (e.g. to account for regional variations such as 'center' vs 'centre'). The following example has two orthographic forms for the same word:

```
<?xml version="1.0" encoding="UTF-8"?>
<lexicon version="1.0"
          xmlns="http://www.w3.org/2005/01/pronunciation-lexicon"
          alphabet="ipa"
          xml:lang="en-GB">
   <lexeme>
       <grapheme>analogue</grapheme>
       <grapheme>analog</grapheme>
       <phoneme>'anəlɒg</phoneme>
   </lexeme>
</lexicon>
```

For the case of homophones, that is words with a different origin and meaning but the same pronunciation, one may insert multiple `<grapheme>` elements in the same `<lexeme>`, as illustrated in the previous example, or alternatively use a separate `<lexeme>` for each distinct word. For example:

```
<?xml version="1.0" encoding="UTF-8"?>
<lexicon version="1.0"
          xmlns="http://www.w3.org/2005/01/pronunciation-lexicon"
          alphabet="ipa"
          xml:lang="en-GB">
```

```
<lexeme>
    <grapheme>blue</grapheme>
    <phoneme>blu:</phoneme>
</lexeme>
<lexeme>
    <grapheme>blew</grapheme>
    <phoneme>blu:</phoneme>
</lexeme>
</lexicon>
```

11.6 Multiple pronunciations

We have shown in the previous section how a lexical entry can have multiple orthographies by simply inserting multiple <grapheme> elements. It is also possible for a single lexical entry to have multiple pronunciations by inserting multiple <phoneme> elements (or <alias> elements). However, with multiple pronunciations comes a subtle asymmetry between speech synthesis and speech recognition usage. When a pronunciation lexicon contains multiple <phoneme> elements (or <alias> elements), a speech recogniser using this lexicon will consider each of the pronunciations as equally valid for that word or phrase. However, the current PLS standard only allows a single pronunciation for speech synthesis. A speech synthesiser selects the first pronunciation in document order from the pronunciation lexicon and ignores the others. As an authoring convenience, an attribute called prefer (with value true) may be added to the <phoneme> or <alias> element, to pick that pronunciation as the one to use for speech synthesis (if more than one prefer attribute is present then the first element in document order with this attribute present is selected). The following example illustrates a very simple PLS document with a lexical entry for a homograph:

```
<?xml version="1.0" encoding="UTF-8"?>
<lexicon version="1.0"
         xmlns="http://www.w3.org/2005/01/pronunciation-lexicon"
         alphabet="ipa"
         xml:lang="en-GB">
   <lexeme>
       <grapheme>read</grapheme>
       <phoneme>rɛd</phoneme>
       <phoneme prefer="true">riːd</phoneme>
   </lexeme>
</lexicon>
```

In this example, the speech recogniser will use both pronunciations while a speech synthesiser will only use the one with the prefer attribute set to true). Observe that had we omitted the prefer attribute, then a speech synthesiser using this pronunciation lexicon would have used the pronunciation contained in the first <phoneme> element.

It is possible that a future version of the PLS standard will permit multiple pronunciations for speech synthesis. To make this feature useful, it may be necessary to associate additional information with each pronunciation to provide further information to the speech synthesiser such as the applicable part of speech or tense for that pronunciation. Standardising that additional information in a language independent way is a challenge, and one reason why the first version of PLS errs on the side of simplicity by omitting it.

11.7 Summary

In this chapter, we introduced the W3C's syntax for pronunciation lexicons called PLS for use with speech synthesisers and speech recognisers. Pronunciation lexicons provide 'dictionaries' for synthesisers and recognisers: they map the written form of words and phrases into a phonetic representation expressed in a phonetic alphabet. The PLS syntax is sufficiently flexible to allow a single dual-purpose document to be written (i.e. specifying pronunciations for both synthesis and recognition) or alternatively a separate PLS document may be authored for synthesis and recognition.

This chapter concludes Part III of the book, which focuses on data representation formats used within the MRCP protocol. Part IV, delves into the details of each of the different MRCP resource types, starting with the speech synthesiser resource.

Part IV

Media Resources

12

Speech synthesiser resource

The MRCP speech synthesiser resource processes plain or marked-up text on behalf of the MRCP client and delivers synthesised speech in real time over the media session. MRCP defines two speech synthesiser resource types: the *basicsynth* resource operates on concatenated audio clips and hence can only provide limited synthesis capabilities based on a restricted vocabulary; the *speechsynth* resource, on the other hand, is a fully featured speech synthesiser capable of rendering any text into speech. This chapter delves into the details of both resource types by studying the methods, events and header fields that apply to them. At this point, the reader may wish to review Chapter 8, which discusses the Speech Synthesis Markup Language (SSML) used for providing marked-up text for synthesis.

12.1 Overview

The two speech synthesiser resource types, basicsynth and speechsynth, may be defined in terms of the SSML features they support. Table 12.1 lists the media resource type and the corresponding SSML elements and input formats supported.

All resources must support input in the form inline SSML content and a list of URIs identifying the SSML markup. The speechsynth resource also accepts input to be synthesised in plain text, supplied either inline or via a URI.

The speech synthesiser media resource supports a total of seven request messages and two event messages, which are summarised in Tables 12.2 and 12.3 respectively.

The speech synthesiser resource has a state machine that is driven by requests from the MRCP client and events generated by the media resource itself. The state machine is illustrated in Figure 12.1. The state machine reflects the state of the resource for the request at the head of the speech synthesiser resource queue.

Table 12.4 summarises the resource-specific header fields for the basicsynth and speechsynth resource types. See Table 7.5 for a list of generic headers that also apply.

Speech Processing for IP Networks   Dave Burke
© 2007 John Wiley & Sons, Ltd

Table 12.1 Definition of the basicsynth and speechsynth resource types

Resource type	Supported SSML	Supported input formats
basicsynth	Restricted SSML: `<speak>`, `<audio>`, `<mark>`, `<say-as>`	`text/uri-list` `application/ssml+xml`
speechsynth	All SSML elements and plain text	`text/uri-list` `application/ssml+xml` `text/plain`

Table 12.2 Request messages (methods)

Message name	Description
`SPEAK`	Provides the speech synthesiser with input to synthesise and initiates speech synthesis and audio streaming.
`PAUSE`	Pauses speech synthesis.
`RESUME`	Resumes a paused speech synthesiser.
`STOP`	Stops speech synthesis.
`BARGE-IN-OCCURRED`	Communicates a barge-in event to the speech synthesiser.
`CONTROL`	Modifies the ongoing speech synthesis action.
`DEFINE-LEXICON`	Defines pronunciation lexicons for the session.

Table 12.3 Event messages

Message Name	Description
`SPEECH-MARKER`	Generated when the speech synthesiser encounters a `<mark>` element.
`SPEAK-COMPLETE`	Generated when the corresponding `SPEAK` request completes.

12.2 Methods

12.2.1 SPEAK

The `SPEAK` request method serves two purposes: (i) to provide the speech synthesiser with input to synthesise, and simultaneously (ii) to initiate speech synthesis and audio streaming. The `SPEAK` request body can contain:

- inline SSML markup (`application/ssml+xml`);
- a list of external SSML or plain text URIs (`text/uri-list`);
- inline plain text (`text/plain`) – speechsynth resource only, or
- a multipart message combining any of the above.

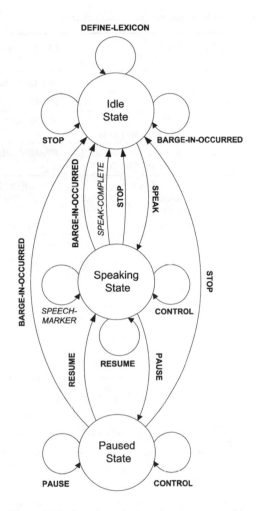

Figure 12.1 Speech synthesiser resource state machine.

The synthesiser resource implements a first-in–first-out queue and thus each SPEAK request is processed serially and in order of receipt. If the resource is idle when it receives a SPEAK request, the corresponding response will specify a request-state of IN-PROGRESS. If the resource is in the speaking or paused state when a SPEAK request arrives, the corresponding response will contain a request state of PENDING indicating that the request is queued.

The voice and prosodic parameters that apply to the synthesis request can be specified at multiple levels, each with a different precedence. The highest precedence applies to the SSML mark-up itself. The next level of precedence applies to the headers in the SPEAK request, i.e. the Voice-, Prosody-, Speaker-Profile and Speech-Language headers. The lowest level of precedence applies to the session default values, which are set using the SET-PARAMS request and apply for subsequent methods invoked during the session.

Table 12.4 Resource-specific header fields for the speech synthesiser resource

Header name	Request	Response	Event	SET-PARAMS / GET-PARAMS
Completion-Cause	—	—	SPEAK-COMPLETE	No
Completion-Reason	—	—	SPEAK-COMPLETE	No
Failed-URI	—	SPEAK DEFINE-LEXICON	SPEAK-COMPLETE	No
Failed-URI-Cause	—	SPEAK DEFINE-LEXICON	SPEAK-COMPLETE	No
Speech-Marker	—	SPEAK BARGE-IN-OCCURRED STOP CONTROL	SPEECH-MARKER	No
Voice-Gender Voice-Age Voice-Variant Voice-Name	SPEAK CONTROL	—	—	Yes
Prosody-Pitch Prosody-Contour Prosody-Range Prosody-Rate Prosody-Duration Prosody-Volume	SPEAK CONTROL	—	—	Yes
Speaker-Profile	SPEAK	—	—	Yes
Speech-Language	SPEAK	—	—	Yes
Kill-On-Barge-In	SPEAK	—	—	Yes
Fetch-Hint	SPEAK	—	—	Yes
Audio-Fetch-Hint	SPEAK	—	—	Yes
Jump-Size	CONTROL SPEAK	—	—	No
Speak-Restart	—	CONTROL	—	No
Speak-Length	SPEAK CONTROL	—	—	No
Load-Lexicon	DEFINE-LEXICON	—	—	No
Lexicon-Search-Order	SPEAK	—	—	Yes

Figure 12.2 illustrates an example of a SPEAK request. The corresponding messages follow.
F1 (client → speechsynth):

```
MRCP/2.0 338 SPEAK 3214
Channel-Identifier: 23eb10a@speechsynth
```

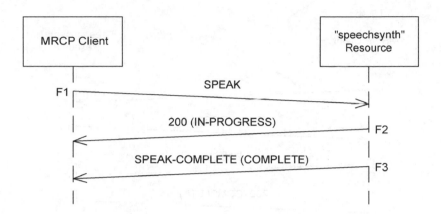

Figure 12.2   SPEAK request example.

```
Content-Type: application/ssml+xml
Content-Length: 213

<?xml version="1.0" encoding="UTF-8"?>
<speak version="1.0" xmlns="http://www.w3.org/2001/10/synthesis"
     xml:lang="en-US">
  Please leave a message after the beep.
  <audio src="beep.wav"/>
</speak>
```

F2 (speechsynth → client):

```
MRCP/2.0 119 3214 200 IN-PROGRESS
Channel-Identifier: 23eb10a@speechsynth
Speech-Marker: timestamp=857206027059
```

F3 (speechsynth → client):

```
MRCP/2.0 156 SPEAK-COMPLETE 3214 COMPLETE
Channel-Identifier: 23eb10a@speechsynth
Speech-Marker: timestamp=861500994355
Completion-Cause: 000 normal
```

12.2.2 PAUSE

The PAUSE request method instructs the speech synthesiser resource to pause speech output if it is currently speaking something. When the speech output is paused, the speech synthesiser resource responds with a status code of 200. The accompanying Active-Request-Id-List header field specifies the request-id of the SPEAK request that was paused. If the speech synthesiser is not currently processing a SPEAK request, it responds with a 402 Method not valid in this state status. If the speech synthesiser resource receives a PAUSE request while already in the paused state, it issues a response with status code 200 Success.

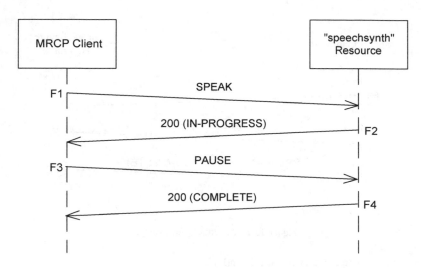

Figure 12.3 PAUSE request example.

Figure 12.3 illustrates an example of a PAUSE request. The corresponding messages follow.

F1 (client → speechsynth):

```
MRCP/2.0 187 SPEAK 3215
Channel-Identifier: 23eb10a@speechsynth
Content-Type: text/uri-list
Content-Length: 70

http://www.example.com/file01.ssml
http://www.example.com/file02.ssml
```

F2 (speechsynth → client):

```
MRCP/2.0 118 3215 200 IN-PROGRESS
Channel-Identifier: 23eb10a@speechsynth
Speech-Marker: timestamp=857206027059
```

F3 (client → speechsynth):

```
MRCP/2.0 67 PAUSE 3216
Channel-Identifier: 23eb10a@speechsynth
```

F4 (speechsynth → client):

```
MRCP/2.0 105 3216 200 COMPLETE
Channel-Identifier: 23eb10a@speechsynth
Active-Request-Id-List: 3215
```

12.2.3 RESUME

The RESUME request method instructs a paused speech synthesiser resource to resume speech output.
When the speech output is resumed, the speech synthesiser resource responds with a status code of 200
success. The accompanying Active-Request-Id-List header field specifies the request-id
of the SPEAK request that was resumed. If the speech synthesiser is in the idle state when it receives a
RESUME request, it responds with a 402 Method not valid in this state status. If the speech
synthesiser resource receives a RESUME request while it is in the speaking state, it issues a response
with status code 200 Success.

 Figure 12.4 illustrates an example of a RESUME request following on from the example shown in
Figure 12.3. The corresponding messages follow.

F5 (client → speechsynth):

```
MRCP/2.0 68 RESUME 3217
Channel-Identifier: 23eb10a@speechsynth
```

F6 (speechsynth → client):

```
MRCP/2.0 105 3217 200 COMPLETE
Channel-Identifier: 23eb10a@speechsynth
Active-Request-Id-List: 3215
```

F7 (speechsynth → client):

```
MRCP/2.0 155 SPEAK-COMPLETE 3215 COMPLETE
Channel-Identifier: 23eb10a@speechsynth
Speech-Marker: timestamp=861500994355
Completion-Cause: 000 normal
```

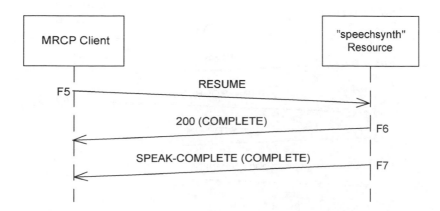

Figure 12.4 RESUME request example.

12.2.4 STOP

The STOP request method instructs the speech synthesiser resource to stop speech output. If the Active-Request-Id-List header field is omitted from the STOP request, the SPEAK request in progress is stopped and any SPEAK requests pending in the queue are removed. If a STOP request successfully terminates one or more PENDING or IN-PROGRESS SPEAK requests, the corresponding response, with status code 200 Success, will include an Active-Request-Id-List header field listing the request-ids of the requests that were terminated. The SPEAK-COMPLETE event is not sent for a terminated request. If no SPEAK requests were terminated, the response status code is still 200 Success but the Active-Request-Id-List header field is omitted.

It is possible to target particular SPEAK requests from the set of PENDING SPEAK requests and the current IN-PROGRESS request by specifying the Active-Request-Id-List header field in the STOP request. If the current IN-PROGRESS request is terminated, then the next available PENDING request in the queue is processed. Note that if the current IN-PROGRESS state is in the paused state and the speech synthesiser receives a STOP request, then the next available PENDING request in the queue becomes IN-PROGRESS but the resource remains in the paused state.

Figure 12.5 illustrates an example of a STOP request. The corresponding messages follow.

F1 (client → speechsynth):

```
MRCP/2.0 143 SPEAK 4000
Channel-Identifier: 23eb10a@speechsynth
Content-Type: text/plain
Content-Length: 28
```

Thank you. Please call again.

F2 (speechsynth → client):

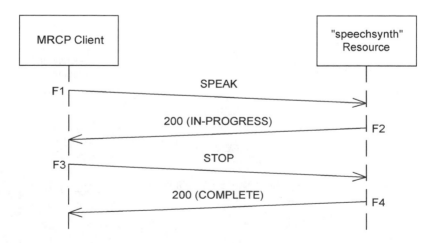

Figure 12.5 STOP request example.

```
MRCP/2.0 122 4000 200 IN-PROGRESS
Channel-Identifier: 23eb10a@speechsynth
Speech-Marker: timestamp=857206027059
```

F3 (client → speechsynth):

```
MRCP/2.0 66 STOP 4001
Channel-Identifier: 23eb10a@speechsynth
```

F4 (speechsynth → client):

```
MRCP/2.0 145 4001 200 COMPLETE
Channel-Identifier: 23eb10a@speechsynth
Speech-Marker: timestamp=861500994355
Active-Request-Id-List: 4000
```

12.2.5 BARGE-IN-OCCURRED

The BARGE-IN-OCCURRED request method operates identically to the STOP request if the active SPEAK request was issued with the header field Kill-On-Barge-In set to true; otherwise the method has no effect. The purpose of this request method is to allow the MRCP client to communicate to the speech synthesiser resource that it has been made aware of an input event, such as a DTMF key press or spoken input. The speech synthesiser resource can then decide whether or not to terminate the ongoing spoken output depending on the current prompts being barge-in-able.

Within the MRCP framework, there are two common deployment scenarios viz-a-viz orchestration of barge-in (here we take the example of a speech recogniser as the input device):

(i) The speech recogniser and speech synthesiser resources are independent of each other.
(ii) The speech recogniser and speech synthesiser resources are tightly coupled (and usually provided by the same vendor). The speech recogniser, by some proprietary means, can internally and directly signal a barge-in event to the speech synthesiser resource, resulting in a snappier barge-in experience.

For either scenario, the MRCP client acts as a proxy: it receives a START-OF-INPUT event from the speech recogniser resource and immediately issues a BARGE-IN-OCCURRED event to the speech synthesiser resource. In the case of tight coupling, the START-OF-INPUT event will include a header field called Proxy-Sync-Id that is placed by the MRCP client in the subsequent BARGE-IN-OCCURRED request. This header field allows the speech synthesiser resource to determine if it has already (internally) processed this barge-in event. MRCP is designed this way so that the client does not have to be aware of whether the media resource server implements barge-in directly on the server.

If the BARGE-IN-OCCURRED request results in one or more SPEAK requests being terminated, the response with status code 200 Success will include an Active-Request-Id-List enumerating the request-ids of the SPEAK requests terminated. If no SPEAK requests were terminated, the response status code is still 200 Success but the Active-Request-Id-List header field is omitted.

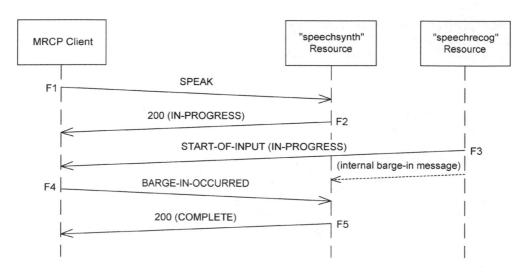

Figure 12.6   BARGE-IN-OCCURRED request example.

Figure 12.6 illustrates an example of a BARGE-IN-OCCURRED request assuming recognition has already commenced via a RECOGNIZE request (not shown). The 'internal barge-in message' only applies if the speech synthesiser and speech recogniser are tightly coupled. The corresponding messages follow.

F1 (client → speechsynth):

```
MRCP/2.0 376 SPEAK 9200
Channel-Identifier: 23eb10a@speechsynth
Content-Type: application/ssml+xml
Content-Length: 251

<?xml version="1.0" encoding="UTF-8"?>
<speak version="1.0" xmlns="http://www.w3.org/2001/10/synthesis"
      xml:lang="en-US">
   Here is your track. You may pause at anytime by simply
   saying pause.
   <audio src="track01.wav"/>
</speak>
```

F2 (speechsynth → client):

```
MRCP/2.0 118 9200 200 IN-PROGRESS
Channel-Identifier: 23eb10a@speechsynth
Speech-Marker: timestamp=857206027059
```

F3 (speechrecog → client):

```
MRCP/2.0 134 START-OF-INPUT 10001 IN-PROGRESS
```

```
Channel-Identifier: a123c31f@speechrecog
Input-Type: speech
Proxy-Sync-Id: 392812
```

F4 (client → speechsynth):

```
MRCP/2.0 103 BARGE-IN-OCCURRED 9201
Channel-Identifier: 23eb10a@speechsynth
Proxy-Sync-Id: 392812
```

F5 (speechsynth → client):

```
MRCP/2.0 144 9201 200 COMPLETE
Channel-Identifier: 23eb10a@speechsynth
Speech-Marker: timestamp=861500994355
Active-Request-Id-List: 9200
```

12.2.6 CONTROL

The CONTROL request method instructs the speech synthesiser resource to change how it is rendering the current SPEAK request. The CONTROL request can be used to jump forward or backwards (by specifying the Jump-Size header), or change the voice or prosodic parameters (by specifying a Voice- or Prosody- header). The CONTROL request is ideal for providing fast-forward and rewind capability, or to modify the speed or volume of the rendered audio. If a CONTROL request is applied to a SPEAK request, the response, with status code 200 Success, will include an Active-Request-Id-List header field with the request-id of the active SPEAK request, to which the CONTROL request was applied. If no SPEAK requests were IN-PROGRESS when the CONTROL request was issued, the response status code is 402 Method not valid in this state.

The behaviour resulting from specifying the Jump-Size header needs special consideration. If the client issues a CONTROL request with a Jump-Size header that requests the speech synthesiser resource to jump beyond the beginning of the current SPEAK request, the active request simply restarts from the beginning and the response includes a Speak-Restart header with a value of true. If the client issues a CONTROL request with a Jump-Size header that requests the speech synthesiser resource to jump beyond the end of the current SPEAK request, the active request simply ends with a SPEAK-COMPLETE event following the response to the CONTROL request. If there are more SPEAK requests in the queue, these are played as normal.

Figure 12.7 illustrates an example of a CONTROL request. The corresponding messages follow.

F1 (client → speechsynth):

```
MRCP/2.0 529 SPEAK 5000
Channel-Identifier: 23eb10a@speechsynth
Content-Type: multipart/mixed; boundary=0a23bf1020
Content-Length: 388

--0a23bf1020
```

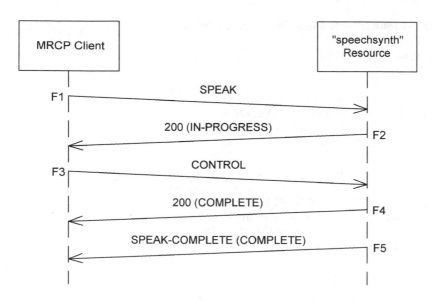

Figure 12.7   CONTROL request example.

```
Content-Type: text/uri-list
Content-Length: 32

http://www.example.com/menu.ssml

--0a23bf1020
Content-Type: application/ssml+xml
Content-Length: 198

<?xml version="1.0" encoding="UTF-8"?>
<speak version="1.0" xmlns="http://www.w3.org/2001/10/synthesis"
       xml:lang="en-US">
   Say operator to speak with a customer representative
</speak>
--0a23bf1020--
```

F2 (speechsynth → client):

```
MRCP/2.0 118 5000 200 IN-PROGRESS
Channel-Identifier: 23eb10a@speechsynth
Speech-Marker: timestamp=857206027059
```

F3 (client → speechsynth):

```
MRCP/2.0 114 CONTROL 5001
Channel-Identifier: 23eb10a@speechsynth
Jump-Size: +3 Second
```

```
Prosody-volume: +20%
```

F4 (speechsynth → client):

```
MRCP/2.0 145 5001 200 COMPLETE
Channel-Identifier: 23eb10a@speechsynth
Active-Request-Id-List: 5000
Speech-Marker: timestamp=861500994355
```

F5 (speechsynth → client):

```
MRCP/2.0 156 SPEAK-COMPLETE 5000 COMPLETE
Channel-Identifier: 23eb10a@speechsynth
Speech-Marker: timestamp=865795961651
Completion-Cause: 000 normal
```

12.2.7 DEFINE-LEXICON

The DEFINE-LEXICON request method instructs the speech synthesiser resource to load or unload one or more lexicon documents for the session. The DEFINE-LEXICON request is issued when the speech synthesiser resource is in the idle state to allow potentially large lexicon documents to be loaded ahead of time (issuing this request in any other state results in a response message with status code 402 Method not valid in this state). The Load-Lexicon header field value is used to indicate whether to load the lexicon (true) or unload the lexicon (false) – the default is true.

The lexicon data is either supplied inline in the message body or as an external URI reference using the text/uri-list. When the lexicon data is supplied inline, a Content-ID header field must also be specified, so that the lexicon data can be later referenced using the session: URI scheme (see Section 7.3.5). The MRCP specification does not mandate a particular lexicon data format but the Pronunciation Lexicon Specification (PLS) described in Chapter 11 is expected to become the predominant standard (corresponding to the application/pls+xml type). Many vendors also support their own proprietary formats for lexicons. If the DEFINE-LEXICON request fails, for example, because the lexicon format is not supported or the access did not succeed, a response with status code 407 Method or operation failed is returned.

The MRCP client can specify one or more lexicon documents to use in a SPEAK request by specifying the Lexicon-Search-Order header field. Typically, the value for this header will contain URIs (both external URIs and the session: URIs). Note that a lexicon referenced in the SSML document takes precedence over those specified in the Lexicon-Search-Order header field. Figure 12.8 illustrates an example of a DEFINE-LEXICON request. The corresponding messages follow.

F1 (client → speechsynth):

```
MRCP/2.0 468 DEFINE-LEXICON 6999
Channel-Identifier: 23eb10a@speechsynth
Content-ID: names@example.com
Content-Type: application/pls+xml
Content-Length: 302

<?xml version="1.0" encoding="UTF-8"?>
```

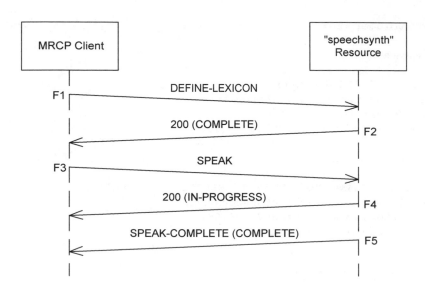

Figure 12.8 DEFINE-LEXICON request example.

```
<lexicon version="1.0"
         xmlns="http://www.w3.org/2005/01/pronunciation-lexicon"
         alphabet="ipa"
         xml:lang="en-US">
     <lexeme>
         <grapheme>Francesca</grapheme>
         <phoneme>Frantʃɛscɔ</phoneme>
     </lexeme>
</lexicon>
```

F2 (speechsynth → client):

```
MRCP/2.0 74 6999 200 COMPLETE
Channel-Identifier: 23eb10a@speechsynth
```

F3 (client → speechsynth):

```
MRCP/2.0 362 SPEAK 7000
Channel-Identifier: 23eb10a@speechsynth
Lexicon-Search-Order: <session:names@example.com>
Content-Type: application/ssml+xml
Content-Length: 184

<?xml version="1.0" encoding="UTF-8"?>
<speak version="1.0" xmlns="http://www.w3.org/2001/10/synthesis"
       xml:lang="en-US">
```

```
    You have a new message from Francesca.
</speak>
```

F4 (speechsynth → client):

```
MRCP/2.0 118 7000 200 IN-PROGRESS
Channel-Identifier: 23eb10a@speechsynth
Speech-Marker: timestamp=857206027059
```

F5 (speechsynth → client):

```
MRCP/2.0 156 SPEAK-COMPLETE 7000 COMPLETE
Channel-Identifier: 23eb10a@speechsynth
Speech-Marker: timestamp=861500994355
Completion-Cause: 000 normal
```

12.3 Events

12.3.1 SPEECH-MARKER

The SPEECH-MARKER event message is generated when the speech synthesiser resource encounters a marker tag in the speech mark-up it is currently rendering. Within SSML, the <mark> element with its associated name attribute is used to insert a marker into the speech markup (see Chapter 8). The SPEECH-MARKER event message is also sent when a SPEAK request in the PENDING state enters the IN-PROGRESS state (i.e. as the speech synthesiser resource pulls the next request from its queue). The purpose of the SPEECH-MARKER in this case is to indicate the time at which the request went IN-PROGRESS.

The SPEECH-MARKER event message includes a header field called Speech-Marker that specifies a timestamp and the name of the mark encountered (if any). The timestamp designates the absolute time measured in seconds and fractions of a second since 0h UTC 1st January 1900 at which the text corresponding to the marker was emitted as speech by the resource. The timestamp is formally an NTP timestamp [35] encoded as a 64-bit decimal number. To correlate the NTP timestamp with a timestamp received in an RTP packet, the MRCP client must inspect the RTCP sender reports to infer the offset between the RTP timestamp and the NTP wallclock time (recall from Chapter 6 that the RTCP sender report indicates the RTP timestamp of the packet at a given NTP timestamp).

Figure 12.9 illustrates an example of the SPEECH-MARKER event. The corresponding messages follow.

F1 (client → speechsynth):

```
MRCP/2.0 330 SPEAK 8000
Channel-Identifier: 23eb10a@speechsynth
Content-Type: application/ssml+xml
Content-Length: 205

<?xml version="1.0" encoding="UTF-8"?>
<speak version="1.0" xmlns="http://www.w3.org/2001/10/synthesis"
        xml:lang="en-US">
```

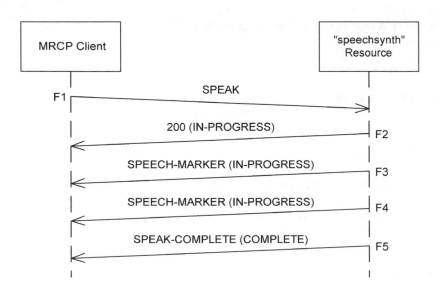

Figure 12.9 SPEECH–MARKER event example.

From <mark name="here"/> here to <mark name="there"/> there
</speak>

F2 (speechsynth → client):

```
MRCP/2.0 118 8000 200 IN-PROGRESS
Channel-Identifier: 23eb10a@speechsynth
Speech-Marker: timestamp=857206027059
```

F3 (speechsynth → client):

```
MRCP/2.0 132 SPEECH-MARKER 8000 IN-PROGRESS
Channel-Identifier: 23eb10a@speechsynth
Speech-Marker: timestamp=861500994355;here
```

 F4 (speechsynth → client):

```
MRCP/2.0 133 SPEECH-MARKER 8000 IN-PROGRESS
Channel-Identifier: 23eb10a@speechsynth
Speech-Marker: timestamp=865795961651;there
```

F5 (speechsynth → client):

```
MRCP/2.0 161 SPEAK-COMPLETE 8000 COMPLETE
Channel-Identifier: 23eb10a@speechsynth
Speech-Marker: timestamp=865795961652;there
Completion-Cause: 000 normal
```

12.3.2 SPEAK-COMPLETE

The SPEAK-COMPLETE event message is generated when the speech synthesiser resource finishes processing the corresponding SPEAK request (identified by the request-id included in the event). The request state contained in a SPEAK-COMPLETE event message is always COMPLETE, indicating that no further events will be generated for the original request. The SPEAK-COMPLETE event message includes the Completion-Cause header field, which indicates the reason the SPEAK request completed. Many of the examples in the preceding sections illustrate the SPEAK-COMPLETE event.

12.4 Headers

12.4.1 Completion-Cause

The Completion-Cause header is always present in the SPEAK-COMPLETE response to indicate why the SPEAK request has completed. The value for the header field includes both a cause code and corresponding cause name. Table 12.5 summaries the different cause codes and cause names.
   Example:

```
Completion-Cause: 002 parse-failure
```

12.4.2 Completion-Reason

The Completion-Reason header is optionally present in the SPEAK-COMPLETE response to supply more information on the reason why the SPEAK request terminated. This header field is intended for use by the MRCP client for logging purposes.
   Example:

```
Completion-Reason: SSML not well-formed
```

Table 12.5 Cause codes and cause names for the Completion-Cause header field

Cause code	Cause name	Description
000	normal	SPEAK request completed normally.
001	barge-in	SPEAK request terminated because of a barge-in. This value can only be returned when the speech recogniser and speech synthesiser resources are tightly coupled.
002	parse-failure	The markup in the SPEAK request failed to parse.
003	uri-failure	The SPEAK request terminated because access to a URI failed. Note that a failed <audio> does not trigger this cause code, rather fallback audio is played (see Section 8.3).
004	error	An error occurred and the SPEAK request terminated prematurely.
005	language-unsupported	The requested language is not supported.
006	lexicon-load-failure	A lexicon document failed to load.
007	cancelled	The queued SPEAK request was cancelled due to an error encountered processing a request before it.

12.4.3 *Failed-URI*

The `Failed-URI` header field indicates a URI to which access failed. This applies to SSML and text documents for synthesising and also to lexicon files. The header field may be returned in a response to the `SPEAK` or `DEFINE-LEXICON` requests or in the `SPEAK-COMPLETE` event. See also the `Failed-URI-Cause` header field.
   Example:

```
Failed-URI: http://www.example.com/dictionary.pls
```

12.4.4 *Failed-URI-Cause*

The `Failed-URI-Cause` header field is used in conjunction with the `Failed-URI` header field to provide protocol specific response code information. For HTTP URIs, the value will typically be a HTTP *4XX* or *5XX* response code.
   Example:

```
Failed-URI-Cause: 404 Not Found
```

12.4.5 *Speech-Marker*

The `Speech-Marker` header field has a dual purpose: (i) to report the last encountered mark, and (ii) to report timestamp information corresponding to certain actions performed by the speech synthesiser resource. The `Speech-Marker` header field is present in the `SPEECH-MARKER` event message, where its value includes a `timestamp` parameter to indicate when the event occurred and the name of the mark that was encountered, if any. The mark name may be omitted, for example, if the `SPEECH-MARKER` event is emitted as a result to a `SPEAK` request transitioning from `PENDING` to `IN-PROGRESS`. The `Speech-Marker` header field also appears in several response messages to indicate timestamp information. The `Speech-Marker` header field is present in the response to an `IN-PROGRESS SPEAK` request to indicate the timestamp for the start of playback, and in the responses to `BARGE-IN-OCCURRED`, `STOP` and `CONTROL` requests to indicate when the corresponding action was taken. In these cases, the last mark name encountered (if any) is inserted into the `Speech-Marker` header field. The `timestamp` parameter value is the unsigned decimal value of the 64-bit NTP timestamp. Implementations are not required to report a value for the `timestamp` parameter (in this case the value to the right of the equal sign is omitted).
   Example:

```
Speech-Marker: timestamp=857206027059;my_mark_name
```

12.4.6 *Voice- parameters*

There are four header fields derived from the attributes of SSML's `<voice>` element (see Chapter 8): `Voice-Gender`, `Voice-Age`, `Voice-Variant`, and `Voice-Name`. The headers apply the same meaning to the entire `SPEAK` request as do their SSML equivalents for mark-up enclosed within the

<voice> element.[1] The Voice- parameters may be set for the session using the SET-PARAMS request method. The Voice- parameters may also be set for the current SPEAK request. Note that the <voice> element within SSML mark-up takes precedence over the Voice- parameter header fields. Finally, some implementations also support the ability to change the Voice- parameters 'on the fly' by specifying them in the CONTROL request method (an implementation that does not support this feature will return a response with status code 403 Unsupported header).
Example:

```
Voice-Gender: female
```

12.4.7 Prosody- parameters

There are six header fields derived from the attributes of SSML's <prosody> element in total (see Chapter 8): Prosody-Pitch, Prosody-Contour, Prosody-Range, Prosody-Rate, Prosody-Duration, and Prosody-Volume. The header fields are applied using the same rules as the Voice- parameter header fields described in the preceding section.
Example:

```
Prosody-Volume: soft
```

12.4.8 Speaker-Profile

The Speaker-Profile header field is used to supply a URI which references the profile of the speaker used for synthesis. Speaker profiles are collections of voice parameters such as voice name, gender, age, etc. MRCP does not specify a document format for a speaker profile document but rather leaves that as an implementation specific detail. The Speaker-Profile header may be used on a per-SPEAK request basis or set for the session via the SET-PARAMS request.
Example:

```
Speaker-Profile: http://www.example.com/speaker01.pfl
```

12.4.9 Speech-Language

The Speech-Language header field is used to indicate the default language of the speech data if it is not otherwise specified in the markup (recall from Chapter 8 that SSML allows the language to be specified via the xml:lang attribute on several elements including <speak>). The format for the value is the same as that used for SSML's xml:lang attribute and is specified in RFC 3066 [55]. The Speech-Language header may be used on a per-SPEAK request basis or set for the session via the SET-PARAMS request.
Example:

```
Speech-Language: en-GB
```

---

[1] The Voice- header fields also allow voice control when plain text is being synthesised.

12.4.10 Kill-On-Barge-In

The Kill-On-Barge-In header field takes a boolean value to indicate whether or not a SPEAK request can be interrupted by barge-in from DTMF or speech input. If set to true, the SPEAK request stops immediately on barge-in and any queued SPEAK requests in the PENDING state are discarded. If set to false, the barge-in event is simply ignored. The Kill-On-Barge-In header may be used on a per-SPEAK request basis or set for the session via the SET-PARAMS request. The default value if omitted is true.

If the speech synthesiser resource and speech recogniser resource are part of the same session (i.e. the control channels are part of the same control session – see Chapter 5), then barge-in can be communicated internally. Otherwise, the MRCP client, on receipt of the START-OF-INPUT event from the speech recogniser resource (or indeed on detecting input via some other proprietary means if an MRCP speechrecog resource is not used), can issue a BARGE-IN-OCCURRED request to the speech synthesiser resource. Note that MRCP requires that the BARGE-IN-OCCURRED request be issued regardless of whether the event is communicated internally in the media resource server.

Example:

```
Kill-On-Barge-In: false
```

12.4.11 Fetch-Hint

This Fetch-Hint header field determines when the speech synthesiser resource should make a request for a resource identified by a URI, such as SSML and text documents to synthesise, audio files, and lexicon documents. A value of prefetch indicates that the content is to be downloaded when the SPEAK request is received. A value of safe indicates that the content may be downloaded when the content is actually needed. The Fetch-Hint header may be used on a per-SPEAK request basis or set for the session via the SET-PARAMS request. The default value if omitted is prefetch.

Example:

```
Fetch-Hint: safe
```

12.4.12 Audio-Fetch-Hint

The Audio-Fetch-Hint header field only applies to audio files and overrides the Fetch-Hint header field described in the preceding section. In addition to the prefetch and safe values, it also accepts a value of stream that indicates that the content can be streamed as it is being downloaded. This assumes the download speed is as fast or faster than the rate at which the audio is transmitted over RTP and is typically the case. The Audio-Fetch-Hint header may be used on a per-SPEAK request basis or set for the session via the SET-PARAMS request. The default value if omitted is prefetch.

Example:

```
Audio-Fetch-Hint: stream
```

12.4.13 *Jump-Size*

The Jump-Size[2] header field is used to jump forward or backwards in the speech. The value for this header is either a marker name to jump to (i.e. corresponding to the name attribute of an SSML <mark>) or otherwise consists of three parts: a '+' or '−' to indicate a forward or backward jump respectively, a number to indicate the size of the jump, and the units of the jump. The units can either be one of Second, Word, Sentence, or Paragraph. Not all speech synthesiser resources are expected to support all units; if a speech synthesiser does not support a unit it returns a response with status code 409 Unsupported header value. The Jump-Size header may be used with a SPEAK request basis to specify an initial offset, or on CONTROL to effect an offset of an ongoing SPEAK request.

Example:

```
Jump-Size: -5 Second
```

12.4.14 *Speak-Restart*

The Speak-Restart header field is used to inform the MRCP client that a CONTROL request with a negative Jump-Size has resulted in the speech output jumping to the start of the current SPEAK request. The header field takes a boolean value and is specified in the response to the CONTROL request (otherwise it is simply omitted). No special treatment is required for the converse case where a Jump-Size results in the speech output jumping to or past the end of the SPEAK request as the request will complete and the MRCP client will receive a SPEAK-COMPLETE event message as normal.

Example:

```
Speak-Restart: true
```

12.4.15 *Speak-Length*

The Speak-Length header field is used to specify the length of speech to speak. It can either be specified in the SPEAK request, in which case the length is taken relative to the beginning of the speech, or in a CONTROL request, in which case the length is taken relative to the current speaking point. The legal values for this header are identical to those of Jump-Size (except that negative values are not allowed). If a speech synthesiser does not support a unit it returns a response with status code 409 Unsupported header value.

Example:

```
Speak-Length: 25 Second
```

12.4.16 *Load-Lexicon*

The Load-Lexicon header field may be specified in the DEFINE-LEXICON method to indicate whether the lexicon is to be loaded or unloaded. The default value is true.

---

[2] This header field is called Jump-Target in MRCPv1 (see Appendix A, located on the Web at http://www.daveburke.org/speechprocessing/).

Example:

```
Load-Lexicon: false
```

12.4.17 Lexicon-Search-Order

The `Lexicon-Search-Order` header field specifies an ordered list of URIs for lexicons to search through. Lexicons specified in the SSML markup take precedence. The `Lexicon-Search-Order` header may be used on a per-`SPEAK` request basis or set for the session via the `SET-PARAMS` request. URIs are placed within angle brackets. Multiple URIs are delimited by whitespace.

Example:

```
Lexicon-Search-Order: <http://example.com/nouns.pls>
                      <http://10.0.0.1/places.pls>
```

12.5 Summary

This chapter provides an in-depth look at the speech synthesiser resource. MRCP defines two subtypes that differ by capability. The first, the basicsynth, is ideal for simple prompt playback based on concatenated audio prompts. The second, the speechsynth, is a fully fledged speech synthesiser that can synthesis arbitrary text. Both resources accept SSML mark-up, albeit with a limited number of XML elements supported for the basicsynth. The speechsynth can also accept plain text to synthesise. The MRCP speech synthesiser resource is sufficiently rich to provide all the underlying capabilities required by a variety of media processing devices, ranging from simple media servers to VoiceXML IVR service nodes.

In the next chapter, we continue our in-depth look at the MRCP media resources by studying the speech recogniser resource.

13

Speech recogniser resource

The MRCP speech recogniser analyses audio and DTMF events carried over the media session and returns text-based recognition results. MRCP specifies two recogniser resource types. The *dtmfrecog* resource only supports recognition of DTMF digits and is therefore restricted to operating with DTMF grammars. The *speechrecog* resource supports recognition of speech audio and DTMF and hence supports both speech grammars and DTMF grammars.[1] This chapter delves into the details of both resource types by studying the methods, events and header fields that apply to them. At this point, the reader may wish to review Chapter 9, which discusses the Speech Recognition Grammar Specification (SRGS). SRGS is a data format that specifies the words and phrases (and DTMF sequences) a speech recogniser may recognise. In addition, it may be advantageous to review Chapter 10, which discusses the Natural Language Semantics Markup Language (NLSML) used to format recognition results returned to the MRCP client.

13.1 Overview

The dtmfrecog resource provides a subset of functionality of the speechrecog resource by only supporting DTMF input and corresponding DTMF grammars. DTMF digits are recognised via RTP events using the mechanism described in RFC 2833 [44] or, optionally, by analysing the input audio stream for DTMF tones–see Chapter 6. In every other way, the dtmfrecog resource behaves as a speechrecog resource thus allowing us to concentrate on the general speechrecog resource for the remainder of this chapter (which we will generically refer to as a 'recogniser resource').

MRCP recogniser resources are required to support the XML form of the SRGS [6] described in Chapter 9 and may also support the equivalent ABNF form of SRGS. With this approach, grammars written in textual form specify the words and phrases (and DTMF sequences) that can be recognised for a given recognition request. In addition, semantic interpretation tags may be added to the grammars to annotate the semantics or meaning of different utterances – the semantics corresponding to one

---

[1] Some speech recogniser implementations may choose not to support DTMF recognition, although the MRCP specification recommends that they do.

or more matched utterances are subsequently returned in the recognition results (see Chapter 9). Since the MRCP protocol carries grammars opaquely, it is straightforward for recogniser resource implementations to support their own proprietary grammar formats.

MRCP speech recognisers may optionally support *voice enrolled* grammars. These grammars differ fundamentally from other grammar types in that they are created at run-time by the speaker's voice, usually by repeating a word or phrase several times to train the speech recogniser. The grammars are naturally speaker dependent. A voice enrolled grammar is conceptually equivalent to an SRGS grammar containing a list of alternatives specified by the `<one-of>` element (see Chapter 9) – the difference being that each alternative is trained from the speaker's voice and the actual grammar format is implementation specific. While the use of a voice enrolled grammar in the basic recognition use-case is identical to that of say an SRGS grammar, a series of new methods and header fields are required specifically for training and managing the voice enrolled grammars. Since, in practice, voice enrolled grammars are typically used in a minority of applications, we separate out the discussion of those methods and header fields related to voice enrolment in the discussion both for clarity and to allow the reader the option of skipping the voice enrollment mechanisms on first reading.

The recogniser resource may operate in one of two modes:

- normal mode recognition;
- hotword mode recognition.

Normal mode recognition refers to the usual case where the speech recogniser tries to match the entire speech utterance or DTMF string against the active grammars and returns a `RECOGNITION-COMPLETE` event indicating success (in which case the recognition results are contained in the message body) or some failure status (e.g. `001 no-match` if the input failed to match an active grammar. Hotword mode recognition, on the other hand, is where the speech recogniser looks for a specific speech phrase(s) or DTMF sequence(s) and ignores speech or DTMF that does not match. The recognition action only completes for a successful match of the grammar, if the MRCP client cancels it, or if the `No-Input-Timeout` or `Recognition-Timeout` timer expires. As a consequence, a hotword recognition never terminates with a `Completion-Cause` value of `001 no-match`. Hotword is useful in situations where a special 'keyword phrase' is being listened for. For example, during a call transfer, the phrase 'main menu' could be used to terminate the transfer and return the caller to a menu.

The recogniser resource also supports a text-based 'interpretation' mode of operation. A text utterance sent to the recogniser over the control session (i.e. instead of speech over the media session) will be matched against the active grammars and, assuming a match, will result in recognition results being returned to the MRCP client along with the semantic interpretation (if the grammar contained semantic tags – see Section 9.8 for more information on semantic interpretation). This feature allows text-based tools to be created for testing grammar coverage and correct operation of semantic interpretation tags, for example.

The recognition resource supports a total of six request messages and three event messages. Voice enrolment adds a further five request methods specifically for creating and managing voice enrolled grammars. The methods and events are summarised in Tables 13.1–13.3.

The recogniser resource has a state machine that is driven by requests from the MRCP client and events generated by the media resource itself. The state machine is illustrated in Figure 13.1. A much simpler state machine applies to the case of training voice enrolled grammars. In this case, the `START-SESSION` and `END-SESSION` methods are not considered to change the state of the

Table 13.1 Recognition request messages (methods) not specific to voice enrollment

Message name	Description
RECOGNIZE	Requests that the recogniser resource commence recognition and specifies one or more grammars to activate.
DEFINE-GRAMMAR	Provides one or more grammars to the recogniser resource and requests that it access, fetch, and compile the grammars as needed in preparation for recognition.
START-INPUT-TIMERS	Starts the no input timer.
GET-RESULT	Retrieves recognition results for a completed recognition. This method can be used by the MRCP client to specify a different confidence level or *n*-best list length to retrieve different result alternatives, for example.
STOP	Stops an ongoing recognition.
INTERPRET	Requests that the recogniser resource perform an interpretation of supplied text by matching it against the active grammar(s).

Table 13.2 Recognition request messages (methods) specific to voice enrollment

Message name	Description
START-PHRASE-ENROLLMENT	Starts a new phrase enrollment session. The MRCP client subsequently issues multiple RECOGNIZE requests to enrol a new utterance in the grammar.
ENROLLMENT-ROLLBACK	Discards the utterance captured from the previous RECOGNIZE request.
END-PHRASE-ENROLLMENT	Ends a phrase enrolment session.
MODIFY-PHRASE	Used to modify an enrolled phrase.
DELETE-PHRASE	Used to delete a phrase in a voice enrolled grammar.

Table 13.3 Event messages

Message name	Description
START-OF-INPUT	Generated when the recogniser resource first detects speech or DTMF.
RECOGNITION-COMPLETE	Generated when the recognition request completes.
INTERPRETATION-COMPLETE	Generated when the interpretation request completes.

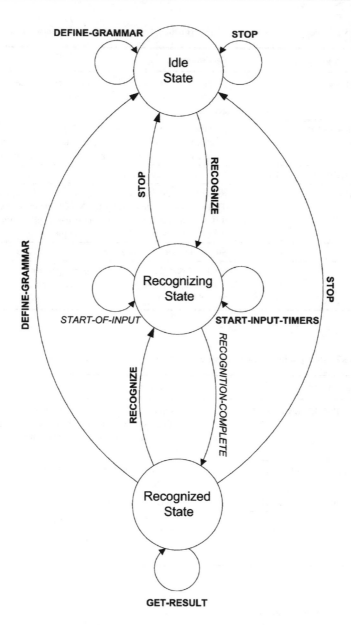

Figure 13.1 Recogniser resource state machine.

resource, and each utterance capture (triggered by calling RECOGNIZE) results in the state machine transitioning from idle to the recognising state and back to idle when the RECOGNITION-COMPLETE event fires.

Tables 13.4 and 13.5 summarises the resource-specific header fields for the recogniser resource. See Table 7.5 for a list of generic headers that also apply.

Table 13.4 Recogniser resource header fields (not specific to voice enrollment)

Header name	Request	Response	Event	GET-PARAMS / SET-PARAMS
Completion-Cause	—	RECOGNIZE DEFINE-GRAMMAR	RECOGNITION-COMPLETE	No
Completion-Reason	—	RECOGNIZE DEFINE-GRAMMAR	RECOGNITION-COMPLETE	No
Failed-URI	—	RECOGNIZE DEFINE-GRAMMAR	RECOGNITION-COMPLETE	No
Failed-URI-Cause	—	RECOGNIZE DEFINE-GRAMMAR	RECOGNITION-COMPLETE	No
Recognition-Mode	RECOGNIZE	—	—	No
Input-Type	—	—	START-OF-INPUT	No
Confidence-Threshold	RECOGNIZE GET-RESULT	—	—	Yes
Sensitivity-Level	RECOGNIZE	—	—	Yes
Speed-Vs-Accuracy	RECOGNIZE	—	—	Yes
N-Best-List-Length	RECOGNIZE GET-RESULT	—	—	Yes
No-Input-Timeout	RECOGNIZE	—	—	Yes
Recognition-Timeout	RECOGNIZE	—	—	Yes
Speech-Complete-Timeout	RECOGNIZE	—	—	Yes
Speech-Incomplete-Timeout	RECOGNIZE	—	—	Yes
Hotword-Max-Duration	RECOGNIZE	—	—	Yes
Hotword-Min-Duration	RECOGNIZE	—	—	Yes
DTMF-Interdigit-Timeout	RECOGNIZE	—	—	Yes
DTMF-Term-Timeout	RECOGNIZE	—	—	Yes
DTMF-Term-Char	RECOGNIZE	—	—	Yes
DTMF-Buffer-Time	—	—	—	Yes
Clear-DTMF-Buffer	RECOGNIZE	—	—	No
Save-Waveform	RECOGNIZE	—	—	Yes
Waveform-URI	—	END-PHRASE-ENROLLMENT	RECOGNITION-COMPLETE	No
Input-Waveform-URI	RECOGNIZE	—	—	No
Media-Type	RECOGNIZE	—	—	Yes
Start-Input-Timers	RECOGNIZE	—	—	No
Speech-Language	RECOGNIZE DEFINE-GRAMMAR	—	—	Yes
Cancel-If-Queue	RECOGNIZE	—	—	No
New-Audio-Channel	RECOGNIZE	—	—	No
Ver-Buffer-Utterance	RECOGNIZE	—	—	No
Early-No-Match	RECOGNIZE	—	—	Yes
Interpret-Text	RECOGNIZE	—	—	No
Recognizer-Context-Block	—	—	—	Yes

Table 13.5   Recogniser resource header fields (specific to voice enrollment)

Header name	Request	Response	Event	GET-PARAMS / SET-PARAMS
Enroll-Utterance	RECOGNIZE	—	—	No
Num-Min-Consistent-Pronunciations	START-PHRASE-ENROLLMENT	—	—	Yes
Consistency-Threshold	START-PHRASE-ENROLLMENT	—	—	Yes
Clash-Threshold	START-PHRASE-ENROLLMENT	—	—	Yes
Personal-Grammar-URI	START-PHRASE-ENROLLMENT MODIFY-PHRASE DELETE-PHRASE	—	—	No
Phrase-ID	START-PHRASE-ENROLLMENT MODIFY-PHRASE DELETE-PHRASE	—	—	No
New-Phrase-ID	MODIFY-PHRASE	—	—	No
Phrase-NL	START-PHRASE-ENROLLMENT MODIFY-PHRASE	—	—	No
Weight	START-PHRASE-ENROLLMENT MODIFY-PHRASE	—	—	No
Save-Best-Waveform	START-PHRASE-ENROLLMENT	—	—	No
Confusable-Phrases-URI	RECOGNIZE	—	—	No
Abort-Phrase-Enrollment	END-PHRASE-ENROLLMENT	—	—	No

13.2 Recognition methods

13.2.1 RECOGNIZE

The RECOGNIZE request method starts the recognition process and simultaneously specifies the active grammars specifying the words and phrases to which the input audio stream is to be matched against. The recognition results are returned in the message body of the RECOGNITION-COMPLETE event. Hotword recognition is requested by specifying the Recognition-Mode header field with a value of hotword (the default is normal). The RECOGNIZE request may also contain header fields to tune the recognition request such as language, confidence thresholds, timer settings, etc. A separate use of the RECOGNIZE request method is to enrol an utterance into a personal grammar and is described in Section 13.3.1.

The active grammars for the recognition action are specified in the body of the RECOGNIZE request. Either inline grammar documents or URI references to grammar documents may be specified (or indeed a combination of both if a multipart message body is used – see Section 7.1.4.1). In the former case, a complete SRGS grammar may be specified in the RECOGNIZE request message body identified by the Content-Type header field value of application/srgs+xml (assuming the XML form of SRGS). When the grammar document is contained in the message body, a Content-ID header field is required to be present. This allows subsequent RECOGNIZE requests to supply a URI for that grammar in the form of a session: URI (see Section 7.3.5) instead of having to re-send the (potentially large) grammar document for each recognition request. To supply grammar references, a list of URIs may be specified either using the text/uri-list or[2] text/grammar-ref-list (see Section 7.1.4.2). Since a voice enrolled grammar is identified by a grammar URI, no special considerations are required to recognise against such a grammar. It is possible that more than one active grammar could match the input utterance, and therefore the order of inclusion of the grammars within the RECOGNIZE request method is used to determine the precedence. This is important for VoiceXML interpreters built upon MRCP where VoiceXML-document ordering of grammars are used by application authors to specify grammar precedence (see Chapter 16).

The speech recognition resource supports the ability either to queue multiple RECOGNIZE requests or, alternatively, operate without a queue. The Cancel-If-Queue header field, taking a Boolean value, is used to select the desired behaviour. A value of false implies that multiple RECOGNIZE requests are queued in the same way as SPEAK requests are for the speech synthesis resource, i.e. if the resource is idle or in the recognized state when it receives a RECOGNIZE request, the corresponding response will contain a request-state of IN-PROGRESS. If the resource is in the recognizing state when a RECOGNIZE request arrives, the corresponding response will indicate a request-state of PENDING. Each request is processed in a first-in–first-out manner and if an error occurs with one request, all queued requests are cancelled. On the other hand, a value of true for Cancel-If-Queue means that if the MRCP client issues another RECOGNIZE request while one is already IN-PROGRESS, that request will be cancelled (specifically returning a RECOGNITION-COMPLETE with Completion-Cause 011 cancelled). Queuing of recognition requests was introduced to facilitate use-cases where a hotword recognition is followed immediately by a normal recognition. For example, 'Computer, call Rob please' where 'Computer' is the hotword and 'call Rob please' matches the grammar of a normal recognition. Recognition accuracy is improved when the hotword phrase is short and queuing avoids losing audio between the two recognition requests.

MRCP defines a number of timers for controlling recognition behaviour that differ depending on whether a normal or hotword recognition is being performed. Tables 13.6 and 13.7 summarise the timers for normal and hotword recognitions respectively.

Figure 13.2 illustrates an example of a RECOGNIZE request. The corresponding messages follow.

F1 (client → speechrecog):

```
MRCP/2.0 439 RECOGNIZE 00001
Channel-Identifier: 23af1e13@speechrecog
Content-ID: <grammar1@form-level.store>
Content-Type: application/srgs+xml
Content-Length: 265
```

---

[2] The text/grammar-ref-list provides the same functionality as text/uri-list but with the added advantage that individual grammar weights may be specified per URI.

Table 13.6 Timers applicable to normal recognition

Timer name (header field)	Purpose	Timer started when?	Timer expiry (`Completion-Cause`)
`No-Input-Timeout`	Handles the case when the user does not respond.	On receipt of the `RECOGNIZE` request (if `Start-Input Timers: true`) otherwise when the `START-INPUT -TIMERS` request is issued.	`002 no-input- timeout`
`Recognition- Timeout`	Limits the duration of recognition. Similar to VoiceXML's `maxspeechtimeout` property.	When the `START-OF-INPUT` event is generated	• `008 success- maxtime` if there is a full match • `015 no-match- maxtime` if there is a no match • `014 partial-match- maxtime`[a] if there is a no match
`Speech-Complete- Timeout`	Tunes the trade-off between fast system response and breaking an utterance inappropriately.	Silence after an utterance that fully matches an active grammar.	`000 success`
`Speech- Incomplete- Timeout`	Same as `Speech- Complete-Timeout`. Usually `Speech- Incomplete- Timeout` is longer than the `Speech- Complete-Timeout` to allow users to pause/breathe mid-utterance.	Silence after an utterance that partially matches an active grammar.	`013 partial-match`
`DTMF-Interdigit- Timeout`	Specifies the amount of time between DTMF key presses to wait before finalising a result.	Re-started each time a DTMF digit is received.	• `000 success` if there is a full match • `013 partial match` if there is a partial match
`DTMF-Term-Timeout`	Specifies the amount of time to wait for the termchar when no more input is allowed by the grammar.	When the grammar reaches a termination point.	`000 success`

[a] In general, a recogniser resource that cannot differentiate between a partial match and a no match can return a `001 no-match` in place of `013 partial-match` or `014 partial-match-maxtime`

Table 13.7   Timers applicable to hotword recognition

Timer name (header field)	Purpose	Timer started when?	Timer expiry (Completion-Cause)
`No-Input-Timeout`	Handles the case when the user does not respond.	On receipt of the `RECOGNIZE` request (if `Start-Input Timers: true`) otherwise when the `START-INPUT-TIMERS` request is issued. Note, for hotword, the timer is only stopped when an utterance is matched.	`002 no-input-timeout`
`Recognition-Timeout`	Limits the duration of recognition. Similar to VoiceXML's `maxspeech timeout` property.	When speech is detected. The timer is reset on detection of silence (i.e. after `Speech-Incomplete-Timeout` expires).	• `008 success-maxtime` if there is a full match • `003 hotword-maxtime` if there is a no match
`Hotword-Max-Duration`	Specifies the maximum length of an utterance to be considered for hotword recognition	Start of utterance	Do not consider current utterance for hotword
`Hotword-Min-Duration`	Specifies the minimum length of an utterance to be considered for hotword recognition	Start of utterance	Consider current utterance for hotword
`Speech-Complete-Timeout`	Tunes the tradeoff between fast system response and breaking an utterance inappropriately	Silence after an utterance that fully matches an active grammar	`000 success`
`Speech-Incomplete-Timeout`	Same as `Speech-Complete-Timeout`. Usually `Speech-Incomplete-Timeout` is longer than the `Speech-Complete-Timeout` to allow users to pause/breathe mid-utterance	Silence after an utterance that partially matches an active grammar.	Ignore utterance if there is no match.

Table 13.7   (*continued*)

Timer name (header field)	Purpose	Timer started when?	Timer expiry (completion-cause)
`DTMF-Interdigit-Timeout`	Specifies the amount of time between DTMF key presses to wait before finaliaing a result	Restarted each time a DTMF digit is received	• `000 success` if there is a full match • ignore DTMF sequence if there is a partial match or no match
`DTMF-Term-Timeout`	Specifies the amount of time to wait for the termchar when no more input is allowed by the grammar	When the grammar reaches a termination point	`000 success`

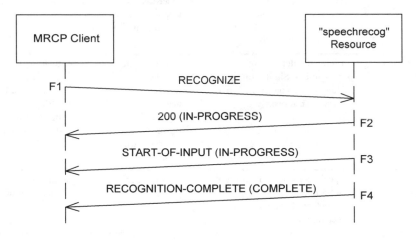

Figure 13.2   `RECOGNIZE` request example.

```
<?xml version="1.0" encoding="UTF-8"?>
<grammar version="1.0" xmlns="http://www.w3.org/2001/06/grammar"
      xml:lang="en-IE">
   <rule id="yesno">
    <one-of>
       <item>yes</item>
       <item>no</item>
    </one-of>
   </rule>
</grammar>
```

F2 (speechrecog → client):

```
MRCP/2.0 79 00001 200 IN-PROGRESS
Channel-Identifier: 23af1e13@speechrecog
```

F3 (speechrecog → client):

```
MRCP/2.0 111 START-OF-INPUT 00001 IN-PROGRESS
Channel-Identifier: 23af1e13@speechrecog
Input-Type: speech
```

F4 (speechrecog → client):

```
MRCP/2.0 474 RECOGNITION-COMPLETE 00001 COMPLETE
Channel-Identifier: 23af1e13@speechrecog
Completion-Cause: 000 success
Content-Type: application/nlsml+xml
Content-Length: 291

<?xml version="1.0" encoding="UTF-8"?>
<result grammar="session:grammar1@form-level.store"
        xmlns="http://www.ietf.org/xml/ns/mrcpv2">
    <interpretation confidence="0.9">
     <instance>
        yes
     </instance>
     <input>yes</input>
    </interpretation>
</result>
```

13.2.2 DEFINE-GRAMMAR

The purpose of the DEFINE-GRAMMAR method is to request the recogniser resource to access, fetch and compile the grammars contained in the message body (grammars are contained in the body in the same way as they are for RECOGNIZE requests, i.e. as inline grammar documents or grammar URIs). The MRCP client does not necessarily need to issue the DEFINE-GRAMMAR method to perform recognition – rather this method simply gives the client an option to prepare potentially large grammars ahead of recognition time. Inline grammars and their corresponding session: URIs are only valid for the session that defined them (either through a DEFINE-GRAMMAR or RECOGNIZE request). However, since recogniser resource implementations usually include a global grammar cache, external grammars prepared in a previous session may be served (subject to freshness criteria) from the cache thus optimising the grammar preparation step for subsequent uses.

The response to the DEFINE-GRAMMAR request always carries a request-state of COMPLETE. A successful response carries the status code of 200 Success and a Completion-Cause header field with a value of 000 success. If there is an error (the grammar could not be fetched or compiled or the recogniser is in the recognising state, for example), the response carries the status code of 407 Method or Operation Failed and an appropriate Completion-Cause header field value.

Figure 13.3 illustrates an example of a DEFINE-GRAMMAR request. The corresponding messages follow.

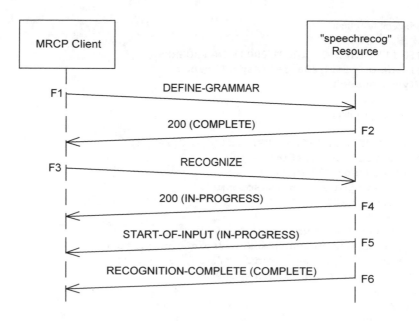

Figure 13.3 DEFINE-GRAMMAR request example.

F1 (client → speechrecog):

```
MRCP/2.0 365 DEFINE-GRAMMAR 10000
Channel-Identifier: 23af1e13@speechrecog
Content-ID: <menu@example.com>
Content-Type: application/srgs+xml
Content-Length: 197

<?xml version="1.0" encoding="UTF-8"?>
<grammar version="1.0" xmlns="http://www.w3.org/2001/06/grammar"
        xml:lang="en-US">
    <rule id="menu">
        main menu
    </rule>
</grammar>
```

F2 (speechrecog → client):

```
MRCP/2.0 108 10000 200 COMPLETE
Channel-Identifier: 23af1e13@speechrecog
Completion-Cause: 000 success
```

F3 (client → speechrecog):

```
MRCP/2.0 223 RECOGNIZE 10001
Channel-Identifier: 23af1e13@speechrecog
```

```
No-Input-Timeout: 10000
Recognition-Timeout: 30000
Content-Type: text/grammar-ref-list
Content-Length: 39

<session:menu@example.com>;weight="1.0"
```

F4 (speechrecog → client):

```
MRCP/2.0 79 10001 200 IN-PROGRESS
Channel-Identifier: 23af1e13@speechrecog
```

F5 (specchrecog → client):

```
MRCP/2.0 111 START-OF-INPUT 10001 IN-PROGRESS
Channel-Identifier: 23af1e13@speechrecog
Input-Type: speech
```

F6 (speechrecog → client):

```
MRCP/2.0 466 RECOGNITION-COMPLETE 10001 COMPLETE
Channel-Identifier: 23af1e13@speechrecog
Completion-Cause: 000 success
Content-Type: application/nlsml+xml
Content-Length: 283

<?xml version="1.0" encoding="UTF-8"?>
<result grammar="session:menu@example.com"
        xmlns="http://www.ietf.org/xml/ns/mrcpv2">
   <interpretation confidence="0.7">
      <instance>
        menu
      </instance>
      <input>menu</input>
   </interpretation>
</result>
```

13.2.3 *START-INPUT-TIMERS*[3]

By default, a No-Input-Timeout timer is started when recognition starts. If no input is detected before the timer expires, the recognition terminates with a RECOGNITION-COMPLETE message with Completion-Cause set to 002 noinput-timeout.

Typically, the MRCP client will issue the RECOGNIZE request immediately after a prompt soliciting the user for input completes. For some applications, it is desirable to start the recognition at the

---

[3] This method is called RECOGNITION-START-TIMERS in MRCPv1 (see Appendix A, located on the Web at http://www.daveburke.org/speechprocessing/).

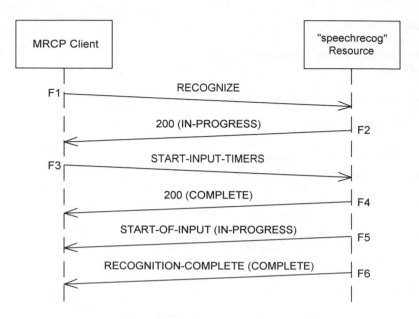

Figure 13.4 START-INPUT-TIMERS request example.

same time as the prompt, thus allowing familiar users to barge-in over the prompt and perform the recognition earlier. When the recognition is commenced at the same time as a prompt playback, one does not want to start the No-Input-Timeout timer until after the playback completes.[4] This is achieved by issuing the RECOGNIZE message with a header field called Start-Input-Timers with a value of false. Later, when the MRCP client learns that the prompt playback has completed (e.g. it receives a SPEAK-COMPLETE event from a speech synthesiser resource), it can issue the START-INPUT-TIMERS method to the recogniser resource to start the No-Input-Timeout timer.

Figure 13.4 illustrates an example of a START-INPUT-TIMERS request. The corresponding messages follow.

F1 (client → speechrecog):

```
MRCP/2.0 286 RECOGNIZE 20000
Channel-Identifier: 23af1e13@speechrecog
No-Input-Timeout: 2000
Start-Input-Timers: false
Content-Type: text/grammar-ref-list
Content-Length: 103

<http://example.com/popularnames.grxml>;weight="1.0"
<http://example.com/rarenames.grxml>;weight="0.5"
```

---

[4] If one always knew the length of the prompt playback, the No-Input-Timeout could be set to this value plus the time to wait for input. However, in general, the MRCP client will not know the duration of the prompt playback ahead of time.

F2 (speechrecog → client):

```
MRCP/2.0 79 20000 200 IN-PROGRESS
Channel-Identifier: 23af1e13@speechrecog
```

F3 (client → speechrecog):

```
MRCP/2.0 82 START-INPUT-TIMERS 20001
Channel-Identifier: 23af1e13@speechrecog
```

F4 (speechrecog → client):

```
MRCP/2.0 77 20001 200 COMPLETE
Channel-Identifier: 23af1e13@speechrecog
```

F5 (speechrecog → client):

```
MRCP/2.0 111 START-OF-INPUT 20000 IN-PROGRESS
Channel-Identifier: 23af1e13@speechrecog
Input-Type: speech
```

F6 (speechrecog → client):

```
MRCP/2.0 492 RECOGNITION-COMPLETE 20000 COMPLETE
Channel-Identifier: 23af1e13@speechrecog
Completion-Cause: 000 success
Content-Type: application/nlsml+xml
Content-Length: 309

<?xml version="1.0" encoding="UTF-8"?>
<result grammar="http://example.com/rarenames.grxml"
      xmlns="http://www.ietf.org/xml/ns/mrcpv2">
    <interpretation confidence="0.2">
        <instance>
          Mr. Rabbit
        </instance>
        <input>roger rabbit</input>
    </interpretation>
</result>
```

13.2.4 GET-RESULT

When a recognition request successfully completes, the recognition results are returned in the body of the RECOGNITION-COMPLETE event and the recogniser enters the recognized state. The returned results can depend on the settings for the recognition such as Confidence-Threshold and N-Best-Length-List, for example. The GET-RESULT request method can be used while the recogniser is in the recognized state to request that it recompute and return the results

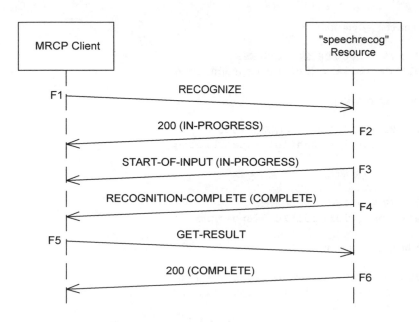

Figure 13.5 GET-RESULT request example.

according to the recognition settings supplied in the GET-RESULT request. For example, the MRCP client could issue a GET-RESULT request with a lower Confidence-Threshold and higher N-Best-List-Length to get more recognition hypotheses.

Figure 13.5 illustrates an example of a GET-RESULT request. The corresponding messages follow.

F1 (client → speechrecog):

```
MRCP/2.0 205 RECOGNIZE 30000
Channel-Identifier: 23af1e13@speechrecog
Confidence-Threshold: 0.9
Content-Type: text/grammar-ref-list
Content-Length: 47

<http://example.com/options.grxml>;weight="1.0"
```

F2 (speechrecog → client):

```
MRCP/2.0 75 30000 IN-PROGRESS
Channel-Identifier: 23af1e13@speechrecog
```

F3 (speechrecog → client):

```
MRCP/2.0 111 START-OF-INPUT 30000 IN-PROGRESS
Channel-Identifier: 23af1e13@speechrecog
Input-Type: speech
```

F4 (speechrecog → client):

```
MRCP/2.0 480 RECOGNITION-COMPLETE 30000 COMPLETE
Channel-Identifier: 23af1e13@speechrecog
Completion-Cause: 000 success
Content-Type: application/nlsml+xml
Content-Length: 297

<?xml version="1.0" encoding="UTF-8"?>
<result grammar="http://example.com/options.grxml"
        xmlns="http://www.ietf.org/xml/ns/mrcpv2">
    <interpretation confidence="0.95">
        <instance>
            sales
        </instance>
        <input>sales</input>
    </interpretation>
</result>
```

F5 (client → speechrecog):

```
MRCP/2.0 102 GET-RESULT 30001
Channel-Identifier: 23af1e13@speechrecog
Confidence-Threshold: 0.5
```

F6 (speechrecog → client):

```
MRCP/2.0 627 RECOGNITION-COMPLETE 30001 COMPLETE
Channel-Identifier: 23af1e13@speechrecog
Completion-Cause: 000 success
Content-Type: application/nlsml+xml
Content-Length: 444

<?xml version="1.0" encoding="UTF-8"?>
<result grammar="http://example.com/options.grxml"
        xmlns="http://www.ietf.org/xml/ns/mrcpv2">
    <interpretation confidence="0.95">
        <instance>
            sales
        </instance>
        <input>sales</input>
    </interpretation>
    <interpretation confidence="0.61">
        <instance>
            presales
        </instance>
        <input>presales</input>
```

```
    </interpretation>
</result>
```

13.2.5 STOP

The STOP request method is used to stop either an active (IN-PROGRESS) or queued (PENDING) RECOGNIZE request. If the Active-Request-Id-List header field is omitted from the STOP request, the RECOGNIZE request in progress is stopped and any RECOGNIZE requests pending in the queue are removed. If a STOP request successfully terminates one or more PENDING or IN-PROGRESS RECOGNIZE requests, the corresponding response, with status code 200 Success, will include an Active-Request-Id-List header field listing the request-ids of the requests that were terminated. The RECOGNITION-COMPLETE event is not sent for a terminated request. If no RECOGNIZE requests were terminated, the response status code would still be 200 Success but the Active-Request-Id-List header field is omitted.

It is possible to target particular RECOGNIZE requests from the set of PENDING RECOGNIZE requests and the current IN-PROGRESS request by specifying the Active-Request-Id-List header field in the STOP request. If the current IN-PROGRESS request is terminated, then the next available PENDING request in the queue is processed.

Figure 13.6 illustrates an example of a STOP request. The corresponding messages follow.

F1 (client → speechrecog):

```
MRCP/2.0 205 RECOGNIZE 40000
Channel-Identifier: 23af1e13@speechrecog
Confidence-Threshold: 0.9
Content-Type: text/grammar-ref-list
Content-Length: 47
```

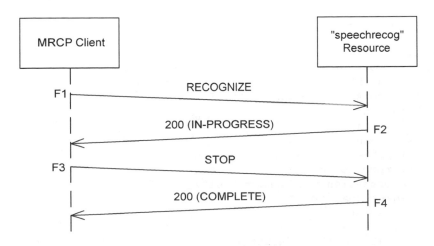

Figure 13.6 STOP request example.

```
<http://example.com/colours.grxml>;weight="1.0"
```

F2 (speechrecog → client):

```
MRCP/2.0 75 40000 IN-PROGRESS
Channel-Identifier: 23af1e13@speechrecog
```

F3 (client → speechrecog):

```
MRCP/2.0 68 STOP 40001
Channel-Identifier: 23af1e13@speechrecog
```

F4 (speechrecog → client):

```
MRCP/2.0 105 40001 COMPLETE
Channel-Identifier: 23af1e13@speechrecog
Active-Request-Id-List: 40000
```

13.2.6 INTERPRET

The INTERPRET request method behaves in much the same way as the RECOGNIZE method except that instead of analysing the input audio stream, the input is given in textual form in a body part of the message body specified by the Interpret-Text header. The interpretation results are returned in the message body of the INTERPRETATION-COMPLETE event. Unlike the RECOGNIZE method, INTERPRET requests cannot be queued – issuing another INTERPRET while one is in progress results in a response with status code 402 Method not valid in this state.

Figure 13.7 illustrates an example of an INTERPRET request. The corresponding messages follow.

F1 (client → speechrecog):

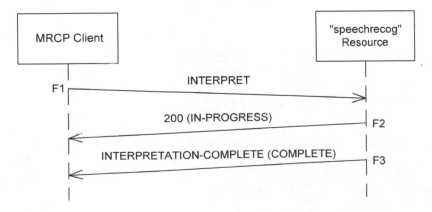

Figure 13.7 INTERPRET request example.

```
MRCP/2.0 740 INTERPRET 50000
Channel-Identifier: 23af1e13@speechrecog
Interpret-Text: text@example.com
Content-Type: multipart/mixed; boundary=a0f2b1e4f9
Content-Length: 559

--a0f2b1e4f9
Content-ID: <text@example.com>
Content-Type: text/plain
Content-Length: 3

yes
--a0f2b1e4f9
Content-ID: <gram@example.com>
Content-Type: application/srgs+xml
Content-Length: 338

<?xml version="1.0" encoding="UTF-8"?>
<grammar version="1.0" xmlns="http://www.w3.org/2001/06/grammar"
        xml:lang="en-IE" tag-format="semantics/1.0">
    <rule id="yesno">
        <one-of>
            <item>yes<tag>out.result=1</tag></item>
            <item>no<tag>out.result=0</tag></item>
        </one-of>
    </rule>
</grammar>
--a0f2b1e4f9--
```

F2 (speechrecog → client):

```
MRCP/2.0 79 50000 200 IN-PROGRESS
Channel-Identifier: 23af1e13@speechrecog
```

F3 (speechrecog → client):

```
MRCP/2.0 484 INTERPRETATION-COMPLETE 50000 COMPLETE
Channel-Identifier: 23af1e13@speechrecog
Completion-Cause: 000 success
Content-Type: application/nlsml+xml
Content-Length: 298

<?xml version="1.0" encoding="UTF-8"?>
<result grammar="session:test@example.com"
        xmlns="http://www.ietf.org/xml/ns/mrcpv2">
    <interpretation confidence="1.0">
        <instance>
            <result>1</result>
```

```
      </instance>
      <input>yes</input>
   </interpretation>
</result>
```

13.3 Voice enrolment methods

13.3.1 START-PHRASE-ENROLLMENT

The START-PHRASE-ENROLLMENT request method is used by the MRCP client to start an enrolment session. The purpose of an enrolment session is add a phrase to a speaker-dependent, personal grammar by employing the user's voice. During the enrolment session, the MRCP client repetitively calls RECOGNIZE with the Enroll-Utterance header field set to true.[5] Each call to RECOGNIZE captures a single example of the phrase in question and the progress of the enrolment is conveyed in NLSML contained in the RECOGNITION-COMPLETE event message. Typically, a small number of repetitions is required to train the phrase. The enrolment session is terminated by the MRCP client calling END-PHRASE-ENROLLMENT and is used either to commit the new phrase to the personal grammar or to abort it.

A number of header fields must be specified in the START-PHRASE-ENROLLMENT method to configure the enrolment session. The Personal-Grammar-URI specifies a URI for the speaker-dependent personal grammar. That URI may already exist, in which case an additional phrase is being added to the grammar. Otherwise, a new grammar is created. The format for the Personal-Grammar-URI is platform specific. The Phrase-ID header field specifies a unique identifier for the phrase. The Phrase-NL specifies a natural language or semantic result that will be returned during a later recognition if the phrase is recognised. The set of header fields pertaining to voice enrolment are explained in detail in Section 13.6.

Figure 13.8 illustrates a complete example of the flow required to enrol a phrase. The corresponding messages follow

F1 (client → speechrecog):

```
MRCP/2.0 237 START-ENROLLMENT-SESSION 60000
Channel-Identifier: 23af1e13@speechrecog
Personal-Grammar-URI: http://enrolledgrammars/user01gram02.dat
Phrase-ID: 10231
Phrase-NL: jbloggs@example.com
Num-Min-Consistent-Pronunciations: 2
```

F2 (speechrecog → client):

```
MRCP/2.0 76 60000 200 COMPLETE
Channel-Identifier: 23af1e13@speechrecog
```

F3 (client → speechrecog):

---

[5] During an enrolment session, the MRCP client may legally issue a RECOGNIZE request with Enroll-Utterance set to false to perform a regular recognition.

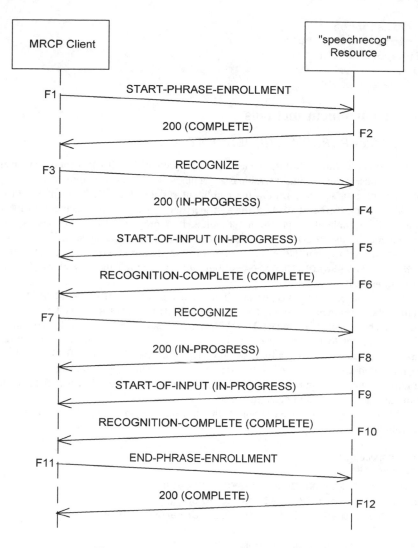

Figure 13.8 Complete voice enrolment example.

```
MRCP/2.0 97 RECOGNIZE 60001
Channel-Identifier: 23af1e13@speechrecog
Enroll-Utterance: true
```

F4 (speechrecog → client):

```
MRCP/2.0 79 60001 200 IN-PROGRESS
Channel-Identifier: 23af1e13@speechrecog
```

F5 (speechrecog → client):

```
MRCP/2.0 111 START-OF-INPUT 60001 IN-PROGRESS
Channel-Identifier: 23af1e13@speechrecog
Input-Type: speech
```

F6 (speechrecog → client):

```
MRCP/2.0 621 RECOGNITION-COMPLETE 60001 COMPLETE
Channel-Identifier: 23af1e13@speechrecog
Completion-Cause: 000 success
Content-Type: application/nlsml+xml
Content-Length: 438

<?xml version="1.0" encoding="UTF-8"?>
<result xmlns="http://www.ietf.org/xml/ns/mrcpv2"
        grammar="http://enrolledgrammars/user01gram02.dat">
    <enrollment-result>
        <num-good-repetitions>1</num-good-repetitions>
        <num-repetitions-still-needed>1</num-repetitions-still-needed>
        <consistency-status>undecided</consistency-status>
        <num-clashes>0</num-clashes>
    </enrollment-result>
</result>
```

F7 (client → speechrecog):

```
MRCP/2.0 97 RECOGNIZE 60002
Channel-Identifier: 23af1e13@speechrecog
Enroll-Utterance: true
```

F8 (speechrecog → client):

```
MRCP/2.0 79 60002 200 IN-PROGRESS
Channel-Identifier: 23af1e13@speechrecog
```

F9 (speechrecog → client):

```
MRCP/2.0 111 START-OF-INPUT 60002 IN-PROGRESS
Channel-Identifier: 23af1e13@speechrecog
Input-Type: speech
```

F10 (speechrecog → client):

```
MRCP/2.0 622 RECOGNITION-COMPLETE 60002 COMPLETE
Channel-Identifier: 23af1e13@speechrecog
Completion-Cause: 000 success
Content-Type: application/nlsml+xml
```

```
Content-Length: 431

<?xml version="1.0" encoding="UTF-8"?>
<result xmlns="http://www.ietf.org/xml/ns/mrcpv2"
        grammar="http://enrolledgrammars/user01gram02.dat">
    <enrollment-result>
        <num-good-repetitions>2</num-good-repetitions>
        <num-repetitions-still-needed>0</num-repetitions-still-needed>
        <consistency-status>consistent</consistency-status>
        <num-clashes>0</num-clashes>
    </enrollment-result>
</result>
```

F11 (client → speechrecog):

```
MRCP/2.0 85 END-PHRASE-ENROLLMENT 60003
Channel-Identifier: 23af1e13@speechrecog
```

F12 (speechrecog → client):

```
MRCP/2.0 76 60003 200 COMPLETE
Channel-Identifier: 23af1e13@speechrecog
```

13.3.2 ENROLLMENT-ROLLBACK

The ENROLLMENT-ROLLBACK request method can be used by the MRCP client to undo the previous RECOGNIZE request. This capability is useful if the utterance is deemed unsuitable for training. Only the audio captured by the previous RECOGNIZE request can be discarded (i.e. invoking ENROLLMENT-ROLLBACK more than once will have no effect).

Figure 13.9 illustrates an example of the ENROLLMENT-ROLLBACK request method. The corresponding messages follow.

F1 (client → speechrecog):

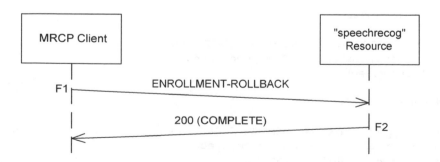

Figure 13.9   ENROLLMENT-ROLLBACK request example.

```
MRCP/2.0 83 ENROLLMENT-ROLLBACK 70000
Channel-Identifier: 23af1e13@speechrecog
```

F2 (speechrecog → client):

```
MRCP/2.0 76 70000 200 COMPLETE
Channel-Identifier: 23af1e13@speechrecog
```

13.3.3 END-PHRASE-ENROLLMENT

The END-PHRASE-ENROLLMENT request method is used by the MRCP client to terminate the enrolment session and either commit the enrolled phrase or abort it. This request is usually called to commit the phrase after the NLSML returned in the RECOGNITION-COMPLETE event indicates that the number of repetitions still needed (<num-repetitions-still-needed>) is 0. To abort the phrase enrolment, the END-PHRASE-ENROLLMENT request can be issued with the header field Abort-Phrase-Enrollment set to true. If the Save-Best-Waveform header field was present and set to true in the START-PHRASE-ENROLLMENT request, the response to the END-PHRASE-ENROLLMENT will contain the Waveform-URI header field indicating the location of the recording containing the best repetition of the learned phrase. One use of the recorded waveform is to allow the MRCP client to enumerate the choices the user may speak corresponding to the personal grammar. An example of the END-PHRASE-ENROLLMENT request method is given in Section 13.3.1.

13.3.4 MODIFY-PHRASE

The MODIFY-PHRASE request method is used to modify certain attributes of a phrase that was previously enrolled in a personal grammar. The Phrase-ID, Phrase-NL, or Weight can be modified. The phrase being modified is identified by the Phrase-ID header field value and new values for Phrase-NL or Weight can be supplied via those header fields. If the Phrase-ID value itself is to be changed, the New-Phrase-ID header field is used to indicate the new ID value.

Figure 13.10 illustrates an example of the MODIFY-PHRASE request method. The corresponding messages follow.

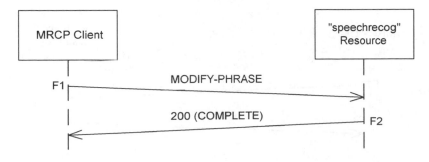

Figure 13.10 MODIFY-PHRASE request example.

F1 (client → speechrecog):

```
MRCP/2.0 133 MODIFY-PHRASE 90000
Channel-Identifier: 23af1e13@speechrecog
Phrase-ID: Name01
New-Phrase-ID: Name03
Weight: 0.5
```

F2 (speechrecog → client):

```
MRCP/2.0 76 90000 200 COMPLETE
Channel-Identifier: 23af1e13@speechrecog
```

13.3.5 DELETE-PHRASE

The DELETE-PHRASE request method is used to delete a phrase (indicated by the Phrase-ID) from a personal grammar (indicated by the Personal-Grammar-URI). Figure 13.11 illustrates an example of the DELETE-PHRASE request method. The corresponding messages follow.

F1 (client → speechrecog):

```
MRCP/2.0 162 DELETE-PHRASE 100000
Channel-Identifier: 23af1e13@speechrecog
Personal-Grammar-URI: http://enrolledgrammars/user01gram02.dat
Phrase-ID: Name01
```

F2 (speechrecog → client):

```
MRCP/2.0 77 100000 200 COMPLETE
Channel-Identifier: 23af1e13@speechrecog
```

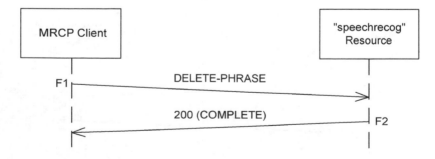

Figure 13.11 DELETE-PHRASE request example.

13.4 Events

13.4.1 START-OF-INPUT [6]

The START-OF-INPUT event message is used by the recogniser resource to inform the MRCP client that it has received either speech or DTMF input (the type of input is indicated by the Input-Type header field). This event message is typically used by the MRCP client to detect barge-in and stop the playback of audio – see Section 12.2.5 for more information. Receipt of the event can also be used by the MRCP client to estimate when barge-in occurred (c.f. VoiceXML's marktime shadow variable discussed in Section 16.6.4.3). The START-OF-INPUT is not generated for hotword recognition. Many of the examples in the preceding sections illustrate the START-OF-INPUT event.

13.4.2 RECOGNITION-COMPLETE

The recogniser resource uses the RECOGNITION-COMPLETE event to indicate to the MRCP client that the recognition action has completed and to convey the recognition results in the message body. The Completion-Cause and optional Completion-Reason header fields indicate the status of the completed recognition. After the recognition is complete, the recogniser is in the recognized state and maintains the recognition results and the audio waveform input for the recognition until the next recognition. The GET-RESULT method can be used by the MRCP client to request that the recogniser resource recomputes the recognition results subject to given parameter settings – see Section 13.2.4. In the context of a voice enrolment session and where RECOGNIZE is issued with the header field Enroll-Utterance set to true, the RECOGNITION-COMPLETE event is also used to indicate ongoing enrolment results – see Section 13.3.1. Many of the examples in the preceding sections illustrate the RECOGNITION-COMPLETE event.

13.4.3 INTERPRETATION-COMPLETE

The INTERPRETATION-COMPLETE event is analogous to the RECOGNITION-COMPLETE event in that it indicates completion of an interpretation action and conveys interpretation results. An example of the INTERPRETATION-COMPLETE event was given in Section 13.2.6.

13.5 Recognition headers

13.5.1 Completion-Cause

The Completion-Cause header is always present in the RECOGNITION-COMPLETE response to indicate the reason why the RECOGNIZE request has completed. If the RECOGNIZE request returns with an error response, the Completion-Cause header field will be present. The Completion-Cause header field also appears in the response to the DEFINE-GRAMMAR request to indicate the result of accessing and compiling the grammar(s) identified in the request. The value for the header field includes both a cause code and corresponding cause name. Table 13.8 summaries the different cause codes and cause names.

---

[6] This event is called START-OF-SPEECH in MRCPv1 (see Appendix A, located on the Web at http://www.daveburke.org/speechprocessing/).

Table 13.8 Cause codes and cause names for the `Completion-Cause` header field

Cause code	Cause name	Description
000	success	The `RECOGNIZE` request completed with a match or the `DEFINE-GRAMMAR` request completed successfully (the grammar was access and compiled correctly).
001	no-match	A no-match occurred, i.e. the input did not match any active grammar.
002	no-input-timeout	The `No-Input-Timeout` timer expired before a match occurred.
003	hotword-maxtime	The `Recognition-Timeout` timer expired during a hotword recognition.
004	grammar-load-failure	The `RECOGNIZE` or `DEFINE-GRAMMAR` request failed due to a grammar load failure.
005	grammar-compilation-failure	The `RECOGNIZE` or `DEFINE-GRAMMAR` request failed due to a grammar compilation failure.
006	recognizer-error	The `RECOGNIZE` or `DEFINE-GRAMMAR` request failed due to a general error in the recogniser.
007	speech-too-early	The recognition failed because the user was already speaking before the `RECOGNIZE` request was issued.
008	success-maxtime	The `Recognition-Timeout` timer expired but the recognition still completed with a match.
009	uri-failure	An error occurred accessing a URI.
010	language-unsupported	The `RECOGNIZE` or `DEFINE-GRAMMAR` request failed because the requested language is not supported by the recogniser.
011	cancelled	A new `RECOGNIZE` cancelled the previous `RECOGNIZE` request.
012	semantics-failure	The recognition succeeded but there was an error computing the semantic interpretation. The results are supplied without the semantic interpretation.
013	partial-match	The `Speech-Incomplete-Timeout` timer expired before there was a complete match. The partial results may be returned in the NLSML.
014	partial-match-maxtime	The `Recognition-Timeout` timer expired before there was a complete match. The partial results may be returned in the NLSML.
015	no-match-maxtime	The `Recognition-Timeout` timer expired before there was even a partial match.
016	grammar-definition-failure	An error occurred defining a grammar other than `004 grammar-load-failure` or `005 grammar-compilation-failure` (e.g. there was insufficient space on the media resource server to store the grammar).

Example:

```
Completion-Cause: 001 no-match
```

13.5.2 Completion-Reason

The Completion-Reason header is optionally present alongside the Completion-Cause header field to supply more information on the reason why the request terminated. This header field is intended for use by the MRCP client for logging purposes.
Example:

```
Completion-Reason: Recursive grammars not supported
```

13.5.3 Failed-URI

The Failed-URI header field indicates a URI for which access to failed. See also the Failed-URI-Cause header field.
Example:

```
Failed-URI: http://192.168.1.1/numbers.grxml
```

13.5.4 Failed-URI-Cause

The Failed-URI-Cause header field is used in conjunction with the Failed-URI header field to provide protocol specific response code information. For HTTP URIs, the value will typically be a HTTP $4xx$ or $5xx$ response code.
Example:

```
Failed-URI-Cause: 404 Not Found
```

13.5.5 Recognition-Mode

The Recognition-Mode header field is used to indicate the mode for recognition and takes the value of either normal or hotword. A value of normal means that the input is matched to the active speech or DTMF grammars, and if any portion of the input does not match the grammar, the recognition terminates with a Completion-Cause header field value of 001 no-match. Hotword recognition operates by instead looking for specific keywords or phrases (or DTMF sequences) in the input, and input that does not match the active grammars is ignored. The Recognition-Mode header may be used on a per-RECOGNIZE request basis and defaults to normal.
Example:

```
Recognition-Mode: hotword
```

13.5.6 Input-Type

The Input-Type header field is present in the START-OF-INPUT event to indicate the type of input detected. Legal values are speech or dtmf.
    Example:

```
Input-Type: speech
```

13.5.7 Confidence-Threshold

The Confidence-Threshold header field is used to set the threshold for the minimum confidence level that a recognition must meet in order be considered a successful recognition. A recognition result with a confidence level value below the Confidence-Threshold setting is returned with a Completion-Cause header field value 001 no-match. The value for the Confidence-Threshold header field is required to be in the range of 0.0–1.0. The Confidence-Threshold header may be used on a per-RECOGNIZE request basis or set for the session via the SET-PARAMS request. The default value is left as implementation specific.
    Example:

```
Confidence-Threshold: 0.5
```

13.5.8 Sensitivity-Level

Speech detection is used by the recognition resource to filter out background noise from the speech that is to be analysed. The Sensitivity-Level controls how sensitive the speech detector is. A low value means the detector is less sensitive to noise and a high value means it is more sensitive to quiet input. The Sensitivity-Level header field may be specified on a per-RECOGNIZE request basis or set for the session via the SET-PARAMS request. The default value is left as implementation specific.
    Example:

```
Sensitivity-Level: 0.65
```

13.5.9 Speed-Vs-Accuracy

Speech recognisers often have the ability to be tuned for performance or accuracy. The Speech-Vs-Accuracy header field value varies from 0.0. to 1.0 where higher values imply more accuracy and CPU processing, and lower values imply lower accuracy but reduced CPU processing (hence a higher number of concurrent recognition requests can be handled by the same resource). The Speech-Vs-Accuracy header field may be specified on a per-RECOGNIZE request basis or set for the session via the SET-PARAMS request. The default value is left as implementation specific.
    Example:

```
Speed-Vs-Accuracy: 0.35
```

13.5.10 N-Best-List-Length

The purpose of the N-Best-List-Length header field is to allow the MRCP client to indicate to the recogniser resource that multiple recognition result alternatives may be returned up to a maximum of N. Different result alternatives correspond to different paths through the grammar, but only those paths with a confidence level greater than the Confidence-Threshold value are returned. Note that the recogniser resource may (and often does) choose to return fewer than N results. A larger value for N implies the recogniser resource must perform increased processing. The N-Best-List-Length header field may be specified on a per-RECOGNIZE request basis or set for the session via the SET-PARAMS request. The default (and minimum) value is 1.
Example:

N-Best-List-Length: 5

13.5.11 No-Input-Timeout

The No-Input-Timeout timer is used to detect the scenario where a recognition is started but the user does not start speaking or enter DTMF within a given time period. If the No-Input-Timeout timer expires before input is detected, the RECOGNITION-COMPLETE event is generated with a Completion-Cause value of 002 no-input-timeout.The usual course of action for the speech application is to prompt the user that the system did not hear anything and restart the recognition. The No-Input-Timeout header field may be specified on a per-RECOGNIZE request basis or set for the session via the SET-PARAMS request. The units for the No-Input-Timeout header field value are in milliseconds. The value can range from 0 to an implementation specific maximum and there is no default value specified.
Example:

No-Input-Timeout: 3000

13.5.12 Recognition-Timeout

The Recognition-Timeout timer is used to limit the overall duration of the recognition. This is useful to prevent background noise from occupying the recogniser for an extended period of time. This header field is equivalent to VoiceXML's maxspeechtimeout (see Section 16.6.3 for more information).

For normal (non-hotword) recognition, the timer is started when the START-OF-INPUT is generated. If a successful match has been reached when the timer expires, a RECOGNITION-COMPLETE event is generated with Completion-Cause value of 008 success-maxtime; if a partial match has been achieved when the timer expires, a Completion-Cause of 014 partial-match-maxtime is returned; if no match has been achieved when the timer expires, a Completion-Cause of 015 no-match-maxtime is returned.

For hotword recognition, the Recognition-Timeout timer starts when the recognition starts (since START-OF-INPUT is not generated for hotword recognition). If there is a successful match reached when the timer expires, a RECOGNITION-COMPLETE event is generated with Completion-Cause value of 008 success-maxtime; otherwise a Completion-Cause of 003 hotword-maxtime is returned. For hotword recognition, the timer is reset when silence is detected.

The Recognition-Timeout header field may be specified on a per-RECOGNIZE request basis or set for the session via the SET-PARAMS request. The units for the Recognition-Timeout

header field value are in milliseconds. The value can range from 0 to an implementation specific maximum. The default value is 10 seconds.
   Example:

```
Recognition-Timeout: 7000
```

13.5.13 Speech-Complete-Timeout

The Speech-Complete-Timeout header field indicates the amount of time that must elapse after the user has stopped speaking before the recogniser returns a successful recognition result. The Speech-Complete-Timeout applies when the spoken utterance is a complete match of an active grammar (i.e. it is not possible to speak further and still match the grammar). Too short a value for the timeout could result in an utterance being broken up inappropriately while too long a timeout will make the system's response appear slow. The Speech-Complete-Timeout header field may be specified on a per-RECOGNIZE request basis or set for the session via the SET-PARAMS request. The units for the Speech-Complete-Timeout header field value are in milliseconds. The value can range from 0 to an implementation specific maximum with typical values in the range of 300 to 1000 ms. There is no specified default.
   Example:

```
Speech-Complete-Timeout: 500
```

13.5.14 Speech-Incomplete-Timeout

The Speech-Incomplete-Timeout header field is similar to the Speech-Complete-Timeout except that it applies in the case where the speech prior to the silence is an incomplete (i.e. prefix) match of all active grammars. If the timer triggers, a RECOGNITION-RESULT event is generated with a Completion-Cause value of 013 partial- match. The Speech-Incomplete-Timeout also applies when there is a complete match of a grammar but it is possible to speak further and still match the grammar. The Speech-Incomplete-Timeout value is usually longer than the Speech-Complete-Timeout value to allow users to pause mid-utterance (for example, to breathe). The Speech-Incomplete-Timeout header field may be specified on a per-RECOGNIZE request basis or set for the session via the SET-PARAMS request. The units for the Speech-Incomplete-Timeout header field value are in milliseconds. The value can range from 0 to an implementation specific maximum. There is no specified default.
   Example:

```
Speech-Incomplete-Timeout: 1000
```

13.5.15 Hotword-Max-Duration

The Hotword-Max-Duration header field only applies to hotword mode recognition and is used to limit the maximum length of an utterance to be considered for recognition. The header field is used in conjunction with the Hotword-Min-Duration header field to tune the recogniser and prevent recognition of utterances that are either too long or too short. The Hotword-Max-Duration header field may be specified on a per-RECOGNIZE request basis or set for the session via the SET-PARAMS request. The units for the `Hotword-Max-Duration` header field value are in milliseconds. There is no specified default.

Example:

```
Hotword-Max-Duration: 5000
```

13.5.16 Hotword-Min-Duration

The `Hotword-Min-Duration` header field only applies to hotword mode recognition and is used to limit the minimum length of an utterance to be considered for recognition. The header field is used in conjunction with the `Hotword-Max-Duration` header field to tune the recogniser and prevent recognition of utterances that are either too long or too short. The `Hotword-Min-Duration` header field may be specified on a per-`RECOGNIZE` request basis or set for the session via the `SET-PARAMS` request. The units for the `Hotword-Min-Duration` header field value are in milliseconds. There is no specified default.

Example:

```
Hotword-Min-Duration: 250
```

13.5.17 DTMF-Interdigit-Timeout

The `DTMF-Interdigit-Timeout` header field specifies the amount of time allowable between successive DTMF key presses before the recogniser finalises a result. The `DTMF-Interdigit-Timeout` header field may be specified on a per-`RECOGNIZE` request basis or set for the session via the `SET-PARAMS` request. The units for the `DTMF-Interdigit-Timeout` header field value are in milliseconds. The value can range from 0 to an implementation specific maximum. The default value is 5000 ms.

Example:

```
DTMF-Interdigit-Timeout: 3000
```

13.5.18 DTMF-Term-Timeout

The `DTMF-Term-Timeout` header field is similar to the `DTMF-Interdigit-Timeout` header field except that it applies when there is a complete match of an active grammar (i.e. it is not possible to enter further digits and still match the grammar). The purpose of the timer is to allow for the case where other grammars allowing longer digit sequences are active and can be further matched or a `DTMF-Term-Char` is specified (see next section). The `DTMF-Term-Timeout` header field may be specified on a per-`RECOGNIZE` request basis or set for the session via the `SET-PARAMS` request. The units for the `DTMF-Term-Timeout` header field value are in milliseconds. The value can range from 0 to an implementation specific maximum. The default value is 10000 ms.

Example:

```
DTMF-Term-Timeout: 2000
```

13.5.19 DTMF-Term-Char

The `DTMF-Term-Char` header field is used to specify the terminating DTMF character for DTMF input (it can be thought of as a kind of 'return key' and is not present in the recognition result). The `DTMF-Term-Char` header field may be specified on a per-RECOGNIZE request basis or set for the session via the `SET-PARAMS` request. The default value is none, indicated by an empty value.
    Example:

```
DTMF-Term-Char: #
```

13.5.20 DTMF-Buffer-Time

The `DTMF-Buffer-Time` header field specifies the size of the DTMF type ahead buffer in milliseconds. Between `RECOGNIZE` requests, the recogniser buffers DTMF key presses. When a new `RECOGNIZE` request is issued, the contents of the buffer can be used for the recognition, and further digits can be collected from the user to match the active grammar(s) as necessary. This allows power users familiar with a service to type ahead. The `DTMF-Buffer-Time` header field can be specified for the session via the `SET-PARAMS` request. There is no specified default value.
    Example:

```
DTMF-Buffer-Time: 10000
```

13.5.21 Clear-DTMF-Buffer

The `Clear-DTMF-Buffer` header field can be specified on a `RECOGNIZE` request with a value of `true` to clear any buffered DTMF in the type ahead buffer (see also the `DTMF-Buffer-Time` header field). The default value is `false`.
    Example:

```
Clear-DTMF-Buffer: true
```

13.5.22 Save-Waveform

The `Save-Waveform` header field takes a Boolean value to indicate to the recogniser whether or not to save the recognised audio (without endpointing). If set to `true`, the captured audio is indicated via the `Waveform-URI` header field in the `RECOGNITION-COMPLETE` event message (thereby allowing the MRCP client to retrieve the audio). The format for the captured audio is specified by the `Media-Type` header field (see Section 13.2.25). This feature is useful for several purposes. For example, if the recogniser fails to match the utterance, the recorded audio can be archived and manually inspected by a human operator to determine if the grammar should be expanded. The recorded audio may also be submitted to a subsequent recognition by specifying a `RECOGNIZE` with the `Input-Waveform-URI` header field (see Section 13.5.24). The `Save-Waveform` header field may be specified on a per-RECOGNIZE request basis or set for the session via the `SET-PARAMS` request. The default value is `false`.
    Example:

```
Save-Waveform: true
```

13.5.23 Waveform-URI[7]

The Waveform-URI header field is used to communicate a URI to the MRCP client for the audio captured during recognition. This header field appears in the RECOGNITION-COMPLETE event if the Save-Waveform header field is specified with a value of true (see preceding section). This header field is also used in the response to the END-PHRASE-ENROLLMENT method to communicate a URI corresponding to the audio for the best repetition of an enrolled phrase if the Save-Best-Waveform header field was previously specified with a value of true. The Waveform-URI header field is post-pended with size and duration parameters for the captured audio, measured in bytes and milliseconds respectively. The URI indicated by the header field must be accessible for the duration of the session. If there is an error recording the audio, the Waveform-URI header field value is left blank.

Example:

```
Waveform-URI: <http://10.0.0.1/utt01.wav>;size=8000;duration=1000
```

13.5.24 Input-Waveform-URI

The Input-Waveform-URI header field is used to supply audio for the RECOGNITION instead of using the media session. This capability is useful for performing a re-recognition against a different grammar employing audio previously captured (e.g. sourced from the Waveform-URI header field), and is also useful for automatic tools that test for grammar coverage, for example.

Example:

```
Input-Waveform-URI: http://10.0.0.1/utt01.wav
```

13.5.25 Media-Type

The Media-Type header field is used to indicate the media type to encode the captured audio referenced by the Waveform-URI header field (see Section 13.5.22). MRCP makes no attempt to mandate particular media types that must be supported. There are some commonly used audio media types but since they are not officially registered with IANA, their definition may differ across implementations – see Table 8.3. The Media-Type header field may be specified on a per-RECOGNIZE request basis or set for the session via the SET-PARAMS request. There is no specified default.

Example:

```
Media-Type: audio/x-wav
```

13.5.26 Start-Input-Timers[8]

The Start-Input-Timers header field may appear in the RECOGNIZE request to tell the recogniser whether or not to start the No-Input-Timeout timer. See the earlier discussion in Section 13.2.3 for more information). The default value is true.

---

[7] This header field is called Waveform-URL in MRCPv1 (see Appendix A, located on the Web at http://www.daveburke.org/speechprocessing/).
[8] This header field is called Recognizer-Start-Timers in MRCPv1 (see Appendix A, located on the Web at http://www.daveburke.org/speechprocessing/).

Example:

```
Start-Input-Timers: false
```

13.5.27 Speech-Language

The Speech-Language header field is used to indicate the default language for the grammar if it is not otherwise specified in the markup (recall from Chapter 9 that SRGS specifies the language via the xml:lang attribute). The format for the value is the same as used for SRGS's xml:lang attribute and is specified in RFC 3066 [55]. The Speaker-Language header may be used on a per-RECOGNIZE or per-DEFINE-GRAMMAR request basis or set for the session via the SET-PARAMS request.
   Example:

```
Speech-Language: de-CH
```

13.5.28 Cancel-If-Queue

The Cancel-If-Queue header field is used to indicate whether multiple RECOGNIZE requests can be queued or not. A value of true implies that no queue is used and an IN-PROGRESS RECOGNIZE request is cancelled on receipt of a new RECOGNIZE request. A cancelled recognition terminates with a RECOGNITION-COMPLETE event message with a Completion-Cause header field value of 011 cancelled. A value of false implies that the RECOGNIZE requests are queued. See also the discussion in Section 13.2.1. The Cancel-If-Queue header field may be specified on a per-RECOGNIZE request basis. There is no default value.
   Example:

```
Cancel-If-Queue: true
```

13.5.29 New-Audio-Channel

The New-Audio-Channel header field can be specified on a RECOGNIZE request with a value of true to indicate to the recogniser resource that from this point on, audio is being received from a new channel. A recogniser resource will usually interpret this as a request to reset its endpointing algorithm, thus discarding any adaption performed to the previous channel. This feature is useful in allowing an existing session to be reused on a new telephone call. The New-Audio-Channel header field may be specified on a per-RECOGNIZE request basis or set for the session via the SET-PARAMS request. The default value for this header field is false.
   Example:

```
New-Audio-Channel: true
```

13.5.30 Ver-Buffer-Utterance

The Ver-Buffer-Utterance header field may be specified in a RECOGNIZE request with a value of true to indicate to the recogniser resource that it should make the audio captured available in a

buffer for a co-resident speaker verification engine to use later (for example, to use the audio to verify the identity of the speaker later on). The buffer in question is allocated when a verification resource is added to the same session and shared across the recogniser resource and the speaker verification and identification resource (see Chapter 15 for more information). The default value for this header field is `false`.
   Example:

```
Ver-Buffer-Utterance: true
```

13.5.31 *Early-No-Match*

The `Early-No-Match` header field is used to indicate whether the recogniser should perform early matching of the grammar – i.e. that it should not wait until the endpointer indicates the end of speech before attempting to match the collected speech against the active grammar(s). Recognisers that do not support early matching silently ignore this header field. The default value for this header field is `false`.
   Example:

```
Early-No-Match: true
```

13.5.32 *Interpret-Text*

The `Interpret-Text` header field is used within the `INTERPRET` request method to reference the text to use for the interpretation. The value for the header field is the `Content-ID` of a `text/plain` body part within a multipart message in the `INTERPRET` request message body (see the example in Section 13.2.6).
   Example:

```
Interpret-Text: text@example.com
```

13.5.33 *Recognizer-Context-Block*

The `Recognizer-Context-Block` header field is used to retrieve (via `GET-PARAMS`) and set (via `SET-PARAMS`) the recogniser context block. The recogniser context block refers to vendor-specific data that contains information relating to the session, such as endpointer settings that have been adapted over the course of the session. This mechanism allows the session to be handed off to a different recogniser resource from the same vendor that, perhaps, is more suitable (e.g. supports the required speech language). The `Recognizer-Context-Block` header field references the `Content-ID` of a binary MIME message of type `application/octets`. This header field can be used to retrieve the block of data from a recogniser resource by issuing a `GET-PARAMs` request and specifying the `Recognizer-Context-Block` header field with a blank value. The response to the `GET-PARAMS` response will contain the `Recognizer-Context-Block` header field with a value of a `Content-ID` referencing the `application/octets` message body (i.e. the value of the `Content-ID` header field and the `Recognizer-Context-Block` header field will be the same in the `GET-PARAMS` response and the recogniser context block will be in the message body). The block

of data can then be sent to a new recogniser resource (from the same vendor) using the SET-PARAMS request method and supplying the Recognizer-Context-Block and Content-ID header fields and the application/octets message body.
   Example:

```
Recognizer-Context-Block: data@vendor-x.com
```

13.6 Voice enrolment headers

13.6.1 Enroll-Utterance

The Enroll-Utterance header field can be specified in the RECOGNIZE request method with a value of true to indicate that a phrase is to be enrolled (as opposed to performing a normal recognition). A RECOGNIZE request with a value of true for Enroll-Utterance can only be issued within an enrolment session (i.e. after a START-PHRASE-ENROLLMENT request was issued). The default value is false.
   Example:

```
Enroll-Utterance: true
```

13.6.2 Num-Min-Consistent-Pronunciations

The Num-Min-Consistent-Pronunciations header field is used to specify the minimum number of consistent pronunciations that must be obtained from the user before the enrolled phrase can be committed to the personal grammar. The NLSML contained in the RECOGNITION-COMPLETE event indicates the number of repetitions still needed in the <num-repetitions-still-needed> element (see Chapter 10). The Num-Min-Consistent-Pronunciations header field may be specified on a per-START-PHRASE-ENROLLMENT request basis or set for the session via the SET-PARAMS request. The minimum value and default value is 1.
   Example:

```
Num-Min-Consistent-Pronunciations: 1
```

13.6.3 Consistency-Threshold

The Consistency-Threshold header field is used to set a measure of how close an utterance of a phrase must be to previous utterances to be considered consistent. The higher the header field value, the closer the match of the utterance must be to the previous ones. The consistency status of an utterance is indicated by the <consistency-status> element of the NLSML contained in the RECOGNITION-COMPLETE event (see Chapter 10). The Consistency-Threshold header field may be specified on a per-START-PHRASE-ENROLLMENT request basis or set for the session via the SET-PARAMS request. The header field value can be specified in the range of 0.0 to 1.0. There is no specified default value.
   Example:

```
Consistency-Threshold: 0.75
```

13.6.4 Clash-Threshold

The Clash-Threshold header field is used to set a measure of how close an utterance of a phrase can be to an existing phrase in the personal grammar. This is to prevent phrases being enrolled that the recogniser finds difficult to distinguish between. A smaller value for the header field reduces the number of clashes detected. The number of clashes is reported via the <num-clashes> element of the NLSML contained in the RECOGNITION-COMPLETE event (see Chapter 10). The Clash-Threshold header field may be specified on a per-START-PHRASE-ENROLLMENT request basis or set for the session via the SET-PARAMS request. The header field value can be specified in the range of 0.0 to 1.0, where a value of 0 switches off clash testing. There is no specified default value.

Example:

```
Consistency-Threshold: 0.75
```

13.6.5 Personal-Grammar-URI

The Personal-Grammar-URI header field is specified in the START-PHRASE-ENROLLMENT, MODIFY-PHRASE, or DELETE-PHRASE request methods to indicate the speaker-dependent personal grammar to be operated upon. The header field value is a URI and the format of the grammar is implementation specific. There is no specified default value.

Example:

```
Personal-Grammar-URI: http://example.com/enroll/user1.dat
```

13.6.6 Phrase-ID

The Phrase-ID header field is specified in the START-PHRASE-ENROLLMENT, MODIFY-PHRASE, or DELETE-PHRASE request methods to indicate the particular phrase ID within the personal grammar to be operated upon. There is no specified default value.

Example:

```
Phrase-ID: name_joe_bloggs
```

13.6.7 New-Phrase-ID

The New-Phrase-ID header field may be specified in the MODIFY-PHRASE request method to indicate a new identifier for a given, existing Phrase-ID. There is no specified default value.

Example:

```
New-Phrase-ID: name_joe_bloggs_02
```

13.6.8 Phrase-NL

The Phrase-NL header field specifies the natural language or semantic interpretation corresponding to the enrolled phrase. This header field is specified in the START-PHRASE-ENROLLMENT request

to indicate the semantic interpretation for the new phrase being enrolled and may also appear in the MODIFY-PHRASE request to change the associated semantic interpretation. The value for the Phrase-NL header field is returned in the contents of the <instance> NLSML element in the RECOGNITION-COMPLETE event for a recognition result.
   Example:

Phrase-NL: item01

13.6.9 Weight

The Weight header field specifies the occurrence likelihood for a phrase. Conceptually, this is equivalent to the weight attribute on an <item> alternative in an SRGS grammar (see Chapter 9). Weight values are normalised when the grammar is compiled and hence only relative values matter. The Weight header field may appear in the START-PHRASE-ENROLLMENT request to specify a weight value for the new phrase being enrolled and may also appear in the MODIFY-PHRASE request to modify the weight value. There is no specified default value.
   Example:

Weight: 2.0

13.6.10 Save-Best-Waveform

The Save-Best-Waveform header field may be specified in the START-PHRASE-ENROLLMENT request method, with a value of true to request that the audio stream captured for the best repetition of the enrolled phrase be saved. The Waveform-URI header field specified in the END-PHRASE-ENROLLMENT response indicates the saved waveform. The default value is false.
   Example:

Save-Best-Waveform: true

13.6.11 Confusable-Phrases-URI

The Confusable-Phrases-URI header field may be placed in the RECOGNIZE request to specify the URI of a grammar that defines phrases that may not be enrolled (such as command phrases). If an attempt is made to enrol a phrase that is confusable, the NLSML returned in the RECOGNITION-COMPLETE event will specify a list of confusable phrases within the <confusable-phrases> element (see Chapter 10).
   Example:

Confusable-Phrase-URI: file://c:\data\commands.dat

13.6.12 Abort-Phrase-Enrollment

The Abort-Phrase-Enrollment header field may be placed in the END-PHRASE-ENROLLMENT request method with a value of true to indicate to the recogniser resource to abort the phrase being enrolled. The default value is false.

Example:

```
Abort-Phrase-Enrollment: true
```

13.7 Summary

In this chapter, we took a close look at the recogniser resource. The speechrecog resource is a fully fledged speech recogniser, usually with additional DTMF recognition capabilities, while the dtmfrecog resource is restricted to operating on DTMF input only. The recogniser resource supports both normal and hotword modes of operation. In addition, the recogniser resource supports the ability to perform interpretation of text input to test grammar coverage, and to determine the associated semantic interpretation. An optional feature for speech recognisers is the ability to support voice enrolled grammars whereby the user can add phrases to a speaker-dependent personal grammar.

In the next chapter, we continue our in-depth look at the MRCP media resources by studying the recorder resource.

14

Recorder resource

The MRCP recorder resource is designed to capture audio and store it at a URI where it can be retrieved later in the same session. The recorder resource is expected to have some basic speech processing capabilities to enable speech detection. This feature allows recordings to be end-pointed – silence can be suppressed at the beginning and end of the recording (and optionally in the middle of the recording). In addition, a recording can be aborted if no audio is received within a predetermined timeout, and automatically terminated on receipt of a certain amount of silence at the end of the recording. The recorder resource is ideal for applications requiring audio capture such as voicemail. This chapter delves into the details of the recorder resource by studying the methods, events, and header fields that apply to it.

14.1 Overview

The recorder resource supports a total of three request messages and two event messages, summarised in Tables 14.1 and 14.2 respectively.

The recorder resource has a state machine that is driven by requests from the MRCP client and events generated by the media resource itself. The state machine is illustrated in Figure 14.1.

Table 14.3 summarises the resource-specific header fields for the recorder resource. See Table 7.5 for a list of generic headers that also apply.

14.2 Methods

14.2.1 RECORD

When the MRCP client issues the RECORD request, it puts the recorder resource into the recording state. By default, the resource starts capturing audio. If, however, the Capture-On-Speech header is present and a value of true is specified, the resource only commences capturing audio when speech

Table 14.1 Request messages (methods)

Message name	Description
RECORD	Puts the resource into the recording state. Depending on headers supplied, recording either starts immediately or when speech is detected.
START-INPUT-TIMERS	Starts the no-input timer.
STOP	Stops the recording.

Table 14.2 Event messages

Message name	Description
START-OF-INPUT	Generated when the recorder resource first detects speech.
RECORD-COMPLETE	Generated when the recording automatically terminates.

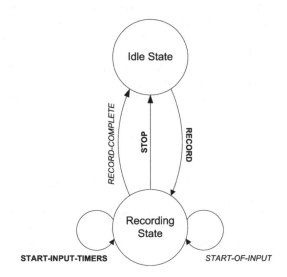

Figure 14.1   Recorder resource state machine.

is detected. The format of the recording is determined by the Media-Type header, which is required
to be present in the RECORD request. The recorder resource saves the recording to the URI indicated
by the Record-URI. If the header field is present but specified with no value, the resource chooses
a convenient location to store the audio and returns the URI in the Record-URI header field in the
RECORD-COMPLETE event message, or in the response to a STOP request if one is issued. If the
Record-URI header field is omitted from the RECORD request, the recorder resource returns the data
in the message body of the RECORD-COMPLETE event or STOP response as appropriate (in both cases
the Record-URI header field value indicates the value of the Content-ID of the message body).
   MRCP requires that the recorder resource at least supports the https:// URI scheme (i.e. HTTP
over SSL or TLS) for the Record-URI header field – necessary for ensuring secure transmission

Table 14.3 Resource-specific header fields for the recorder resource

Header name	Request	Response	Event	SET-PARAMS / GET-PARAMS
Completion-Cause	—	—	RECORD-COMPLETE	No
Completion-Reason	—	—	RECORD-COMPLETE	No
Failed-URI	—	RECORD	RECORD-COMPLETE	No
Failed-URI-Cause	—	RECORD	RECORD-COMPLETE	No
Record-URI	RECORD	STOP	RECORD-COMPLETE	No
Media-Type	RECORD	—	—	No
Capture-On-Speech	RECORD	—	—	Yes
No-Input-Timeout	RECORD	—	—	Yes
Max-Time	RECORD	—	—	Yes
Final-Silence	RECORD	—	—	Yes
Sensitivity-Level	RECORD	—	—	Yes
Trim-Length	STOP	—	—	No
Start-Input-Timers	RECORD START-INPUT-TIMERS	—	—	No
Ver-Buffer-Utterance	RECORD	—	—	No
New-Audio-Channel	RECORD	—	—	No

of potentially sensitive audio recordings. The http:// and file:// URI schemes are also common. If the Record-URI value is not valid, a response with status 404 Illegal value for header is returned. If the recorder resource is unable to create the requested stored content for some reason, a response with status 407 Method or operation failed is returned.

There is no concept of queuing for RECORD requests – if a RECORD request is received while the resource is in the recording state, a 402 Method not valid in this state response is returned. Figure 14.2 illustrates an example of a RECORD request. The corresponding messages follow.

F1 (client → recorder):

```
MRCP/2.0 168 RECORD 1000
Channel-Identifier: 11F018BE6@recorder
Record-URI:
Media-Type: audio/basic
Capture-On-Speech: true
Final-Silence: 2000
Max-Time: 5000
```

F2 (recorder → client):

```
MRCP/2.0 76 1000 200 IN-PROGRESS
Channel-Identifier: 11F018BE6@recorder
```

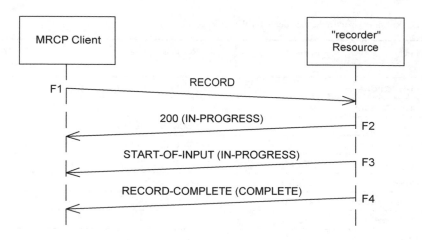

Figure 14.2   RECORD request example.

F3 (recorder → client):

```
MRCP/2.0 87 START-OF-INPUT 1000 IN-PROGRESS
Channel-Identifier: 11F018BE6@recorder
```

F4 (recorder → client):

```
MRCP/2.0 200 RECORD-COMPLETE 1000 COMPLETE
Channel-Identifier: 11F018BE6@recorder
Completion-Cause: 000 success-maxtime
Record-URI: <file://recorder/audio/file01.wav>;
          size=40000;duration=5000
```

14.2.2 START-INPUT-TIMERS

By default, a no-input timer is started when recording starts (its value is set by the No-Input-Timeout header field). If no speech is detected before the timer expires, the recording terminates with a RECORD-COMPLETE message specifying a Completion-Cause value of 002 noinput-timeout.

Typically, the MRCP client will issue the RECORD request immediately after a prompt soliciting the user for input completes. For some applications, it is desirable to start the record operation at the same time as the prompt, thus allowing familiar users to barge-in over the prompt and start the recording operation earlier. When the record operation is commenced at the same time as a prompt playback, one doesn't want to start the no-input timer until after the playback completes. This is achieved by issuing the RECORD request with a header field called Start-Input-Timers with a value of false. Later, when the MRCP client learns that the prompt playback has completed (e.g. it receives a SPEAK-COMPLETE event from a speech synthesiser resource), it can issue the START-INPUT-TIMERS method to the recorder resource to start the no-input timer.

Figure 14.3 illustrates an example of a START-INPUT-TIMERS request. The corresponding

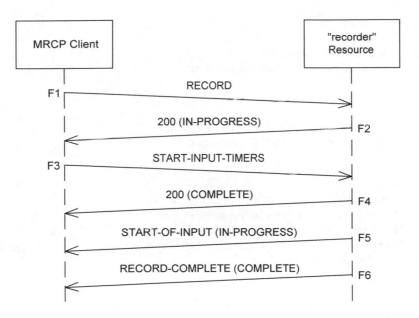

Figure 14.3 START-INPUT-TIMERS request example.

messages follow.

F1 (client → recorder):

```
MRCP/2.0 219 RECORD 1500
Channel-Identifier: 11F018BE6@recorder
Record-URI:
Media-Type: audio/basic
Capture-On-Speech: true
Start-Input-Timers: false
No-Input-Timeout: 2000
Final-Silence: 2000
Max-Time: 5000
```

F2 (recorder → client):

```
MRCP/2.0 76 1500 200 IN-PROGRESS
Channel-Identifier: 11F018BE6@recorder
```

F3 (client → recorder):

```
MRCP/2.0 79 START-INPUT-TIMERS 1501
Channel-Identifier: 11F018BE6@recorder
```

F4 (recorder → client):

```
MRCP/2.0 73 1501 200 COMPLETE
Channel-Identifier: 11F018BE6@recorder
```

F5 (recorder → client):

```
MRCP/2.0 87 START-OF-INPUT 1500 IN-PROGRESS
Channel-Identifier: 11F018BE6@recorder
```

F6 (recorder → client):

```
MRCP/2.0 200 RECORD-COMPLETE 1500 COMPLETE
Channel-Identifier: 11F018BE6@recorder
Completion-Cause: 000 success-silence
Record-URI: <https://10.0.0.1/audio/f1912.wav>;
    size=40000;duration=5000
```

14.2.3 STOP

The STOP request terminates the ongoing recording action and moves the recorder resource's state from the recording state to the idle state. The recording up to the point at which the STOP request is made is available to the MRCP client as indicated by the Record-URI header in the STOP response. The STOP response contains the Active-Request-Id-List specifying the request ID of the request that was stopped and the RECORD-COMPLETE event is not sent in this case. Making the recording up to the point of the STOP request available is crucial for implementing the common voicemail use case where the user simply hangs up to terminate the recording. It is also useful for the case where the recording is stopped by user input detected by, for example, a dtmfrecog resource. The Trim-Length header field may be specified in the STOP request to specify the number of milliseconds to trim from the end of the recording after the stop.

Figure 14.4 illustrates an example of a STOP request. The corresponding messages follow.

F1 (client → recorder):

```
MRCP/2.0 201 RECORD 2000
Channel-Identifier: 11F018BE6@recorder
Record-URI: file://recorder/audio/file04.wav
Media-Type: audio/x-wav
Capture-On-Speech: true
Final-Silence: 2000
Max-Time: 5000
```

F2 (recorder → client):

```
MRCP/2.0 76 2000 200 IN-PROGRESS
Channel-Identifier: 11F018BE6@recorder
```

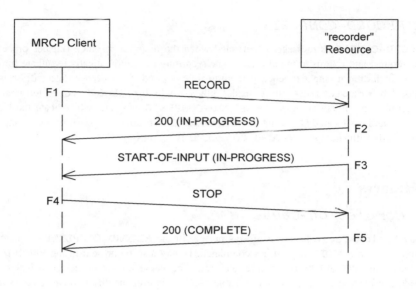

Figure 14.4 STOP request example.

F3 (recorder → client):

```
MRCP/2.0 87 START-OF-INPUT 2000 IN-PROGRESS
Channel-Identifier: 11F018BE6@recorder
```

F4 (client → recorder):

```
MRCP/2.0 65 STOP 2001
Channel-Identifier: 11F018BE6@recorder
```

F5 (recorder → client):

```
MRCP/2.0 179 2001 200 COMPLETE
Channel-Identifier: 11F018BE6@recorder
Active-Request-Id-List: 2000
Record-URI: <file://recorder/audio/file04.wav>;
          size=16000;duration=2000
```

14.3 Events

14.3.1 START-OF-INPUT

The START-OF-INPUT event is generated when the recorder resource first detects speech. The MRCP client may use this event to terminate speech playback by, for example, generating a BARGE-IN-OCCURRED message to an active speech synthesiser resource (see Chapter 12). All of the examples in the preceding section illustrate the START-OF-INPUT event message.

14.3.2 RECORD-COMPLETE

The RECORD-COMPLETE message is generated when the recorder resource finishes processing the RECORD request and returns to the idle state. The recording can automatically terminate for a variety of reasons, including no-input timeout, maximum time expired, final silence detected, or as a result of an error. The Completion-Cause header indicates why the RECORD request terminated. If the request terminated without an error, the RECORD-COMPLETE message will indicate the URI where the captured audio is located (specified in the Record-URI header field). All of the examples in the preceding section illustrate the RECORD-COMPLETE event message.

14.4 Headers

14.4.1 Completion-Cause

The Completion-Cause header is always present in the RECORD-COMPLETE event to indicate the reason why the RECORD request has completed. It may also be present in the RECORD response with request state COMPLETE if a failure occurred. The value for the header field includes both a cause code and corresponding cause name. Table 14.4 summaries the different cause codes and cause names.

Example:

```
Completion-Cause: 002 noinput-timeout
```

14.4.2 Completion-Reason

The Completion-Reason header is optionally present in the RECORD-COMPLETE response to supply more information on the reason why the RECORD request terminated. This header field is intended for use by the MRCP client for logging purposes.

Table 14.4   Cause codes and cause names for the Completion-Cause header field

Cause code	Cause name	Description
000	success-silence	The RECORD request completed due to final silence detection.
001	success-maxtime	The RECORD request completed because the maximum time was reached.
002	noinput-timeout	The RECORD request terminated because no speech was detected.
003	uri-failure	The RECORD request terminated because access to a URI failed.
004	error	The RECORD request terminated because of an error

Example:

```
Completion-Reason: Out of disk space
```

14.4.3 Failed-URI

The `Failed-URI` header field is returned in the RECORD response or RECORD-COMPLETE event when access to the `Record-URI` fails. This header field indicates the URI to which access failed. See also the `Failed-URI-Cause` header field.
   Example:

```
Failed-URI: http://192.168.1.10/audio/mailbox01.wav
```

14.4.4 Failed-URI-Cause

The `Failed-URI-Cause` header field is used in conjunction with the `Failed-URI` header field to provide protocol specific response code information. For HTTP URIs, the value will typically be a HTTP $4xx$ or $5xx$ response code.
   Example:

```
Failed-URI-Cause: 404 Not Found
```

14.4.5 Record-URI

The `Record-URI` header field indicates the location of the captured audio. If present in the RECORD request and with a specified value, the recorder resource stores the audio to that URI. If the header field is present in the RECORD request but specified with an empty value, the recorder resource chooses a convenient URI itself. If the `Record-URI` header field is omitted from the RECORD request, the recorded data is returned in the message body of the RECORD-COMPLETE event or STOP response. The `Record-URI` header field is also present in the RECORD-COMPLETE event message and STOP response. The recorder resource adds the size in bytes and duration in milliseconds to the returned URI (in this case the URI is enclosed in angle brackets so that this information is not misconstrued as being part of the URI).
   Recorder resources are required to support the HTTPS URI scheme for the `Record-URI`. Support of HTTP is also common. Support of HTTP(S) implies the recorder resource implements, or has access to, a web server to serve the content up to the MRCP client. When a HTTP(S) URI is specified in the RECORD request, the recorder resource must upload the recording to that URI; this is usually achieved either by the HTTP POST or PUT verbs (see Appendix C, located on the Web at http://www.daveburke.org/speechprocessing/). Some resources also support the file:// scheme – this works if the MRCP client and media resource server have access to the same network file system.
   Example:

```
Record-URI: <http://10.0.0.2/recordings/audio10.wav>;
            size=40000;duration=5000
```

14.4.6 Media-Type

The Media-Type header field specifies the MIME type of the desired audio recording and is required to be present in the RECORD request message. MRCP makes no attempt to mandate particular media types that must be supported. There are some commonly used audio media types but since they are not officially registered with IANA, their definition may differ across implementations – see Table 8.3. The presence of a header in the file at least allows the receiver of the audio to introspect the content to find out more about the format. The Media-Type header field may be specified on a per-RECORD request basis or set for the session via the SET-PARAMS request. There is no specified default.
   Example:

```
Media-Type: audio/x-wav
```

14.4.7 Capture-on-Speech

The Capture-On-Speech header field takes a Boolean value to indicate whether the audio capture should only start when speech is detected. If omitted, the default value of false is assumed, indicating that audio capture starts immediately on receipt of the RECORD request. The Capture-On-Speech header field may be specified on a per-RECORD request basis or set for the session via the SET-PARAMS request.
   Example:

```
Capture-On-Speech: true
```

14.4.8 No-Input-Timeout

The no-input timer is used to terminate a recording when no speech is detected at the beginning of the recording. The No-Input-Timeout header field specifies the number of milliseconds to wait for speech to be detected before terminating the recording and returning a RECORD-COMPLETE event message with a Completion-Cause value of 002 noinput-timeout. The No-Input-Timeout header field may be specified on a per-RECORD request basis or set for the session via the SET-PARAMS request. The default value is implementation specific.
   Example:

```
No-Input-Timeout: 3000
```

14.4.9 Max-Time

The Max-Time header field is used to specify the maximum duration for the recording in milliseconds from the time the recording starts and not including any silence suppression that may be applied to the recording. When the maximum time is reached, the recording terminates and the RECORD-COMPLETE event is generated with a Completion-Cause of 000 success-silence. The Max-Time header field may be specified on a per-RECORD request basis or set for the session via the SET-PARAMS request. The default value is 0 and implies a maximum time of infinity.
   Example:

```
Max-Time: 10000
```

14.4.10 Final-Silence

The Final-Silence header field specifies the amount of silence in milliseconds that is interpreted as being the end of the recording. A value of 0 implies infinity (i.e. no final silence detection is applied). The Final-Silence header field may be specified on a per-RECORD request basis or set for the session via the SET-PARAMS request. The default value is left as implementation specific.

Example:

```
Final-Silence: 3500
```

14.4.11 Sensitivity-Level

Speech detection is used by the recorder resource for detecting no-input conditions, when to commence audio capture, final silence conditions, and for silence suppression. The Sensitivity-Level controls how sensitive the speech detector is by specifying a float value between 0.0 and 1.0. A low value means it is less sensitive to noise and a high value means it is more sensitive to quiet input. The Sensitivity-Level header field may be specified on a per-RECORD request basis or set for the session via the SET-PARAMS request. The default value is left as implementation specific.

Example:

```
Sensitivity-Level: 0.85
```

14.4.12 Trim-Length

The Trim-Length header field may be specified on a STOP request. It indicates the number of milliseconds to trim from the end of the recording. The default value for this header field is 0.

Example:

```
Trim-Length: 1000
```

14.4.13 Start-Input-Timers

The Start-Input-Timers header field is specified in the RECORD request to delay commencement of the no-input timer until the MRCP client issues a START-INPUT-TIMERS request. The Start-Input-Timers header takes a Boolean value: a value of true (the default) indicates that the no-input timer is started on receipt of the RECORD request, while a value of false means the no-input timer is not started. A value of false is typically used to allow the recorder resource to be put into the recording state at the same time as the prompts begin playback. This allows the audio to be captured if the user barges in over the prompts. If the user does not start speaking during the prompt playback, the no-input timer is started at the end of the playback (via the START-INPUT-TIMERS request) to give the user a finite amount of time to respond.

Example:

```
Start-Input-Timers: false
```

14.4.14 Ver-Buffer-Utterance

The `Ver-Buffer-Utterance` header is used in the RECORD request to indicate that the recorded audio can be later considered for speaker verification, assuming a speaker verification resource has been allocated as part of the same session (see Chapter 15). The header field takes a Boolean value and defaults to `false`.
   Example:

```
Ver-Buffer-Utterance: true
```

14.4.15 New-Audio-Channel

Specifying the header field `New-Audio-Channel` with a value of `true` in the RECORD request indicates to the recorder resource that the audio being sent now comes from a different audio source, channel or speaker. The presence of this header field with a value of `true` is used to reset any adaptation algorithms that might have been running. For example, speech detectors used in recorder resources usually maintain a long-term running average of statistical parameters corresponding to speech and non-speech audio. If an audio source were to be changed without resetting the algorithm, the adapted parameter values for the previous channel could adversely affect the speech detector.
   Example:

```
New-Audio-Channel: false
```

14.5 Summary

This chapter provides a detailed look at the MRCP recorder resource. This resource provides the basic capability of audio recording but notably includes speech detection capabilities resulting in a rich feature set that enables no-input detection, final silence detection, automatic commencement of recording, and the ability to suppress silence at the beginning and end of recording. The recorder resource is ideal for any IVR application requiring capture of audio and is sufficiently complete to provide the underlying functionality required by VoiceXML (see Chapter 16).

In the next chapter, we continue our in-depth look at the MRCP media resources by studying the speaker verification resource.

15

Speaker verification resource

The MRCP speaker verification resource (*speakverify*) provides speaker verification and identification functions to the MRCP client. Speaker verification is concerned with the problem of establishing that the speaker is the person they purport to be (their claimed identity is usually asserted via a customer code, pin number, caller-ID, etc.). A spoken utterance is compared to a previously stored voiceprint, associated with the claimed identity, to determine an accept/reject decision. Speaker identification deals with the problem of determining who is speaking from a set of known speakers. A recorded utterance is compared with a set of previously stored voiceprints to determine the identity of the person most likely to have spoken the utterance (or possibly a 'none-of-the-above'). From the standpoint of using the MRCP speaker verification resource, the speaker identification use-case can be considered 'multi-verification' – verifying against multiple voiceprints. We will therefore use the term verification loosely in this chapter also to mean identification.

The MRCP speaker verification resource makes no distinction between text-dependent and text-independent modes of operation – rather this is a decision for the application designer to take into account, given the capabilities of the particular resource being employed. See Section 2.3 for a general discussion on speaker verification and identification technologies. This chapter delves into the details of the MRCP speaker verification resource by studying the methods, events and header fields that apply to it. The speaker verification resource leverages the Natural Language Semantics Markup Language (NLSML) format for encapsulating verification and identification results – see Chapter 10 for a review of NLSML.

15.1 Overview

The MRCP speaker verification resource can operate independently of other resources or in conjunction with either the recogniser resource or the recorder resource. When operating independently, the speaker verification resource analyses the audio received in real-time over the media session. When operating in conjunction with another resource, the speaker verification resource shares its verification buffer with a speech recogniser or recorder resource. The speech recogniser and recorder resources can write to the buffer by specifying the `Ver-Buffer-Utterance` header field with a value of `true` (see Sections 13.5.30 and 14.4.14). This allows audio collected during a recognition or recording to be analysed

Table 15.1 Speaker verification resource request messages (methods)

Message name	Description
START-SESSION	Starts a training or verification session.
END-SESSION	Terminates a training or verification session.
VERIFY	Trains or verifies a voiceprint from live audio.
VERIFY-FROM-BUFFER	Trains or verifies a voiceprint from buffered audio.
VERIFY-ROLLBACK	Discards the last buffered or collected live utterance.
START-INPUT-TIMERS	Starts the no input timer.
GET-INTERMEDIATE-RESULT	Used to poll for intermediate results from an ongoing verification request.
STOP	Stops the active VERIFY or VERIFY-FROM-BUFFER request.
CLEAR-BUFFER	Clears the verification buffer.
QUERY-VOICEPRINT	Used to get a status on a voiceprint.
DELETE-VOICEPRINT	Used to delete a voiceprint.

Table 15.2 Speaker verification resource event messages

Message name	Description
START-OF-INPUT	Generated when the speaker verification resource first detects speech.
VERIFICATION-COMPLETE	Generated when the speaker verification resource completes a training or verification request.

later in a speaker verification or identification task. For example, a PIN number collected by the speech recogniser could indicate the claimed identity and be used to retrieve the corresponding voiceprint. The audio utterance buffered from the recognition can be used by the speaker verification resource to verify the speaker against the voiceprint. Note that sharing of the audio buffer is only possible when the resources are allocated within the same SIP dialog.

The speaker verification resource supports a total of eleven request messages and two event messages. The methods and events are summarised in Tables 15.1 and 15.2.

The speaker verification resource has a state machine that is driven by requests from the MRCP client and events generated by the media resource itself. The state machine is illustrated in Figure 15.1. Note that the resource operates in one of two modes: training or verifying (the mode is set via the Verification-Mode header field in the START-SESSION request method).

Table 15.3 summarises the resource-specific header fields for the speaker verification resource. See Table 7.5 for a list of generic headers that also apply.

15.2 Methods

15.2.1 START-SESSION

The START-SESSION request method is used to start a training or verification session. The value of the Verification-Mode header field indicates the session type: train indicates a training session and verify indicates a verification session. A verification session is associated with a voiceprint repository indicated by the Repository-URI header field. Each session is associated with one or more voiceprints specified by the Voiceprint-Identifier header field (for training

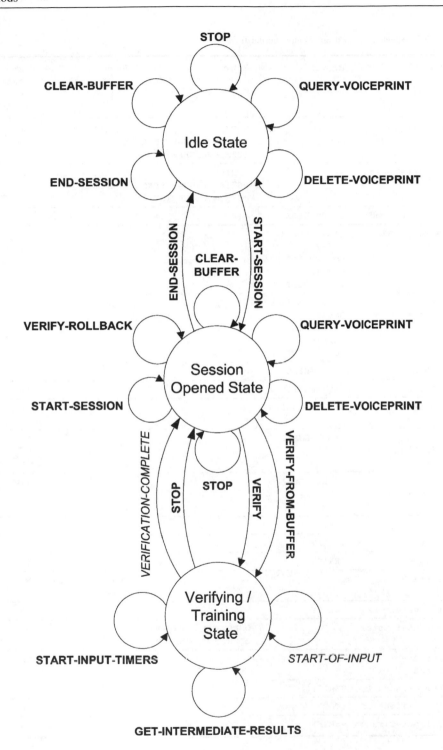

Figure 15.1 Speaker verification resource state machine.

Table 15.3 Speaker verification resource header fields

Header name	Request	Response	Event	GET-PARAMS / SET-PARAMS
Completion-Cause	—	VERIFY VERIFY-FROM-BUFFER QUERY-VOICEPRINT	VERIFICATION-COMPLETE	No
Completion-Reason	—	VERIFY VERIFY-FROM-BUFFER QUERY-VOICEPRINT	VERIFICATION-COMPLETE	No
Verification-Mode	START-SESSION	—	—	No
Repository-URI	START-SESSION QUERY-VOICEPRINT DELETE-VOICEPRINT	—	—	No
Voiceprint-Identifier	START-SESSION QUERY-VOICEPRINT DELETE-VOICEPRINT	—	—	No
Adapt-Model	START-SESSION	—	—	No
Abort-Model	END-SESSION	—	—	No
Min-Verification-Score	START-SESSION	—	—	Yes
Num-Min-Verification-Phrases	START-SESSION	—	—	Yes
Num-Max-Verification-Phrases	START-SESSION	—	—	Yes
No-Input-Timeout	VERIFY	—	—	Yes
Save-Waveform	VERIFY	—	—	No
Media-Type	VERIFY	—	—	Yes
Waveform-URI		—	VERIFICATION-COMPLETE	No
Input-Waveform-URI	VERIFY	—	—	No
Voiceprint-Exists		QUERY-VOICEPRINT DELETE-VOICEPRINT	—	No
Ver-Buffer-Utterance	VERIFY	—	—	No
New-Audio-Channel	VERIFY	—	—	No
Abort-Verification	STOP	—	—	No
Start-Input-Timers	VERIFY	—	—	No

and verification only one voiceprint is specified; for identification there is a semi-colon delimited list of voiceprints). During training, a voiceprint that does not exist will be automatically created. Figure 15.2 illustrates an example of a START-SESSION request. The corresponding messages follow.

F1 (client → speakverify):

```
MRCP/2.0 194 START-SESSION 00001
Channel-Identifier: 1cd3ee59@speakverify
Verification-Mode: train
Repository-URI: http://example.com/voiceprintdatabase
Voiceprint-Identifier: dave.burke
```

F2 (speakverify → client):

```
MRCP/2.0 76 00001 200 COMPLETE
Channel-Identifier: 1cd3ee59@speakverify
```

15.2.2 END-SESSION

The END-SESSION request method terminates a training or verification session. The Abort-Model header field establishes whether or not the voiceprint specified in START-SESSION will be updated. A voiceprint is updated either because a training session was active or a verification session was active and the voiceprint was being adapted (adaption is enabled by setting Adapt-Model: true in the START-SESSION request). A value of false for Abort-Model will result in any changes to the voiceprint being committed to the database, while a value of true will discard any changes. Figure 15.3 illustrates an example of a START-SESSION request. The corresponding messages follow.

F1 (client → speakverify):

```
MRCP/2.0 95 END-SESSION 10000
Channel-Identifier: 1cd3ee59@speakverify
Abort-Model: false
```

F2 (speakverify → client):

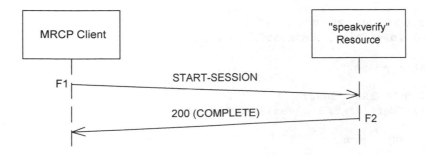

Figure 15.2 START-SESSION request example.

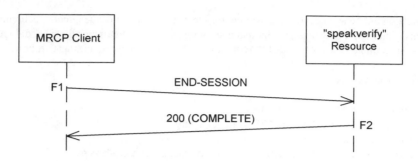

Figure 15.3 END-SESSION request example.

```
MRCP/2.0 76 10000 200 COMPLETE
Channel-Identifier: 1cd3ee59@speakverify
```

15.2.3 VERIFY

The VERIFY method requests that the speaker verification resource operate on the live audio carried over the media session to start training the voiceprint or verify the utterance against one or more voiceprints (the mode of operation is specified via the Verification-Mode header field in the START-SESSION request). A VERIFICATION-COMPLETE event message is generated when the VERIFY request completes. NLSML data within the VERIFICATION-COMPLETE message returns the status of the training or verification attempt. Figure 15.4 illustrates a complete example of training a voiceprint including use of the VERIFY request method. The corresponding messages follow.

F1 (client → speakverify):

```
MRCP/2.0 194 START-SESSION 20000
Channel-Identifier: 1cd3ee59@speakverify
Verification-Mode: train
Repository-URI: http://example.com/voiceprintdatabase
Voiceprint-Identifier: dave.burke
```

F2 (speakverify → client):

```
MRCP/2.0 76 20000 200 COMPLETE
Channel-Identifier: 1cd3ee59@speakverify
```

F3 (client → speakverify):

```
MRCP/2.0 70 VERIFY 20001
Channel-Identifier: 1cd3ee59@speakverify
```

F4 (speakverify → client):

```
MRCP/2.0 79 20001 200 IN-PROGRESS
```

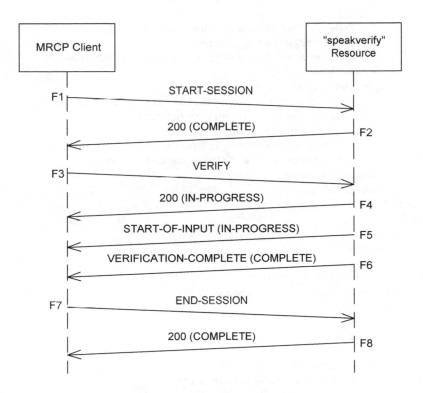

Figure 15.4   Complete voiceprint training example.

```
Channel-Identifier: 1cd3ee59@speakverify
```

F5 (speakverify → client):

```
MRCP/2.0 90 START-OF-INPUT 20001 IN-PROGRESS
Channel-Identifier: 1cd3ee59@speakverify
```

F6 (speakverify → client):

```
MRCP/2.0 921 VERIFICATION-COMPLETE 20001 COMPLETE
Channel-Identifier: 1cd3ee59@speakverify
Completion-Cause: 000 success
Content-Type: application/nlsml+xml
Content-Length: 737

<?xml version="1.0" encoding="UTF-8"?>
<result xmlns="http://www.ietf.org/xml/ns/mrcpv2">
    <verification-result>
        <voiceprint id="dave.burke">
            <incremental>
```

```
               <device>cellular-phone</device>
               <gender>male</gender>
               <utterance-length>751</utterance-length>
          </incremental>
          <cumulative>
               <verification-score>0.93</verification-score>
               <device>cellular-phone</device>
               <gender>male</gender>
               <utterance-length>1522</utterance-length>
               <need-more-data>false</need-more-data>
          </cumulative>
     </voiceprint>
   </verification-result>
</result>
```

F7 (client → speakverify):

```
MRCP/2.0 95 END-SESSION 20002
Channel-Identifier: 1cd3ee59@speakverify
Abort-Model: false
```

F8 (speakverify → client):

```
MRCP/2.0 76 20002 200 COMPLETE
Channel-Identifier: 1cd3ee59@speakverify
```

15.2.4 VERIFY-FROM-BUFFER

The VERIFY-FROM-BUFFER request method performs the same action as the VERIFY request method except that the audio data is sourced from the verification buffer instead of the live media stream. The verification buffer can be written to by the speaker verification resource (by issuing a VERIFY request with Ver-Buffer-Utterance: true), by a recogniser resource allocated in the same SIP dialog (by issuing a RECOGNIZE request with Ver-Buffer-Utterance: true), or by a recorder resource allocated in the same SIP dialog (by issuing a RECORD request with Ver-Buffer-Utterance: true). Note that a START-OF-INPUT event is not generated for the VERIFY-FROM-BUFFER request. Verifying from a buffer is useful for several reasons. A recogniser can be used to ascertain the claimed identity of the user (by recognising an account number for example), and that audio can be subsequently used to verify the user against the corresponding voiceprint. Another example is where the VERIFY request fails to verify the user but the MRCP client wishes to use the same audio to attempt a verification against a different voiceprint. Assuming the initial VERIFY carried the Ver-Buffer-Utterance: true header, then the verification buffer will have been written to and a new session can be started with a different voiceprint (using START-SESSION), and the VERIFY-FROM-BUFFER request can be invoked. Figure 15.5 illustrates an example of the VERIFY-FROM-BUFFER request method. The corresponding messages follow.

F1 (client → speakverify):

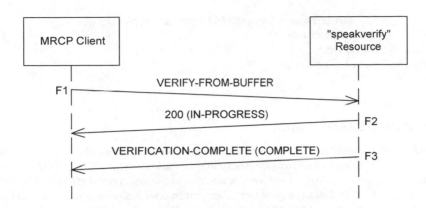

Figure 15.5 VERIFY-FROM-BUFFER request example.

```
MRCP/2.0 82 VERIFY-FROM-BUFFER 30000
Channel-Identifier: 1cd3ee59@speakverify
```

F2 (speakverify → client):

```
MRCP/2.0 79 30000 200 IN-PROGRESS
Channel-Identifier: 1cd3ee59@speakverify
```

F3 (speakverify → client):

```
MRCP/2.0 986 VERIFICATION-COMPLETE 30000 COMPLETE
Channel-Identifier: 1cd3ee59@speakverify
Completion-Cause: 000 success
Content-Type: application/nlsml+xml
Content-Length: 802

<?xml version="1.0" encoding="UTF-8"?>
<result xmlns="http://www.ietf.org/xml/ns/mrcpv2">
    <verification result>
        <voiceprint id="dave.burke">
            <incremental>
                <verification-score>0.85</verification-score>
                <device>carbon-button-phone</device>
                <gender>male</gender>
                <utterance-length>751</utterance-length>
            </incremental>
            <cumulative>
                <verification-score>0.81</verification-score>
                <device>carbon-button-phone</device>
                <gender>male</gender>
                <decision>accepted</decision>
```

```
            <utterance-length>801</utterance-length>
         </cumulative>
      </voiceprint>
   </verification-result>
</result>
```

15.2.5 *VERIFY-ROLLBACK*

The `VERIFY-ROLLBACK` request method undoes the previous `VERIFY` or `VERIFY-FROM-BUFFER` request. As a result, the previously analysed audio will not contribute to the training of the voiceprint (for a training session) or to the analysis and possible adaption of the voiceprint (for a verification session). Issuing the `VERIFY-ROLLBACK` request method more than once has no effect. Figure 15.6 illustrates an example of the `VERIFY-ROLLBACK` request method. The corresponding messages follow.

F1 (client → speakverify):

```
MRCP/2.0 79 VERIFY-ROLLBACK 40000
Channel-Identifier: 1cd3ee59@speakverify
```

F2 (speakverify → client):

```
MRCP/2.0 76 40000 200 COMPLETE
Channel-Identifier: 1cd3ee59@speakverify
```

15.2.6 *START-INPUT-TIMERS*

By default, a `No-Input-Timeout` timer is started when a `VERIFY` request is initiated. If no input is detected before the timer expires, the `VERIFY` request terminates with a `VERIFICATION-COMPLETE` message with `Completion-Cause` set to `002 no-input-timeout`.

Typically, the MRCP client will issue the `VERIFY` request immediately after a prompt soliciting the user for input completes. For some applications, it is desirable to start the training/verification at the same time as the prompt, thus allowing familiar users to barge-in over the prompt and start the training/verification earlier. When the training/verification is commenced at the same time as a

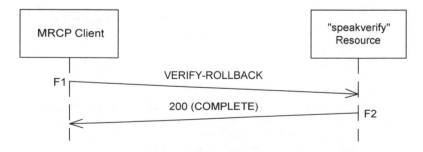

Figure 15.6 `VERIFY-ROLLBACK` request example.

prompt playback, one does not want to start the `No-Input-Timeout` timer until after the playback completes. This is achieved by issuing the `VERIFY` message with the `Start-Input-Timers` header field set to `false`. Later, when the MRCP client learns that the prompt playback has completed (e.g. it receives a `SPEAK-COMPLETE` event from a speech synthesiser resource), it can issue the `START-INPUT-TIMERS` method to the speaker verification resource to start the `No-Input-Timeout` timer. Figure 15.7 illustrates an example of a `START-INPUT-TIMERS` request. The corresponding messages follow.

F1 (client → speakverify):

```
MRCP/2.0 70 VERIFY 50000
Channel-Identifier: 1cd3ee59@speakverify
```

F2 (speakverify → client):

```
MRCP/2.0 79 50000 200 IN-PROGRESS
Channel-Identifier: 1cd3ee59@speakverify
```

F3 (client → speakverify):

```
MRCP/2.0 82 START-INPUT-TIMERS 50001
Channel-Identifier: 1cd3ee59@speakverify
```

F4 (speakverify → client):

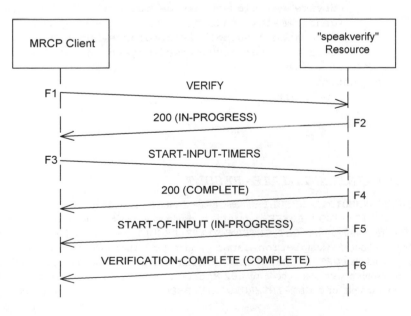

Figure 15.7 `START-INPUT-TIMERS` request example.

```
MRCP/2.0 76 50001 200 COMPLETE
Channel-Identifier: 1cd3ee59@speakverify
```

F5 (speakverify → client):

```
MRCP/2.0 90 START-OF-INPUT 50000 IN-PROGRESS
Channel-Identifier: 1cd3ee59@speakverify
```

F6 (speakverify → client):

```
MRCP/2.0 920 VERIFICATION-COMPLETE 50000 COMPLETE
Channel-Identifier: 1cd3ee59@speakverify
Completion-Cause: 000 success
Content-Type: application/nlsml+xml
Content-Length: 736

<?xml version="1.0" encoding="UTF-8"?>
<result xmlns="http://www.ietf.org/xml/ns/mrcpv2">
    <verification-result>
        <voiceprint id="dave.burke">
            <incremental>
                <device>cellular-phone</device>
                <gender>male</gender>
                <utterance-length>751</utterance-length>
            </incremental>
            <cumulative>
                <verification-score>0.53</verification-score>
                <device>cellular-phone</device>
                <gender>male</gender>
                <utterance-length>1522</utterance-length>
                <need-more-data>true</need-more-data>
            </cumulative>
        </voiceprint>
    </verification-result>
</result>
```

15.2.7 GET-INTERMEDIATE-RESULT

The GET-INTERMEDIATE-RESULT request method can be issued by the MRCP client while a VERIFY or VERIFY-FROM-BUFFER request is in progress to retrieve the accumulated verification results. The response to the GET-INTERMEDIATE-RESULT request method contains NLSML results but without a Completion-Cause header field (because the request has not yet completed). Sending a GET-INTERMEDIATE-RESULT request when the resource is in the idle state results in a response with status code of 402 Method not valid in this state. Figure 15.8 illustrates an example of a GET-INTERMEDIATE-RESULT request. The corresponding messages follow.

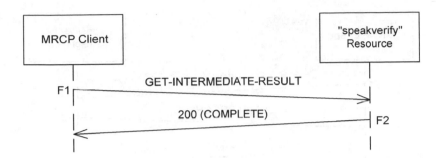

Figure 15.8 GET-INTERMEDIATE-RESULT request example.

F1 (client → speakverify):

```
MRCP/2.0 87 GET-INTERMEDIATE-RESULT 60000
Channel-Identifier: 1cd3ee59@speakverify
```

F2 (speakverify → client):

```
MRCP/2.0 934 60000 200 COMPLETE
Channel-Identifier: 1cd3ee59@speakverify
Content-Type: application/nlsml+xml
Content-Length: 799

<?xml version="1.0" encoding="UTF-8"?>
<result xmlns="http://www.ietf.org/xml/ns/mrcpv2">
    <verification-result>
        <voiceprint id="dave.burke">
            <incremental>
                <verification-score>0.53</verification-score>
                <device>cellular-phone</device>
                <gender>male</gender>
                <utterance-length>751</utterance-length>
            </incremental>
            <cumulative>
                <verification-score>0.67</verification-score>
                <device>cellular-phone</device>
                <gender>male</gender>
                <utterance-length>1522</utterance-length>
                <need-more-data>true</need-more-data>
            </cumulative>
        </voiceprint>
    </verification-result>
</result>
```

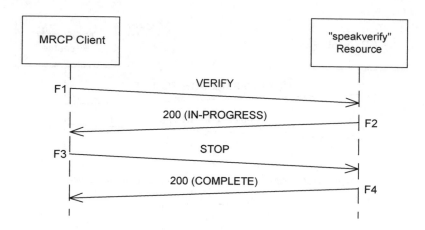

Figure 15.9 STOP request example.

15.2.8 STOP

The STOP request method terminates the ongoing training/verification action and moves the resource's state to idle. The STOP response contains the Active-Request-Id-List specifying the request ID of the request that was stopped. The VERIFICATION-COMPLETE event is not sent in this case. A STOP request may also include an Abort-Verification header field to indicate whether the training/verification result thus far should be discarded (true) or retained (false). A value of false will cause the verification results to be returned in the STOP response. Figure 15.9 illustrates an example of a STOP request. The corresponding messages follow.

F1 (client → speakverify):

```
MRCP/2.0 70 VERIFY 70000
Channel-Identifier: 1cd3ee59@speakverify
```

F2 (speakverify → client):

```
MRCP/2.0 79 70000 200 IN-PROGRESS
Channel-Identifier: 1cd3ee59@speakverify
```

F3 (client → speakverify):

```
MRCP/2.0 95 STOP 70001
Channel-Identifier: 1cd3ee59@speakverify
Abort-Verification: true
```

F4 (speakverify → client):

```
MRCP/2.0 108 70001 200 COMPLETE
Channel-Identifier: 1cd3ee59@speakverify
Active-Request-Id-List: 70000
```

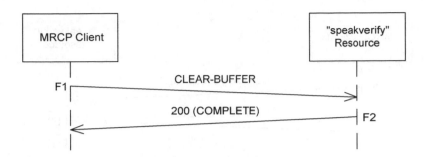

Figure 15.10   CLEAR-BUFFER request example.

15.2.9 CLEAR-BUFFER

The CLEAR-BUFFER request method is used to clear the verification buffer. As already discussed, the verification buffer can be written to via a VERIFY, RECOGNIZE or RECORD request with the Ver-Buffer-Utterance header field set to true and later used via the VERIFY-FROM-BUFFER request method to perform training or verification. Figure 15.10 illustrates an example of a CLEAR-BUFFER request. The corresponding messages follow.

F1 (client → speakverify):

```
MRCP/2.0 76 CLEAR-BUFFER 80000
Channel-Identifier: 1cd3ee59@speakverify
```

F2 (speakverify → client):

```
MRCP/2.0 76 80000 200 COMPLETE
Channel-Identifier: 1cd3ee59@speakverify
```

15.2.10 QUERY-VOICEPRINT

The QUERY-VOICEPRINT request method allows the MRCP client to query whether a voiceprint exists in a repository. A voiceprint is fully specified by a combination of the Repository-URI and the Voiceprint-Identifier and each Voiceprint-Identifier specifies a unique voiceprint within a given Repository-URI. The response to the QUERY-VOICEPRINT request contains a Voiceprint-Exists header field specifying a Boolean value to indicate whether or not the voiceprint was found in the repository. If the Repository-URI itself does not exist, the response includes a Completion-Cause: 007 repository-uri-failure header. Figure 15.11 illustrates an example of a QUERY-VOICEPRINT request. The corresponding messages follow.

F1 (client → speakverify):

```
MRCP/2.0 171 QUERY-VOICEPRINT 90000
Channel-Identifier: 1cd3ee59@speakverify
Repository-URI: http://example.com/voiceprintdatabase
Voiceprint-Identifier: dave.burke
```

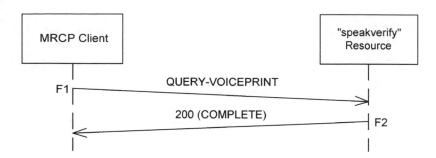

Figure 15.11 QUERY-VOICEPRINT request example.

F2 (speakverify → client):

```
MRCP/2.0 102 90000 200 COMPLETE
Channel-Identifier: 1cd3ee59@speakverify
Voiceprint-Exists: true
```

15.2.11 DELETE-VOICEPRINT

The DELETE-VOICEPRINT request method allows the MRCP client to delete a voiceprint uniquely specified by the combination of the Repository-URI and the Voiceprint-Identifier. If the voiceprint does not exist, no error is reported. Figure 15.12 illustrates an example of a DELETE-VOICEPRINT request. The corresponding messages follow.

F1 (client → speakverify):

```
MRCP/2.0 173 DELETE-VOICEPRINT 100000
Channel-Identifier: 1cd3ee59@speakverify
Repository-URI: http://example.com/voiceprintdatabase
Voiceprint-Identifier: dave.burke
```

F2 (speakverify → client):

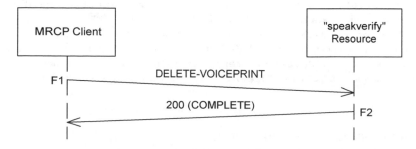

Figure 15.12 DELETE-VOICEPRINT request example.

```
MRCP/2.0 77 100000 200 COMPLETE
Channel-Identifier: 1cd3ee59@speakverify
```

15.3 Events

15.3.1 START-OF-INPUT

The START-OF-INPUT event message is generated by the speaker verification resource to inform the MRCP client that it has received speech input. This event message is typically used by the MRCP client to detect barge-in and stop the playback of audio – see Section 12.2.5 for more information. Receipt of the event can also be used by the MRCP client to estimate the time when barge-in occurred. Section 15.2.3 illustrates an example of the START-OF-INPUT event.

15.3.2 VERIFICATION-COMPLETE

The speaker verification resource generates the VERIFICATION-COMPLETE event to indicate to the MRCP client that the training/verification action has completed and to convey the training/verification results in the message body. The Completion-Cause and optional Completion-Reason header fields indicate the status of the completed training/verification. Section 15.2.3 illustrates an example of the VERIFICATION-COMPLETE event.

15.4 Headers

15.4.1 Completion-Cause

The Completion-Cause header field is always present in the VERIFICATION-COMPLETE response to indicate the reason why the VERIFY or VERIFY-FROM-BUFFER request has completed. The Completion-Cause header field also appears in the response to a VERIFY, VERIFY-FROM-BUFFER or QUERY-VOICEPRINT request to specify more information when a failure status is returned. The value for the header field includes both a cause code and corresponding cause name. Table 15.4 summarises the different cause codes and cause names.
   Example:

```
Completion-Cause: 005 buffer-empty
```

15.4.2 Completion-Reason

The Completion-Reason header is optionally present alongside the Completion-Cause header field to provide more information. This header field is intended for use by the MRCP client for logging purposes.
   Example:

```
Completion-Reason: Out of memory
```

Table 15.4 Cause codes and cause names for the `Completion-Cause` header field

Cause code	Cause name	Description
000	success	The VERIFY or VERIFY-FROM-BUFFER request completed successfully. For verification/identification the decision is encoded within the NLSML `<decision>` element with a value of `accepted`, `rejected`, or `undecided`.
001	error	The VERIFY or VERIFY-FROM-BUFFER request completed prematurely due to a system error.
002	no-input-timeout	The VERIFY request completed due to a No-Input-Timeout timer expiry.
003	too-much-speech-timeout	The VERIFY request completed prematurely due to too much speech.
004	speech-too-early	The VERIFY request completed prematurely because the caller started speaking too soon.
005	buffer-empty	The VERIFY-FROM-BUFFER request completed prematurely because the verification buffer was empty.
006	out-of-sequence	A request method failed because it was out of sequence (e.g. if the MRCP client issues a VERIFY request before a START-SESSION request).
007	repository-uri-failure	A failure occurred accessing the Repository-URI.
008	repository-uri-missing	The Repository-URI header field was not specified.
009	voiceprint-id-missing	The Voiceprint-Identifier header field was not specified.
010	voiceprint-id-not-exist	The voiceprint specified in the Voiceprint-Identifier header field does not exist in the repository.

15.4.3 Verification-Mode

The `Verification-Mode` header field is specified in the `START-SESSION` request method to indicate whether a training session (`train`) or verification session (`verify`) is to be started. There is no default value specified for this header.

Example:

```
Verification-Mode: train
```

15.4.4 Repository-URI

The `Repository-URI` header field is specified in the `START-SESSION`, `QUERY-VOICEPRINT` or `DELETE-VOICEPRINT` request methods to indicate the URI for the database containing the voiceprints. The format for the voiceprint database is implementation specific.

Example:

```
Repository-URI: file://host/voiceprintdb/
```

15.4.5 Voiceprint-Identifier

The Voiceprint-Identifier header field indicates the claimed identity for speaker verification. For speaker identification, the header field value consists of a semi-colon delimited set of voiceprints or a single voiceprint representing a group of speakers. The Voiceprint-Identifier header field is specified in the START-SESSION, QUERY-VOICEPRINT or DELETE-VOICEPRINT request methods. The format for the header field value must be in the format of two character strings separated by a stop (period).
   Example:

```
Voiceprint-Identifier: joe.bloggs
```

15.4.6 Adapt-Model

The Adapt-Model header field is specified in the START-SESSION request method for verification sessions to indicate whether the voiceprint can be adapted when a successful verification is performed. Adapting to ongoing changes in a speaker's speech can improve verification accuracy. The default value for this header field is false.
   Example:

```
Adapt-Model: true
```

15.4.7 Abort-Model

The Abort-Model header field can be specified in the END-SESSION request to indicate whether any changes to a voiceprint (occurring due to a training session or adaption as a result of a verification session) are to be aborted or committed. Setting the header field value to true results in changes to the voiceprint being discarded (note that a value of true for Abort-Model overrides a value of true for Adapt-Model). The default value for this header field is false.
   Example:

```
Abort-Model: true
```

15.4.8 Min-Verification-Score

The Min-Verification-Score header field specifies the minimum value for the verification score for which a decision of accepted may be returned. The value of the header field can vary between -1.0 and 1.0. The Min-Verification-Score header field may be specified on a per-START-SESSION request basis or set for the session via the SET-PARAMS request. There is no specified default value.
   Example:

```
Min-Verification-Score: 0.75
```

15.4.9 Num-Min-Verification-Phrases

The `Num-Min-Verification-Phrases` header field specifies the minimum number of utterances required before a positive decision (i.e. `accepted`) is given for verification. The `Num-Min-Verification-Phrases` header field may be specified on a per-`START-SESSION` request basis or set for the session via the `SET-PARAMS` request. The default and minimum value is 1.

Example:

```
Num-Min-Verification-Phrases: 2
```

15.4.10 Num-Max-Verification-Phrases

The `Num-Max-Verification-Phrases` header field specifies the number of utterances before a decision is forced (either `accepted` or `rejected` – a result of `undecided` cannot be returned after `Num-Max-Verification-Phrases` of utterances have been analysed). The `Num-Max-Verification-Phrases` header field may be specified on a per-`START-SESSION` request basis or set for the session via the `SET-PARAMS` request.

Example:

```
Num-Max-Verification-Phrases: 3
```

15.4.11 No-Input-Timeout

The `No-Input-Timeout` timer is used to detect the scenario where a training or verification is started but the user does not start speaking within a given time period. If the `No-Input-Timeout` timer expires before input is detected, the `VERIFICATION-COMPLETE` event is generated with a `Completion-Cause` value of `002 no-input-timeout`. The usual course of action for the speech application is to prompt the user that the system did not hear anything and restart the training/verification. The `No-Input-Timeout` header field may be specified on a per-`VERIFY` request basis or set for the session via the `SET-PARAMS` request. The units for the `No-Input-Timeout` header field value are in milliseconds. The value can range from 0 to an implementation specific maximum and there is no default value specified.

Example:

```
No-Input-Timeout: 3000
```

15.4.12 Save-Waveform

The `Save-Waveform` header field takes a Boolean value to indicate to the speaker verification resource whether or not to save the audio captured during verification. If set to `true`, the captured audio is indicated via the `Waveform-URI` header field in the `VERIFICATION-COMPLETE` event message (thereby allowing the MRCP client to retrieve the audio). The `Media-Type` header field specifies the format for the captured audio (see Section 15.4.13). The `Save-Waveform` header field may be specified on a per-`VERIFY` request basis (but not `VERIFY-FROM-BUFFER`) or set for the session via the `SET-PARAMS` request. The default value is `false`.

Example:

```
Save-Waveform: true
```

15.4.13 Media-Type

The Media-Type header field is used to indicate the media type for the captured audio referenced by the Waveform-URI header field (see Section 15.4.14). MRCP makes no attempt to mandate particular media types that must be supported. There are some commonly used audio media types but since they are not officially registered with IANA, their definition may differ across implementations – see Table 8.3. The Media-Type header field may be specified on a per-VERIFY request basis or set for the session via the SET-PARAMS request. There is no specified default.
  Example:

```
Media-Type: audio/x-wav
```

15.4.14 Waveform-URI

The Waveform-URI header field is used to communicate a URI to the MRCP client for the audio captured during live verification (i.e. resulting from a VERIFY request). This header field appears in the VERIFICATION-COMPLETE event if the Save-Waveform header field is specified with a value of true. The Waveform-URI header field is post-pended with size and duration parameters for the captured audio measured in bytes and milliseconds respectively. The URI indicated by the header field must be accessible for the duration of the session. If there is an error recording the audio, the Waveform-URI header field value is left blank.
  Example:

```
Waveform-URI: <http://10.0.0.1/utt01.wav>;size=8000;duration=1000
```

15.4.15 Input-Waveform-URI

The Input-Waveform-URI header field is used to supply audio for the VERIFY request instead of using the media session. This capability is useful for sourcing audio from a previous recognition when the speech recogniser resource is separate from the speaker verification resource. It is also useful for automatic tools for batch processing, for example.
  Example:

```
Input-Waveform-URI: http://10.0.0.1/utt01.wav
```

15.4.16 Voiceprint-Exists

The Voiceprint-Exists header field takes a Boolean value and is returned in the response to the QUERY-VOICEPRINT and DELETE-VOICEPRINT requests. For the DELETE-VOICEPRINT request, the header field value indicates the status of the voiceprint just before a deletion was requested.
  Example:

```
Voiceprint-Exists: false
```

15.4.17 Ver-Buffer-Utterance

The Ver-Buffer-Utterance header field may be specified in a VERIFY request, with a value of
true to indicate to the speaker verification resource that it should make the audio captured available
in its verification buffer for later use (via VERIFY-FROM-BUFFER). Note that a speech recogniser
or recorder resource declared in the same SIP dialog can specify this header field in the RECOGNIZE
and RECORD request methods respectively to write to the verification buffer. The default value for this
header field is false.
  Example:

```
Ver-Buffer-Utterance: true
```

15.4.18 New-Audio-Channel

The New-Audio-Channel header field can be specified on a VERIFY request with a value of
true to indicate to the speaker verification resource that from this point on, audio is being received
from a new channel. A speaker verification resource will usually interpret this as a request to reset its
endpointing algorithm, thus discarding any adaption performed to the previous channel. This feature is
useful for allowing an existing session to be reused on a new telephone call, for example. The default
value for this header field is false.
  Example:

```
New-Audio-Channel: true
```

15.4.19 Abort-Verification

The Abort-Verification header field is required on the STOP request to indicate whether the
training/verification result thus far should be discarded (true) or retained (false). A value of false
will cause the verification results to be returned in the STOP response.
  Example:

```
Abort-Verification: true
```

15.4.20 Start-Input-Timers

The Start-Input-Timers header field may appear in the VERIFY request to tell the speaker
verification resource whether or not to start the No-Input-Timeout timer. See the earlier discussion
in Section 15.2.6 for more information). The default value is true.
  Example:

```
Start-Input-Timers: false
```

15.5 Summary

In this chapter, we take a close look at the speaker verification resource. The speaker verification
resource operates in one of two modes: training or verification. The training mode is used to perform

training of a voiceprint by capturing one or more examples of a user speaking an utterance. The verification mode is concerned with matching audio from a speaker with one or more voiceprints. Matching against more than one voiceprint enables the MRCP client to perform speaker identification. The speaker verification resource can operate directly on live audio received over the media stream or on previously recorded audio stored in the verification buffer. The speaker verification, speech recognition and recorder resources can all write to the verification buffer.

This chapter concludes Part IV of the book, which provides an in-depth look at the four MRCP resources. Part V, deals with programming speech applications through VoiceXML and presents some examples of mapping VoiceXML to MRCP.

Part V

Programming Speech Applications

16

Voice Extensible Markup Language (VoiceXML)

Voice Extensible Markup Language (VoiceXML) [4, 5] is a markup-based, declarative, programming language for creating speech-based telephony applications. VoiceXML supports dialogs that feature synthesised speech, digitised audio, recognition of spoken and DTMF key input, recording of audio, mixed initiative conversations, and basic telephony call control. In particular, VoiceXML enables Web-based development and content delivery paradigms to be used with interactive voice response applications. VoiceXML platforms often employ MRCP to provide the underlying speech processing functions required by the language. This chapter provides an introduction to the VoiceXML language. In the next chapter, we discuss the details of how VoiceXML and MRCP work together.

16.1 Introduction

AT&T, IBM, Lucent and Motorola created VoiceXML as part of a joint effort and released version 1.0 of the specification in March 2000 under the auspices of the VoiceXML Forum – an industry body focused on promoting the adoption of VoiceXML technology worldwide. This version was later submitted to the W3C, which has subsequently taken on the responsibility of developing VoiceXML into a series of W3C Recommendations, notably with the publication of the 2.0 and 2.1 versions [4, 5]. In a nutshell, VoiceXML:

- delivers an easy to use language that also permits the creation of complex dialogs;
- promotes application portability across platforms by providing a common language for platform, tool and application providers alike;
- shields application developers from low-level, platform-specific, details;
- separates user interaction code (i.e. VoiceXML) from service logic (e.g. PHP, Java), and
- reuses existing web back-end infrastructure.

The standard VoiceXML architectural model is illustrated in Figure 16.1. The web server, together with the application server, processes HTTP requests issued from the VoiceXML interpreter for VoiceXML documents and other resources such as grammar files and audio files. The application logic

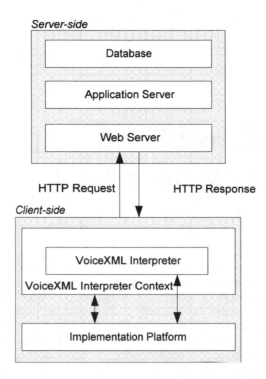

Figure 16.1 VoiceXML architectural model.

resides on the application server implemented in a programming or scripting language such as PHP, ASP .NET, or Java. The application server can dynamically create VoiceXML documents and other resources 'on-the-fly' based on business rules. The web server 'front-ends' the application server and is used to serve up static content. This architecture facilitates a clean separation of the application from the implementation platform.

The VoiceXML interpreter context provides management functions outside of the running VoiceXML interpreter. For example, the VoiceXML interpreter context is responsible for detecting inbound calls, acquiring the initial VoiceXML document and answering the call, while the VoiceXML interpreter conducts the dialog after the answer. The VoiceXML interpreter context also handles events not processed by the VoiceXML document, e.g. an uncaught error event will propagate to the VoiceXML interpreter context resulting in a standard error message and termination of the call. The VoiceXML interpreter context and the VoiceXML interpreter control the implementation platform, which provides services such as speech synthesis, speech and DTMF recognition, audio recording, and call control. An MRCP client can be used within the implementation platform to provide the requisite media processing in a vendor independent, network-distributed fashion.

16.2 Document structure

The basic structure of a VoiceXML document is illustrated below:

```
<?xml version="1.0" encoding="UTF-8"?>
<vxml version="2.1" xmlns="http://www.w3.org/2001/vxml"
```

```
      xml:lang="en-IE">
   <form>
        <block>Welcome to the world of VoiceXML!</block>
   </form>
</vxml>
```

All VoiceXML documents include the root element `<vxml>`. The `version` attribute indicates the version of VoiceXML (the examples in this chapter use Version 2.1). The default namespace for VoiceXML (indicated by the `xmlns` attribute) is defined as `http://www.w3.org/2001/vxml`. The `xml:lang` attribute specifies the primary language identifier. The language identifier can also be specified on the `<prompt>` and `<grammar>` elements to override the language for individual prompts and recognition tasks. The `<form>` element represents a dialog that performs a series of prompt and collect tasks – soliciting the user for information and collecting that information. In the example above, the minimalist form contains a block of executable code indicated by the `<block>` element. This VoiceXML application simply synthesises the prompt 'Welcome to the world of VoiceXML!' before exiting.

VoiceXML relies on ECMAScript[1] [57] for scripting, allowing logical computations to be performed 'client-side', that is, running within the VoiceXML interpreter. Client-side processing also reduces the dependency on network chatter between the VoiceXML interpreter and the Web server: computations required local to the voice user interface can be performed within the VoiceXML dialog and only finalised results need be returned server-side.

16.2.1 Applications and dialogs

Core to VoiceXML are the concepts of *sessions*, *applications* and *dialogs*. A session begins when the user commences interaction with the interpreter context, continues as documents are loaded, executed and unloaded, and terminates when requested to either by the user (e.g. a hang-up), the document (e.g. through an explicit `<exit>`), or by the interpreter context (e.g. because of an unhandled error).

An application is a set of documents that share the same application root document. The application root document remains loaded as long as its URI is specified in the `application` attribute on the `<vxml>` element and becomes unloaded when a transition to a document that does not specify this application root document occurs. While it is loaded, the application root document is used for storing data during the lifetime of the application. The application root document can also specify grammars that will be active throughout the lifetime of the application (e.g. to ensure that a menu option is always available to the user). Figure 16.2 illustrates the concept.

In this example, transitions are made using the `<goto>` element from Doc.1 through to Doc. 3. Each document references the same application root document and hence it remains loaded for the duration of the session. Input received in Doc. 1 could, for example, be stored in the application root document and later retrieved by Doc. 2.

A VoiceXML document contains one or more dialogs. There are two kinds of dialog: forms and menus. A form is the more powerful construct of the two and is represented by the element `<form>` as we have already seen. Forms define the interactions that collect input for its set of form items. Menus, on the other hand, are a more simple but restrictive construct. Menus are represented by the element `<menu>` and contain a number of alternative choices (represented by `<choice>` elements) – a matched choice results in a corresponding transition to a new dialog.

---

[1] ECMAScript is commonly referred to as JavaScript

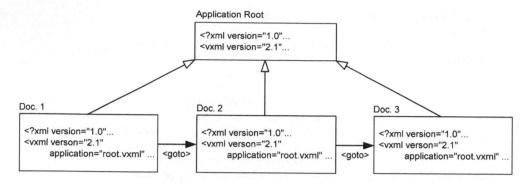

Figure 16.2 Document transitions within an application.

VoiceXML also supports the concept of subdialogs. A subdialog is invoked via the element `<subdialog>` and may be thought of as a kind of function call. A subdialog allows a new VoiceXML dialog to be invoked, during which time the calling context is suspended. When the invoked dialog returns, the original variable instances, grammar and state information are preserved. Subdialogs allow the developer to decompose the application into logical modules and, in particular, to create reusable libraries of dialogs that can be shared across different applications.

16.3 Dialogs

In this section, we take a closer look at the two dialog mechanisms defined in VoiceXML. Here we will focus most of the discussion around the form since the menu is really just syntactic shorthand for a form containing a single field (i.e. a menu just collects a single input from the user).

16.3.1 Forms

A form comprises the following:

- a set of *form items* that are sequentially visited;
- blocks of linearly executable code, called *filled actions*, which are triggered when certain combinations of form items are filled (represented by the `<filled>` element);
- blocks of linearly executable code called *event handlers* which are triggered when certain events occur (represented by the `<catch>` element);
- variable declarations scoped to the form (represented by the `<var>` element).

Form items are further subdivided into *input items* and *control items*. Input items can be filled by user input (i.e. speech or DTMF input) or the result of an action while control items cannot. Each form item has an associated ECMAScript form item variable which can be explicitly named through the `name` attribute on the corresponding element. Table 16.1 summarises the form items and their purpose.

Dialogs in VoiceXML are declarative in nature, that is, one defines the set of input items that can be filled, a set of grammars to match against, the prompts to play at different points, and a set of event handlers and filled actions. The Form Interpretation Algorithm (FIA) is at the core of the

Table 16.1 Form items and their purpose

Form item	Type	Purpose
`<field>`	Input item	Prompts and collects user information.
`<record>`	Input item	Records audio from the caller.
`<transfer>`	Input item	Transfers the caller to a new line.
`<subdialog>`	Input item	Invokes a new dialog and suspends the current one.
`<object>`	Input item	Accesses custom platform extensions.
`<block>`	Control item	Demarcates a set of linearly executable code.
`<initial>`	Control item	Controls the initial interaction in a mixed initiative form.

VoiceXML execution model. The FIA has a main loop that repeatedly selects a form item and visits it. The goal of the FIA is to drive the solicitation and collection of information from the user repeatedly and sequentially until all the input has been received. The main loop of the FIA consists of three phases:

- The *select* phase: This phase selects the next unfilled form item (that is, its form item variable is `undefined`) to visit.
- The *collect* phase: When the form item is visited, any corresponding prompts are played, the appropriate grammars are activated, and the form item waits for input from the user or for an event to occur.
- The *process* phase: The process phase involves filling the appropriate form item variables with information received during the collect phase and executing any corresponding filled actions. If an event was thrown in the previous step, it is processed in this step.

As a consequence of the FIA, at any given time the VoiceXML interpreter can be considered to be in one of two states:

- The *waiting* state: The interpreter is waiting for input from the user. The waiting state corresponds to the FIA's collect phase.
- The *transitioning* state: The interpreter is between input items, e.g. transitioning from one document to another via `<goto>`. The transitioning state corresponds to the FIA's process and select phases.

An important concept to understand in VoiceXML is the difference between prompt *queueing* and prompt *flushing*. While in the transitioning state, prompts are queued for playback at a later stage. The prompts are not flushed (i.e. played) until the interpreter reaches a waiting state, the execution of the document terminates (e.g. on encountering an `<exit>`), or a page transition is called for that specifies audio to play during the fetch (using the `fetchaudio` attribute).

At this point, it is instructive to consider a simple form and discuss how the mechanics of the FIA apply to it. This simple dialog might appear as part of a cinema booking service to ask the user which movie they would like to see and how many tickets they would like to purchase.

```
<?xml version="1.0" encoding="UTF-8"?>
<vxml version="2.1" xmlns="http://www.w3.org/2001/vxml">
    <form id="tickets">
        <block>Welcome to EasyTicket!</block>
```

```
        <field name="movie">
            <prompt>Which movie would you like tickets for?</prompt>
            <grammar src="movie.grxml"/>
        </field>
        <field name="number">
            <prompt>How many tickets would you like?</prompt>
            <grammar src="number.grxml"/>
        </field>
        <filled namelist="movie number" mode="all">
            <submit next="book.vxml" namelist="movie number"/>
        </filled>
        <catch event="noinput">
            Sorry, I did not hear you. Please try again.
        </catch>
        <catch event="nomatch">
            Sorry, I did not understand. Please try again.
        </catch>
    </form>
</vxml>
```

On entering the form, the select phase picks the block since its hidden form item variable is not set (to access the block's form item variable, a `name` attribute can be added to give it an explicit name). The form item variable is automatically set to `true` just before the block is entered and hence a block is typically only executed once. The block is executed during the collect phase – in this example the block simply queues a prompt (the prompt is not yet flushed since the interpreter is still in the transitioning state). On the FIA's second iteration, the block is skipped and the first field is selected since its form item variable, `movie`, is not filled (in ECMAScript terms, the variable evaluates to `undefined`). The collect phase activates the field's grammar, and simultaneously starts the playback of the queued prompts and recognition. The interpreter is now in the waiting state. When the prompt and recognition action completes, the process phase is entered. If the user matches the grammar, the form item variable is filled. The FIA then loops, selecting the next unfilled field – in this case the second field. The process phase for this iteration will execute the filled section since the condition for its execution are met – namely the variables specified in its `namelist` attribute are filled. The `mode` value of `all` (the default) means that the filled section is only invoked when all of the `namelist` variables are filled. In contrast, if the `mode` value was set to `any`, the filled section would have been invoked if any of the variables in the `namelist` were filled. The filled section in this example submits the values obtained from the form to the Web server, which can store this data in a database and use it to determine the next dialog to perform. A `<submit>` element will result in a new VoiceXML page being returned and the execution continues in the new document.

If the recogniser detects no input during the collect phase, a `noinput` event is thrown and the process phase invokes the first catch handler, which simply plays a prompt. The FIA loops around again and selects the first unfilled form item. Similarly, if the recogniser detected input but did not recognise it, a `nomatch` event is thrown and caught by the second catch handler. Note that in both these cases, the selected field's prompt is not repeated – this is the default behaviour. To force a repeat of the field's prompt, a `<reprompt>` element can be added to the end of the catch handler. Below is a sample transcript for the dialog:

Platform: Welcome to EasyTicket! Which movie would you like tickets for?
Caller: *Vampire Movie II.*

Platform: How many tickets would you like?
Caller: <silence>
Platform: Sorry, I did not hear you. Please try again.
Caller: Four.

Note that if there is repeated silence, a caller may soon forget what the original question being asked of them is. We can rectify this situation by adding a `<reprompt>` element to the catch handler for noinput, i.e.

```
<catch event="noinput">
      Sorry, I did not hear you. Please try again.
      <reprompt/>
</catch>
```

In this case, silence from the user would result in the prompt: 'Sorry, I did not hear you. Please try again. How many tickets would you like?'

VoiceXML supports the concept of a prompt counter to permit greater control within repeated prompts. For example, an application developer may wish to insert some variability in repeated prompts. Another common strategy is to use tapered prompts, whereby subsequent repetitions are refined to be more specific and terse. To illustrate the mechanism, we can rewrite the first field in the preceding example thus:

```
<field name="movie">
    <prompt count="1">
        Which movie would you like tickets for?
    </prompt>
    <prompt count="2">
        Please choose from the following movies: Vampire Movie II,
        A Day in the Life, or My Favourite Story.
    </prompt>
    <grammar src="movie.grxml"/>
</field>
```

An example of a transcript using this feature (and assuming the `<reprompt>` element is used in the noinput catch handler) is:

Platform: Welcome to EasyTicket! Which movie would you like tickets for?
Caller: <silence>
Platform: Sorry, I did not hear you. Please try again. Please choose from the following movies: *Vampire Movie II*, *A Day in the Life*, or *My Favourite Story*.
Caller: *A Day in the Life*.
Platform: How many tickets would you like?
Caller: Five.

16.3.2 Menus

The VoiceXML menu is a simpler alternative to the form for specifying a dialog and, as a consequence, is more restrictive. A menu is composed of a number of choices, each represented by the `<choice>`

element. The `<choice>` element contains a word or phrase that is used to prompt the user for that choice and also automatically to generate a grammar for the menu (the word or phrase contained in a `<choice>` element becomes an alternative in a simple grammar). The next attribute on `<choice>` specifies the URI of a dialog that is transitioned to when the choice is matched. A `<menu>` element usually contains a `<prompt>` element to prompt the user of their choices. The `<enumerate>` element works within `<menu>` to list the choices automatically. An example of a simple menu is:

```
<?xml version="1.0" encoding="UTF-8"?>
<vxml version="2.1" xmlns="http://www.w3.org/2001/vxml">
    <menu accept="approximate">
        <prompt>
            Welcome to Cinema One. Please say one of: <enumerate/>
        </prompt>
        <choice next="showtimes.vxml">
            Show times
        </choice>
        <choice name="easyticket.vxml">
            Ticket bookings
        </choice>
        <catch event="noinput">
            Sorry, I did not hear you.
            Please say one of: <enumerate/>
        </catch>
        <catch event="nomatch">
            Sorry, I did not understand.
            Please say one of: <enumerate/>
        </catch>
    </menu>
</vxml>
```

In this example, the initial prompt will say: 'Welcome to Cinema One. Please say one of show times, ticket bookings'. The accept attribute on `<menu>` specifies a value of approximate, which indicates that an approximate recognition phrase can be used. For example, the user can say 'show times', 'show', or 'times' to select the first choice or 'ticket bookings', 'ticket', or 'bookings' to select the second. If the accept attribute value was exact (the default), only one of 'show times' or 'ticket bookings' can be matched. The exact algorithm used to generate a grammar with approximate phrases is platform specific but is usually based on a subset of the choice phrases. For more refined control, application developers are recommended to use the `<form>` with explicit grammars.

16.3.3 Mixed initiative dialogs

The dialogs illustrated thus far fall into the category of *machine directed* dialogs. In a machine directed dialog, the computer determines the interaction and form items are typically executed in a rigid sequential order. In contrast, a *mixed initiative* dialog is one where both the machine and human alternate in directing the dialog. The key to creating a mixed initiative dialog in VoiceXML is to specify a form-level grammar, i.e. a `<grammar>` element whose immediate parent is a `<form>` element. When a form-level grammar is active:

- input items can be matched in any order;
- more than one input item can be filled from a single utterance.

The `<initial>` element is used in mixed initiative dialogs to solicit general, form-wide information from the user. The `<initial>` element does not contain any grammars or filled actions because its purpose is to prompt the user to say information that will fill one or more of the input items, such as fields within the form. The `<initial>` element continues to be visited until a successful recognition occurs, after which the FIA continues by visiting the next unfilled form item. In this way, the mixed initiative dialog robustly 'falls back' to directed dialog as the FIA endeavours to get the remaining necessary input from the user. Let us consider an example of a mixed initiative dialog:

```
<?xml version="1.0" encoding="UTF-8"?>
<vxml version="2.1" xmlns="http://www.w3.org/2001/vxml">
   <form id="tickets">
       <grammar src="cities.grxml"/>
       <initial name="init">
           Let's start the ticket booking process.
           Where would you like to fly from and to?
           <catch event="nomatch" count="1">
               I did not catch that. Please say something like:
               I would like to fly from Dublin to Paris.
           </catch>
           <catch event="nomatch" count="2">
               Sorry, I still do not understand. Let's try by asking
               for one piece of information at a time.
               <assign name="init" expr="true"/>
           </catch>
       </initial>
       <field name="origin">
           <prompt>Which city are you flying from?</prompt>
       </field>
       <field name="dest">
           <prompt>Which city are you flying to?</prompt>
       </field>
       <filled namelist="origin dest" mode="all">
           Checking availability...
           <submit next="check_avail.vxml" namelist="origin dest"
                   fetchaudio="ticktock.wav"/>
       </filled>
   </form>
</vxml>
```

In this example, the user is given the opportunity to fill the two fields origin and dest in one go. If a nomatch occurs, the user is given a second chance to speak all the information. The count attribute on `<catch>` is used to indicate which catch handler to use depending on how many times the event has been thrown – see Section 16.7.4 for more information. After a second chance fails, the `<initial>`'s form item variable is manually set so that it will not be visited. The FIA continues in machine directed fashion by selecting the next unfilled form item, i.e. the `origin` field. A transcription of an efficient exchange is:

> *Platform*: Let's start the ticket booking process.
> Where would you like to fly from and to?
> *Caller*: I would like to fly from Berlin to Rome
> *Platform*: Checking availability...

Now consider a transcription of a less efficient exchange, where the platform receives only the origin city and then reverts to a machine directed dialog:

> *Platform*: Let's start the ticket booking process.
> Where would you like to fly from and to?
> *Caller*: I would like to fly from Berlin
> *Platform*: Which city are you flying to?
> *Caller*: Rome
> *Platform*: Checking availability...

16.4 Media playback

VoiceXML supports media playback through the use of SSML (see Chapter 8). SSML allows application authors to use both prerecorded audio and text-to-speech in their applications. The VoiceXML element `<prompt>` essentially plays the role of the SSML `<speak>` element as a container for SSML (note that the SSML `<speak>` element itself is not allowed in VoiceXML). The `<prompt>` element is legal inside any form item, executable content or `<menu>`. VoiceXML also allows SSML to appear in the `<enumerate>` element, which is legal inside the same elements that `<prompt>` is. As a developer convenience, the `<audio>` element, `<value>` element and plain text to be synthesised may appear outside of `<prompt>` and `<enumerate>` in several other locations as indicated in Table 16.2.

Table 16.2 Container elements for audio markup in VoiceXML

Audio markup type	Container elements
SSML elements	`<prompt>`, `<enumerate>`[a]

Restricted markup (`<audio>`, `<value>`, and plain text)	Form items:
	　`<field>`, `<record>`, `<transfer>`,
	　`<subdialog>`, `<object>`,`<block>`,`<initial>`
	Executable content:
	　`<catch>`, `<error>`, `<help>`, `<noinput>`,
	　`<nomatch>`, `<if>`, `<filled>`
	Other:
	　`<menu>`

[a] The SSML elements `<lexicon>`, `<meta>`, and `<metadata>` are not allowed within `<enumerate>`

VoiceXML goes beyond SSML by introducing the `<value>` element for synthesising ECMAScript variable values. The following example shows how the `<value>` element can be used to synthesise a variable's value:

```
<?xml version="1.0" encoding="UTF-8"?>
<vxml version="2.1" xmlns="http://www.w3.org/2001/vxml">
    <form>
        <field name="digit">
            <grammar src="numbers.grxml"/>
            <prompt>Say a number!</prompt>
            <filled>
                I heard <value expr="digit"/>
            </filled>
        </field>
    </form>
</vxml>
```

VoiceXML also adds the `expr` attribute to the `<audio>` element. This attribute is used for two purposes. The first performs the same function of the `src` attribute to specify a URI but allows an ECMAScript variable to be specified. The second use of `expr` is to playback a reference to audio that was previously recorded (described in Section 16.5).

The `<prompt>` element has several attributes that permit the application author to exercise fine-grained control of media playback. The `bargein` attribute takes a Boolean value and determines whether or not the user can interrupt the prompt. The `bargeintype` property can be set to either `speech` or `hotword`. If set to `speech` (the default), the prompt will be stopped as soon as speech or DTMF input is detected. If set to `hotword`, the prompt will not be stopped until a complete match of an active grammar is detected – input that does not match a grammar is simply ignored and hence the `nomatch` event cannot be generated. The `timeout` attribute specifies the amount of time in milliseconds or seconds after the prompt has finished playing that the platform will wait for input before a `noinput` event is thrown. The language of the prompt can be specified with the `xml:lang` attribute. Finally, the `cond` attribute, which evaluates to an ECMAScript Boolean expression, can be used to deactivate a prompt dynamically (if the `cond` attribute expression evaluates to `false`, the `<prompt>` element is ignored).

16.5 Media recording

Media recording in VoiceXML is performed via the input item `<record>`. A reference to the recorded media is stored in the corresponding form item variable and can either be immediately played back to the user (using the `expr` attribute on `<audio>`) or submitted to the web server (via the `<submit>` element) for server-side storage. Below is an example that records a message, plays it back to the user, and then submits the recording to the web server if the user is satisfied with their recording.

```
<?xml version="1.0" encoding="UTF-8"?>
<vxml version="2.1" xmlns="http://www.w3.org/2001/vxml">
    <form>
        <record name="greeting" beep="true" dtmfterm="true"
                maxtime="7s" finalsilence="3000ms"
```

```
            type="audio/x-wav">
        <prompt>
            Please speak your greeting after the beep
        </prompt>
    </record>
    <field name="confirm">
        <grammar src="yesno.grxml"/>
        <prompt>
            Your greeting is <audio expr="greeting"/>. Would
            you like to keep it?
        </prompt>
        <filled>
            <if cond="confirm == true">
                <submit next="store.php" namelist="greeting"
                        enctype="multipart/form-data"
                        method="post"/>
            <else/>
                <clear/>
            </if>
        </filled>
    </field>
</form>
</vxml>
```

The name attribute on the <record> element specifies the form item variable to store the reference
to the audio. The beep attribute with a value of true requests that the implementation platform
play a beep sound just before recording commences (recording starts after the prompt contained in the
<record> element is played). The dtmfterm attribute indicates whether a DTMF key press will
terminate the recording. The maxtime sets an upper bound on the length of the recording – in this
case 7 seconds. The finalsilence attribute indicates the amount of silence that will automatically
terminate the recording. The type attribute specifies the format for the recording – in this case a
WAV file with a RIFF header (see Table 8.3 for a list of common audio formats and their associated
media types).

Once the record input item is filled, the FIA will select the field and process it. The prompt will
playback the recording via the <audio> element. The user then responds with whether or not they
want to keep the message. If they respond affirmatively, a value of true is assigned to the field's form
item variable. The filled action tests the field's form item variable and if true, submits the recording
via HTTP POST to the Web server (the <submit> element is described in Section 16.8). Otherwise,
the <clear> element has the affect of resetting all the form item variables to undefined and will
result in the FIA starting again by selecting and processing the record input item.

After a recording is made, shadow variables are populated with meta information for the
recording – see Table 16.3.

In some applications, it is desirable to allow the user simply to hang up when they finish
recording their message. This use-case is easily implemented in VoiceXML by leveraging the *final
processing state*. When the user hangs up, the connection.disconnect.hangup event is
thrown. A <catch> element for this event can be inserted, which subsequently <submit>s the audio
recording to the web server (the recording up to the point of hangup is maintained in the <record>

Table 16.3 Shadow variables for a record input item named `greeting`

Shadow variable	Description
`$greeting.duration`	The duration of the recording in milliseconds.
`$greeting.size`	The size of the recording in bytes.
`$greeting.termchar`	This variable contains the DTMF key pressed assuming the `dtmfterm` attribute was set to `true` and the recording was terminated by a key press.
`$greeting.maxtime`	This variable is `true` if the recording terminated because the maximum time (specified in the `maxtime` attribute) was reached.

element's form item variable). To handle the hangup case in the previous example, we need only add the following statement to the `<form>`:

```
<catch event="connection.disconnect.hangup">
    <submit next="store.php" namelist="greeting"
            enctype="multipart/form-data"
            method="post"/>
</catch>
```

16.6 Speech and DTMF recognition

User input, in the form of speech or DTMF key presses, is accepted during the collect phase of the FIA when a `<menu>` is executed or when any of the following form items is executed: `<field>`, `<initial>`, `<record>`, `<transfer>`. In this section, we take a closer look at recognition in VoiceXML.

16.6.1 Specifying grammars

Grammars are used within VoiceXML to constrain the permitted range of words and phrases or DTMF key sequences that may be inputted at a given point in the dialog. All VoiceXML platforms are required to support the XML form of the SRGS format and are recommended to support the ABNF form (see Chapter 9). Within VoiceXML, grammars are specified through the `<grammar>` element. A grammar may either be external or inline. External grammars are referenced through a URI via the `src` attribute on `<grammar>`, as we have already seen with previous examples. Inline grammars, on the other hand, are specified as child content of the `<grammar>` element itself as the following example illustrates:

```
<?xml version="1.0" encoding="UTF-8"?>
<vxml version="2.1" xmlns="http://www.w3.org/2001/vxml">
    <form>
        <field name="show">
            <prompt>
                Please select a department or say operator to
                speak to a member of staff.
```

```
            </prompt>
            <grammar version="1.0" xml:lang="en-IE" root="dept">
                <rule id="dept">
                    <one-of>
                        <item>sales</item>
                        <item>support</item>
                        <item>operator</item>
                    </one-of>
                </rule>
            </grammar>
        </field>
    </form>
</vxml>
```

When an utterance matches a grammar in VoiceXML, a corresponding semantic result is returned in the form item variable; that result may be a simple value or a more complex nested structure.

VoiceXML also has the concept of built-in grammars for common tasks such as recognising digits, dates, times, etc. A built-in grammar can be specified using the `type` attribute on `<field>` without requiring a `<grammar>` element to be specified e.g. `<field name="x" type="digits">`. A drawback of using built-in grammars is that the grammar definition can vary significantly between different platforms and hence adversely affects portability of the application. Use of explicit SRGS grammars instead of builtin grammars is recommended.

16.6.2 Grammar scope and activation

A grammar's scope dictates when it is activated. The scope depends on where the `<grammar>` element is located and on its `scope` attribute:

- *Input item grammars*: Grammars contained within an input item are scoped to that input item only.
- *Form and menu grammars*: Grammars contained as direct descendents of a `<form>` or `<menu>` element are by default given dialog scope – they are active while the dialog is being visited. The grammars can be given document scope by setting the `scope` attribute on the `<form>` element to `document`, or the `scope` attribute on the `<grammar>` element to `document`.
- *Link grammars*: Link grammars are given the scope of the element that contains them. A `<link>` may be a child of `<vxml>`, `<form>`, or the form items `<field>` and `<initial>` (see Section 16.7.3.1 for more information).

When the VoiceXML interpreter waits for input during the collect phase of the FIA, grammars are activated in the following order:[2]

  (i) grammars associated with the current input item;
 (ii) form-level grammars and links associated with the current form;

---

[2] In the event that the input matches more than one grammar, the list defines the precedence order and document order determines precedence if two grammars at the same scope are matched.

(iii) links, forms and menus with document scope;
(iv) links, forms and menus with document scope in the application root;
 (v) platform grammars, if any.

We have already seen the actions of matching input items and form-level grammars. Matching a document scoped grammar associated with a form or menu will result in a transition to that dialog. Matching a grammar associated with a link will trigger the link's transition (see Section 16.7.3.1). Note that the <field> and <record> elements also have a modal attribute, which takes a Boolean value and can be used to restrict the activated grammars just to those contained within the corresponding element while it is executing.

16.6.3 Configuring recognition settings

Recognition settings are configured through the use of VoiceXML properties specified via the <property> element. The <property> element has two attributes: name and value. The <property> element may be specified at the application level (child of <vxml> in the application root), at the document level (child of <vxml> of the executing document), at the dialog level (child of <form> or <menu>), or at the form item level (child of the element representing the form item). Properties apply to their parent element and all their descendents. Properties values defined at a lower level override those set at the parent level. For example, speech recognition can be disabled within a form through the use of the inputmodes property:

```
<?xml version="1.0" encoding="UTF-8"?>
<vxml version="2.1" xmlns="http://www.w3.org/2001/vxml">
    <form>
        <property name="inputmodes" value="dtmf"/>
        <field name="password">
            <grammar src="digits"/>
            <prompt>
                Please enter your four digit pin number.
            </prompt>
        </field>
    </form>
</vxml>
```

Table 16.4 summarises the properties for tuning recognition.

16.6.4 Processing recognition results

In this section, we delve a little more into the details of how recognition results are handled in VoiceXML. VoiceXML provides mapping rules that determine how semantic interpretation results generated from matching a grammar are mapped to form item variables (semantic interpretation is discussed in Section 9.8). Since the mapping rules differ slightly depending on whether the input matches a form-level grammar or a field grammar, we will discuss the two cases separately in the next two sections.

Table 16.4 Properties for tuning recognition in VoiceXML

Name	Description
inputmodes	Specifies which input modalities to use. Allowable values are dtmf and voice. This property is used to disable a modality. For example, in a noisy environment, one might disable speech input and restrict input to DTMF only. The default is 'dtmf voice'.
confidencelevel	A value between 0.0 and 1.0 that specifies a threshold for which recognition results below it are rejected (a nomatch event is thrown). The default is 0.5.
sensitivity	A value between 0.0 and 1.0 that specifies the sensitivity level for speech input. A value of 1.0 means the platform is highly sensitive to quiet input. Conversely, a value of 0.0 means the platform is least sensitive to noise. The default is 0.5.
speedvsaccuracy	A value between 0.0 and 1.0 that specifies the desired balance between speed and accuracy. A value of 0.0 means fastest recognition while a value of 1.0 means best accuracy. The default is 0.5.
universals	A platform may optionally provide platform-specific universal command grammars such as 'help', 'exit', and 'cancel' that are always active. When a command grammar is matched, an associated event is thrown, e.g. the help event is thrown if the help command grammar is matched. The property value is a space delimited list of names, e.g. 'help exit'. The default is 'none'.
maxnbest	Controls the maximum number of recognition results returned in the application.lastresult$ array (see Section 16.6.4.3). The default is 1.
completetimeout	Specifies the length of silence in milliseconds or seconds[a] after user speech before the recogniser finalises a result (either accepting it or throwing a nomatch). The completetimeout is used when the speech is a complete match of the grammar. A long completetimeout value delays result of the recognition completion and makes the platform appear slow. A short completetimeout can result in the recogniser finalising a result before the user has finished talking. The default is platform independent but typically in the range of 0.3 seconds to 1.0 seconds.
incompletetimeout	Similar to the completetimout property except that the incompletetimeout value applies when the speech is an incomplete match of the grammar (e.g. only a subset of a minimum sequence of tokens are matched). Usually, the incompletetimeout value is longer than the completetimeout to allow the user to pause mid-utterance, e.g. to breathe.
maxspeechtimeout	This property specifies the maximum duration of speech in milliseconds or seconds. If the user speaks for longer than this time, the maxspeechtimeout event is thrown.
interdigittimeout	Specifies the length of time between DTMF key presses in milliseconds or seconds before the recogniser finalises a result (either accepting it or throwing a nomatch). The interdigittimeout applies when additional input is allowed by the grammar. The default is platform dependent but typically in the region of 3 seconds.
termtimeout	Similar to the interdigittimeout except that it applies instead when no more input can be accepted by the grammar. The default is 0s.
termchar	Specifies the terminating character of DTMF (analogous to the 'return key' on a computer keyboard). The default is '#'.

bargein	Sets the default barge-in behaviour (barge-in means that a user can provide speech or DTMF during prompt playback with the effect of stopping the playback). The default is `true`.
bargeintype	Sets the default barge-in type to `speech` or `hotword`. Hotword recognition implies that prompts are not stopped until a match occurs and input that does not match a grammar is ignored (i.e. a `nomatch` is never thrown). The default is `speech`.
timeout	Specifies the time in seconds or millliseconds to wait for input after which the `noinput` event is thrown. The default is platform specific but typically about 3 seconds.

[a] VoiceXML employs a consistent time designation for property values that indicate durations. The units of time are indicated by `ms` or `s` for milliseconds or seconds respectively.

16.6.4.1 Mapping field-level results

A grammar specified within an input item (e.g. a `<field>`) will produce a field-level result – matching this grammar can only fill the associated form item variable item. This is useful for directed dialogs when individual pieces of information are being solicited from the user. The simplest case occurs for grammars that do not return a semantic interpretation result, that is, the grammar does not contain semantic interpretation tags. In this case, the raw utterance is taken as the result and will be mapped directly into the form item variable as an ECMAScript string.

If the grammar contains semantic interpretation tags and returns a simple result (i.e. the rule variable associated with the root rule is an ECMAScript string, number, or Boolean type after all rules have been matched), then that result is mapped to the corresponding form item variable and the type is preserved. If the grammar returns an ECMAScript object for the semantic result (i.e. the rule variable associated with the root rule is an ECMAScript object after all rules have been matched), then the mapping rules depend on whether the input item's slot name matches a property of the ECMAScript object or not. The slot name is the value of the `slot` attribute (possible on `<field>` only) or otherwise is the value of the `name` attribute. If a property of the returned ECMAScript object matches the slot name, then that property's value is assigned to the form item variable. Otherwise, the entire object is assigned to the form item variable. For example, consider the ECMAScript object with two properties:

```
{
    size: "large",
    sort: 2
}
```

If a field has a slot name called `sort`, e.g. the field is defined as

```
<field name="sort">
```

or

```
<field name="anything" slot="sort">
```

then the form item variable will be assigned the number 2. However, if the slot name is something other than size or sort, e.g.

```
<field name="anything">
```

then the form item variable will be assigned the entire ECMAScript object. In this case `anything.size` would evaluate to the string 'large' and `anything.sort` would evaluate to the number 2.

16.6.4.2 Mapping form-level results

A grammar specified as a direct descendent of `<form>` will produce a form-level result – the recognition result can fill multiple input items simultaneously. This is useful for mixed initiative dialogs when multiple pieces of information are being solicited from the user, as we have already seen. Form-level grammars must return an ECMAScript object result or otherwise it is not possible to fill any input items. The mapping rules for a form-level grammar are the same as for a field-level grammar that returns an ECMAScript object except that only those input items whose slot value matches a property of the returned ECMAScript object will be filled. If there are no matching slot values, the FIA simple iterates. Below is an example of a mixed initiative dialog.

```
<?xml version="1.0" encoding="UTF-8"?>
<vxml version="2.1" xmlns="http://www.w3.org/2001/vxml">
    <form>
        <grammar src="sizeandsort.grxml"/>
        <initial>
            <prompt>
                Please say the size and sort of pizza you want
            </prompt>
        </initial>
        <field name="size">
            <prompt>Please say the size of pizza you want</prompt>
        </field>
        <field name="sort">
            <prompt>Please say the sort of pizza you want</prompt>
        </field>
        <filled>
            <submit next="placeorder.php" namelist="size sort"/>
        </filled>
    </form>
</vxml>
```

Imagine the user says 'I want a small pepperoni pizza' which matches the form-level grammar and returns the following ECMAScript object (such a grammar was presented in Section 9.8.2):

```
{
    size: "small",
    sort: 1
}
```

The two properties of this object match the slot names of the two input items in the form and hence those input items can be filled simultaneously.

16.6.4.3 Shadow variables

Each successfully filled input item also has an associated set of *shadow variables*, which convey more information about the recognition result. A shadow variable is referenced using the scheme `$name.shadowvar` where `name` is the name of the form item variable, and `shadowvar` is the name of the shadow variable. For example, the confidence of the recognition can be accessed for a field whose name is `size` through the variable `$size.confidence`. The defined shadow variables for a field called `example` are given in Table 16.5.

VoiceXML also defines the `application.lastresult$` array that holds information about the last recognition result to occur within the application. Each element of this array `application.lastresult$[i]` has the same properties as defined for the shadow variables, e.g. `application.lastresult$[i].utterance`. The interpretations are ordered based on the confidence property, from highest to lowest. The maximum number of array entries is governed by the `maxnbest` property (defined in Table 16.4). This array allows the application author to access the 'N-best' recognition results and not just simply the 'best' one which was automatically used to fill the form item variable. As a convenience, `application.lastresult$` evaluates to the first element,

Table 16.5 Shadow variables for a field called `example`

Shadow variable	Description
`$example.utterance`	The raw string of words that was recognised.
`$example.inputmode`	The mode used to provide the input, either `dtmf` or `voice`.
`$example.interpretation`	An ECMAScript variable containing the interpretation (same as the form item variable).
`$example.confidence`	The confidence level in the range of 0.0 to 1.0 where 0.0 indicates minimum confidence and 1.0 indicates maximum confidence.
`$example.markname`	The name of the last `<mark>` element executed by the SSML processor before barge-in occurred or the end of audio playback occurred. If no `<mark>` was executed, this variable is undefined.
`$example.marktime`	The number of milliseconds that elapsed since the last mark was executed by the SSML processor until barge-in occurred or the end of audio playback occurred. If no `<mark>` was executed, this variable is undefined.
`$example.recording`	Stores a reference to a recording of the utterance. Utterances during recognition are only recorded if the VoiceXML property `recordutterance` is set to `true`.
`$example.recordingsize`	The size of the recorded utterance in bytes or undefined if recording of utterances during recognition is not enabled.
`$example.recordingduration`	The duration of the recorded utterance in milliseconds or undefined if recording of utterances during recognition is not enabled.

i.e. `application.lastresult$[0]`. Note that immediately after a recognition is performed, the shadow variables associated with an input item (i.e. as described in Table 16.5) coincide exactly with the properties of `application.lastresult$`.

16.7 Flow control

16.7.1 Executable content

In VoiceXML, executable content represents a block of procedural logic. That is, each XML element within the executable content can be likened to a programmatic statement in a conventional programming language and interpretation of statements occurs in document order. Executable content is contained in specific locations in a VoiceXML document, namely:

- within `<block>` control items;
- within `<filled>` actions, and
- within event handlers (`<catch>`, `<noinput>`, `<nomatch>`, `<help>`, `<error>`).

A VoiceXML document is thus part declarative and part procedural. The application author writes a VoiceXML document with one or more declarative dialogs in the structure of a form or menu, populates the dialog with a number of form items, and specifies grammars and prompts that are activated at various stages of the dialog. The FIA then executes the dialog by interpreting this declarative structure, selecting different form items, and executing them. During the execution, the FIA will periodically execute `<block>`s, `<filled>` actions and event handlers. The contained executable content within theses container elements is executed linearly and the FIA only continues when the block of executable content has ran to completion. Table 16.6 enumerates the elements that comprise executable content and their corresponding purpose.

16.7.2 Variables, scopes and expressions

16.7.2.1 Working with variables

VoiceXML's data model is built upon ECMAScript. Each form item has an associated form item variable that can be explicitly named via the `name` attribute on that form item. VoiceXML also allows variables to be declared using the `<var>` element. The `<var>` element has two attributes. The required `name` attribute is the name for the variable. The optional `expr` attribute can be used to initialise the variable to a given value (if `expr` is omitted, the variable is initialised to ECMAScript `undefined`). For example, the following VoiceXML code will initialise the variable `x` to `undefined` and the variable `y` to 6:

```
<var name="x"/>
<var name="y" expr="1+2+3"/>
```

These elements are equivalent to the ECMAScript expressions:

```
var x;
var y = 1+2+3;
```

Once declared, a variable may be assigned to using the `<assign>` element. For example, to assign a string value of 'hello' to `x`, one may write:

Table 16.6 Executable content

Element	Purpose
`<assign>`	Assigns a value to a variable.
`<clear>`	Clears one or more variables.
`<data>`	Sends data to and fetches arbitrary XML from a web server without transitioning to a new VoiceXML document.
`<disconnect>`	Disconnects the session and results in the `connection.disconnect.hangup` event being generated.
`<exit>`	Exits the session.
`<foreach>`	Iterates through an ECMAScript array and executes the content contained within the `<foreach>` element for each item in the array.
`<goto>`	Transition to another dialog in the same or different document.
`<if>`, `<elseif>`, `<else>`	Provides basic conditional logic.
`<log>`	Specifies logging information.
`<prompt>`, `<enumerate>`, and restricted audio markup (`<audio>`, `<value>`, and plain text).	Queue audio for playback.
`<reprompt>`	Play the prompt within a field when the field is re-visited after an event handler.
`<return>`	Return from a subdialog.
`<script>`	Specifies a block of ECMAScript code for client-side scripting.
`<submit>`	Submits data to a Web server.
`<var>`	Declares a variable.

```
<assign name="x" expr="'hello'"/>
```

Note that an `error.semantic` will be thrown if the application author attempts to assign to a variable that has not already been declared.[3] An ECMAScript variable may be cleared, i.e. set to `undefined`, via the `<clear>` element. For example, the variables x and y may be cleared using the statement:

```
<clear namelist="x y"/>
```

The `<clear>` element is usually used to reset one or more form item variables. If a `<clear>` element is executed without its `namelist` attribute, then all form items in the current form are reset.

---

[3] If this were not the case, ECMAScript rules dictate that the variables would be automatically declared in a global scope, thus breaking the partitioning of data into clearly defined scopes within VoiceXML.

VoiceXML provides the `<if>`, `<elseif>`, and `<else>` elements to implement conditional logic. The `cond` attribute contains the ECMAScript condition to test. These elements operate similarly to conventional programming languages and require little explanation. An example follows:

```
<if cond="x > 1000">
    <assign name="result" expr="'large'"/>
<elseif cond="x > 500"/>
    <assign name="result" expr="'medium'"/>
<else/>
    <assign name="result" expr="'small'"/>
</if>
```

16.7.2.2 Variable scopes

VoiceXML variables are declared at different levels of a predefined ECMAScript scope chain – illustrated in Figure 16.3 and described in Table 16.7. Variables in the current scope and upward are visible to that scope. For example, a script executing at the dialog scope can access variables declared at the dialog level through to the session level. Variable references automatically match the closest enclosing scope according to the scope chain in Figure 16.3. For example, if the following element is placed within a `<form>`:

```
<assign name="x" expr="y">
```

and there is a variable y declared at both the dialog level and document level, the value of the variable declared at the dialog level will be used because it is the closest scope. It is possible to indicate explicitly the scope alongside the variable – for example, if we wanted the document scoped variable y, we could have written:

```
<assign name="x" expr="document.y"/>
```

16.7.2.3 Session variables

VoiceXML defines a number of session variables that are read-only and available for the length of the session. Session variables provide information about the active session. Table 16.8 provides a description of each session variable.

16.7.3 Document and dialog transitions

VoiceXML provides three elements that enable explicit transitions between dialogs and documents to be specified: `<link>`, `<goto>` and `<submit>`. These elements are discussed in the following sub-sections.

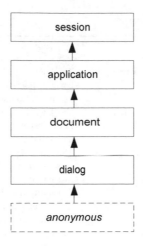

Figure 16.3 ECMAScript scopes in VoiceXML.

Table 16.7 Description of the different ECMAScript scopes in VoiceXML

Scope	Description
session	Read-only variables defined for the entire user session.
application	Variables declared by `<var>` and `<script>`[a] elements which are children of the application root document's `<vxml>` element.
document	Variables declared by `<var>` and `<script>` elements which are children of the executing document's `<vxml>` element.
dialog	Variables declared by `<var>` and `<script>` elements which are children of the executing `<form>` element. Form item variables and shadow variables of both `<form>` and `<menu>` are also declared at the dialog scope.
anonymous	Each `<block>`, `<filled>`, and `<catch>` element defines a new temporary anonymous scope to contain variables declared in that element.

[a] A variable is declared within `<script>` using the keyword var, e.g. var x = 1;.

16.7.3.1 Link

A `<link>` has one or more grammars that are scoped to the element containing the `<link>`. A `<link>` may be placed at the application level (when it is a child of `<vxml>` in the application root document), at the document level (when it is a child of `<vxml>` in the executing document), or at the form level (when it is a child of a `<form>`). When the grammar of a link is matched, the link is activated. Activating a link will cause one of two actions:

- a transition to a new document or dialog, or
- generation of an event.

Table 16.8 Standard session variables

Name	Description
`session.connection.local.uri`	The URI that identifies the local interpreter context. For a PSTN call, this is the dialled number (also called the DNIS).
`session.connection.remote.uri`	The URI that identifies the remote caller device. For a PSTN call, this is the caller ID (also called the ANI).
`session.connection.orignator`	This variable is a reference to either `session.connection.remote.uri` or `session.connection.local uri` depending on whether the caller initiated the call (an inbound call) or the platform initiated the call (an outbound call).
`session.connection.redirect`	An array where each element is a node in the connection path. Each element of the array contains a `uri`, `pi` (presentation information), `si` (screening information), and a `reason` property. The reason property can be either 'unknown', 'user busy', 'no reply', 'deflection during alerting', 'deflection immediate response', or 'mobile subscriber not reachable'.
`session.connection.aai`	Contains application-to-application information passed during call setup. This variable is often used to store user-to-user information obtained from ISDN calls.
`session.connection.protocol.` `name`	Contains the name of the protocol, e.g. 'sip'. This name also identifies a property of `session.connection.protocol` that contains more protocol-specific information. For example, `session.connection.protocol.sip.requesturi` might contain the SIP Request-URI.
`session.connection.protocol.` `version`	Contains the protocol version number.

The `next` attribute on `<link>` is used to specify the document or dialog to transition to (the `expr` attribute performs the same purpose but instead specifies an ECMAScript variable that evaluates to the URI of the document or dialog to transition to). To specify a particular form in a document, one may use an XML fragment, for example `http://www.example.com/doc.vxml#myform` (if the fragment is omitted, the first form in the document will be executed). To specify another form in the same document, one can use a bare fragment, e.g. `#myform`. The following example illustrates a link which causes a transition to a form called `dialog02`:

```
<?xml version="1.0" encoding="UTF-8"?>
<vxml version="2.1" xmlns="http://www.w3.org/2001/vxml">
   <link next="#dialog02">
```

```
        <grammar src="nav.grxml"/>
    </link>
    <form id="dialog01">
        ...
    </form>
    <form id="dialog02">
        ...
    </form>
</vxml>
```

To throw an event instead of making a transition, one uses the `event` attribute to specify the name of the event (or equivalently the `eventexpr` to specify the name of an ECMAScript variable that evaluates to the name of the event to throw). Events are discussed in more detail in Section 16.7.4.

16.7.3.2 Goto

The `<goto>` element is used to:

* transition to another document;
* transition to another dialog in the same document, or
* transition to another form item in the same form.

The `next` and `expr` attributes are used to specify the document or dialog to transition to and follow the same conventions as the identically named attributes on `<link>`. To transition between form items in the same form, the `nextitem` attribute must be used (or equivalently the `expritem` to specify the name of an ECMAScript variable that evaluates to the name of the form item to transition to). For example, to transition to a field whose name is `field01`, one would specify:

```
<goto nextitem="field01">
```

Note that the leading '#' character is not used for transitions between form items because the `nextitem` attribute value takes the name of the form item and not a URI.

16.7.3.3 Submit

The `<submit>` element is used to submit information to the web server using the HTTP methods `GET` or `POST` (see also Appendix C, located on the Web at http://www.daveburke.org/speechprocessing/). In a typical application, information collected from the user is submitted to the application server where server-side logic implementing business rules will act on the data and return a new VoiceXML document. For example, information sent to the application server might be used to lookup information in a database; that information could be inserted into a `<prompt>` in a new VoiceXML document returned in the HTTP response.

The `next` attribute on `<submit>` is used to indicate the URI which will accept the information and return a new VoiceXML document. The `expr` attribute is equivalent to the `next` attribute except that it specifies the name of an ECMAScript variable that evaluates to the URI. The `namelist` attribute on `<submit>` specifies a space-delimited list of variables, the names and values of which are to be sent to the application server via HTTP. How this information is encoded in the HTTP request depends on the value of the `method` attribute on `<submit>`. A value of `get` indicates that the namelist variables are to be encoded in the URI (known as a query string). For example, consider the following snippet of VoiceXML:

```
<submit next="http://example.com/process.asp"
        namelist="field01 field02" method="get"/>
```

Assuming the variable `field01` evaluates to 23 and `field02` evaluates to 'large green', the VoiceXML interpreter will generate the following request:

```
http://example.com/process.asp?field01=23&field02=large%20green
```

The variables that were specified in the `namelist` attribute are inserted in the query string part of the URI. The query string is indicated by the presence the '?' character in the URI and each variable is inserted to the right of this character in the form of `name=value`, separated by the ampersand character '&'. Note that certain characters must be 'escaped' so as not to violate the URI syntax. Special characters are encoded by using a '%' character followed by the ASCII code for that character represented in hexadecimal. In the example above, the space is encoded as %20.

If the method attribute on `<submit>` specifies the value of `post`, the HTTP POST method is used. The POST method encodes `namelist` information in the body of the HTTP request. How that information is encoded in the body further depends on the `enctype` attribute on `<submit>`. If the `enctype` attribute takes the value of `application/x-www-form-urlencoded`, the information is encoded using the same syntax as was used for the query string (but this time it is in the HTTP message body and not part of the URI), i.e. in the example above the body would contain:

```
field01=23&field02=large%20green
```

Using the POST method is preferable to the GET method since there may be restrictions on how large the URI may be allowed to grow (the HTTP message body theoretically has no limit). More importantly, since many Web servers log the URIs used to access them, sensitive information may inadvertently appear in Web server logs. Sending audio information (i.e. audio recorded from the user) to the Web server deserves special consideration. Since the audio is binary data (i.e. a sequence of bytes), the `application/x-www-form-urlencoded` encoding is inappropriate.[4] Instead, the `enctype` value of `multipart/form-data` is used for audio data. Specifying this encoding will insert the raw bytes in a multipart message contained in the HTTP body (see Section 7.1.4.1 for more information on multipart messages).

16.7.4 Event handling

The platform automatically throws events when the user fails to respond, a user utterance does match a grammar, an error occurs, a link is matched that throws an event, etc. Events can be intercepted via catch handlers to allow the insertion of custom handling logic, otherwise a default action will take place. For example, if an `error.semantic` is not caught with a catch handler, the platform will

---

[4] It is actually possible to encode each raw byte using the % notation where the two digit hexadecimal represents the byte value. However, it is inefficient to encode raw bytes in this way since one needs 3 bytes to represent each 1 byte of audio data.

Table 16.9 Predefined general events

Name	Description
`noinput`	The user has not responded within the timeout interval.
`nomatch`	The user's utterance did not match a grammar with sufficient confidence.
`maxspeechtimeout`	The user input was too long.
`connection.disconnect.hangup`	The user has hung up.
`connection.disconnect.transfer`	The user has been transferred to another line and will not return.

`exit`
`help`
`cancel`
	The user has asked to exit, for help, or to cancel by matching a universal grammar (see the `universals` property in Table 16.4).

`error.badfetch`	The fetch of a new document has failed.
`error.badfetch.http.` `responsecode`	This error is thrown for a wide variety of errors including document parse errors. More refined errors may be thrown for HTTP. For example, if the document cannot be found, `error.badfetch.http.404` will be thrown.
`error.semantic`	Indicates a run-time error such as referencing an undefined variable.
`error.noauthorization`	This error is thrown when the application attempts to access a resource for which it does not have permission.
`error.noresource`	The platform requested a resource (for example a speech recogniser) which is not available.

`error.unsupported.builtin`
`error.unsupported.format`
`error.unsupported.language`
`error.unsupported.objectname`
`error.unsupported.element`
	Indicates that the requested built-in, format, language, object, or element is not supported. For the last case, the element is substituted – for example, if the `<transfer>` element is not supported, an `error.unsupported.transfer` event will be thrown when it is encountered.

play a platform-specific prompt and subsequently end the call. Table 16.9 lists the predefined general events in VoiceXML.

As we have already seen, catch handlers are specified using the element `<catch>` with an associated attribute `event` indicating the event to which the handler applies. The exact event name may be specified and a wildcard feature is supported. For example, `<catch event="error.*"/>` will match `error.semantic`, `error.badfetch`, etc. VoiceXML also specifics four shorthand elements for standard events: `<noinput>`, `<nomatch>`, `<help>`, `<error>`. For example, `<noinput>` is equivalent to `<catch event="noinput">`. More information on an event may be available in the `_message` variable accessible in the anonymous scope of the event handler (this variable may evaluate to `undefined` if there is no extra information available).

When an event is thrown, the VoiceXML interpreter selects the best-qualified catch handler – the closest catch handler in terms of scope that matches the event. The `<catch>` element may also specify a `count` attribute. When there are several catch handlers to choose from (at the same scope), the handler selected is the one with the highest count attribute that is less than or equal to the number of times that event has been thrown from the visited form item or form. A catch element can be disabled if an ECMAScript variable is specified in the `<catch>` element's `cond` attribute that evaluates to `false`. This allows catch handlers to be dynamically enabled and disabled through scripting.

An event can be manually thrown via the `<throw>` element. The event to throw is specified in the `event` attribute (or equivalently via `eventexpr` to specify the name of an ECMAScript variable that evaluates to the name of the event to throw). The `_message` property of the event may be specified through the `message` attribute of `<throw>` (or equivalently via `messageexpr` to specify the name of an ECMAScript variable that evaluates to the `_message` value).

16.8 Resource fetching

During the life of an application, the VoiceXML interpreter context must fetch several resources on behalf of the executing application in the VoiceXML interpreter. These resources include:

- VoiceXML documents;
- audio files;
- grammar files;
- script files;
- object code (via the `<object>` element – see [4]);
- XML documents (via the `<data>` element – see [5]).

Typically, the VoiceXML platform will incorporate a cache to optimise resource fetching. In many applications, it is possible to use cached versions of resources to improve performance by obviating the need to contact the Web server. Caching is described further in Appendix C. VoiceXML specifies several attributes to provide fine-grained control over elements that perform resource fetching – see Table 16.10. Elements that fetch VoiceXML documents, i.e. `<goto>`, `<submit>`, `<link>`, `<choice>` and `<subdialog>` also have an additional attribute called `fetchaudio`. This attribute

Table 16.10 Common resource fetching attributes

Attribute	Description
fetchtimeout	The time to wait specified in seconds or milliseconds for the content to be returned before throwing an `error.badfetch` event.
fetchhint	Takes the value of either `prefetch` or `safe` (ultimately defaults to `prefetch`). A value of `safe` indicates that the resource should only be downloaded when actually needed, e.g. an audio file need only be downloaded when it is about to be played. A value of `prefetch` indicates that the resource should be downloaded when the VoiceXML page that references it is downloaded.
maxage	Specifies the maximum acceptable age (in seconds) for a resource returned from the cache. If the age of the resource is older than maxage, the resource must be revalidated against the HTTP server. See Appendix C (located on the Web at http://www.daveburke.org/speechprocessing/) for more information.
maxstale	Specifies the maximum number of seconds the resource can be allowed become stale in the cache and still be used. This value ultimately defaults to 0. See Appendix C (located on the Web at http://www.daveburke.org/speechprocessing/) for more information.

provides the URI of an audio resource to play while the fetching is being performed. This allows any server-side processing delays and network round-trip times to be masked by audio, thereby enhancing the user experience.

16.9 Call transfer

While VoiceXML is first and foremost a dialog language, i.e. it supports conversational dialogs between a human and a computer, the language does provides basic call control features primarily for transferring the caller to a live agent. Call transfer in VoiceXML is designed to be agnostic of the underlying network and hence is specified at a rather high level with a certain degree of abstraction. There are three transfer modes in VoiceXML. Bridge transfer (also called trombone transfer) allows the VoiceXML dialog to resume after the transferred call ends. The blind and consultation transfer types cause the VoiceXML interpreter to drop out of the call and hence do not allow the caller to return to the VoiceXML dialog.

The `<transfer>` element performs the transfer action when it is executed and its form item variable can store the outcome of the transfer. The `dest` attribute (equivalently the `destexpr` for ECMAScript expressions) specifies the destination for the transfer, usually a telephone number or IP telephony address (such as a SIP URI). All VoiceXML platforms must support the `tel:` syntax defined in RFC 2806 [67]. The basic syntax is straightforward, for example an international phone number is preceded by a '+', the country code, and then the rest of the digits, e.g. `tel:+35312345678`. The `type` attribute[5] on `<transfer>` determines the type of transfer to be performed and may be one of `bridge`, `blind` or `consultation`. We consider each transfer type next.

16.9.1 Bridge

With the bridge transfer type, the platform connects the caller to the callee via the platform as illustrated in Figure 16.4. The key point here is that the connection between the original caller and the callee is made inside the VoiceXML platform. Since the caller's media is passing through the VoiceXML platform, the platform may monitor for conditions to terminate the transfer and return the caller to the VoiceXML dialog.

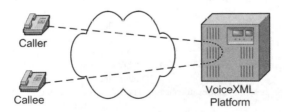

Figure 16.4 Bridge transfer.

---

[5] In VoiceXML 2.0, there was no `type` attribute and instead one chose between bridge and blind using the `bridge` attribute which takes a Boolean value. VoiceXML 2.1 introduced consultation transfer in addition to the more general `type` attribute.

With bridge transfers, the `<transfer>` attribute `connecttimeout` can be used to specify the time in milliseconds or seconds to wait while trying to connect the call before failing with a `noanswer` condition (see below). The `maxtime` attribute on `<transfer>` can be used to specify the time in milliseconds or seconds that the call is allowed to last. A value of 0 s implies indefinite duration. The `transferaudio` attribute on `<transfer>` can be used to specify the URI of an audio source to play to the caller while the transfer attempt is in progress, i.e. before the callee answers.

The `<transfer>` element may contain `<prompt>`s to specify audio to be played to the caller before the transfer attempt is made. With a bridge transfer, it is possible to specify a `<grammar>` element within the `<transfer>` element to recognise against. The `<transfer>`'s grammars are modal (grammars in other scopes are not activated). If the caller's input matches the grammar either during the setup phase or during the call, the callee will be disconnected and VoiceXML execution will continue. Note that hotword recognition is automatically used for grammars specified within the `<transfer>` element and hence an unsuccessful match of the grammar is ignored.

When the transfer input item has finished executing (either with a successful transfer or otherwise), the form item variable is filled with a value depending on the outcome – see Table 16.11.

The following is an example of a VoiceXML application that performs a bridge transfer:

```
<?xml version="1.0" encoding="UTF-8"?>
<vxml version="2.1" xmlns="http://www.w3.org/2001/vxml">
    <form>
        <transfer name="xfer"
                  dest="tel:+353-123-4567"
                  transferaudio="ringback.wav"
                  connecttimeout="60s"
                  type="bridge">
            <prompt>
                Commencing transfer. Say cancel to terminate
```

Table 16.11   `<transfer>` form item variables and associated reasons

Form item variable	Phase	Reason
noanswer	Setup	No answer within the time specified by the `connecttimeout` attribute.
busy	Setup	The callee was busy.
network_busy	Setup	The network refused the call.
unknown	Setup	The transfer ended but the reason is unknown.
near_end_disconnect	Call established	A DTMF or speech recognition from the caller caused the transfer to terminate.
maxtime_disconnect	Call established	The callee was disconnected because the call duration reached the value specified in the `maxtime` attribute.
far_end_disconnect	Call established	The callee hung up.
network_disconnect	Call established	The network disconnected the callee from the platform.

Note: If the caller disconnects during the transfer, a `connection.disconnect.hangup` event is thrown as usual and the call transfer is terminated.

```
                    the transfer at any time.
          </prompt>
          <grammar src="cancel.grxml" type="application/srgs+xml"/>
          <filled>
               <if cond="xfer == 'busy'">
                    <prompt>The line is busy</prompt>
               <elseif cond="xfer == 'noanswer'"/>
                    <prompt>There was no answer</prompt>
               <else/>
                    <prompt>
                         The call terminated with status
                         <value expr="xfer"/>
                    </prompt>
               </if>
          </filled>
     </transfer>
   </form>
</vxml>
```

16.9.2 Blind

Blind transfer differs from bridge transfer in that the caller is redirected to the callee and the platform
drops out of the call – see Figure 16.5. As a consequence, the caller cannot return to the VoiceXML
platform after the transfer has ended. Blind transfer is more efficient than bridge transfer because the
VoiceXML platform is relinquished from the call and does not usurp two separate call legs for the
duration of the transfer. Blind transfer suffers from one major disadvantage: the platform redirects the
call unconditionally and cannot recover the call if the transfer attempt fails.

16.9.3 Consultation

Consultation transfer is similar to blind transfer except that it addresses its shortcomings by avoiding
a drop of the caller in the event of an unsuccessful transfer attempt. Thus, during setup of the transfer,
consultation transfer behaves similarly to the bridge transfer. After the transfer succeeds, the behaviour
is similar to blind transfer – the connection.disconnect.transfer event is thrown and the
VoiceXML platform drops out of the call.

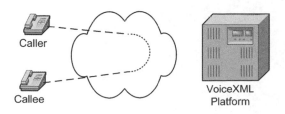

Figure 16.5   Blind transfer.

16.10 Summary

In this chapter, we presented VoiceXML, a popular language for building speech telephony applications that can leverage many of the features that MRCP has to offer. Indeed, many VoiceXML platforms on the market today employ MRCP for providing the necessary speech processing capabilities required of VoiceXML. There are many details in the VoiceXML language itself and, due to space constraints, we only concentrated on the core features. The VoiceXML specifications [4, 5] are eminently accessible and recommended reading for those seeking further information.

The next chapter describes a real world VoiceXML application example that performs basic auto-attendant and voicemail functions. This example also affords us the ability to study realistic MRCP flows derived from a VoiceXML application.

17

VoiceXML – MRCP interworking

VoiceXML and MRCP are synergistic technologies. VoiceXML offers a web-based, declarative programming language for developing speech applications, and MRCP supplies the underlying protocol machinery for accessing the requisite media processing resources. Functionally, the VoiceXML interpreter can be viewed as a media controller sitting atop of an MRCP client, issuing instructions derived from the currently interpreted VoiceXML document.

MRCP is a flexible protocol and there is no single definition of how one maps VoiceXML to MRCP. The approach we take in this chapter is to study some fundamental VoiceXML interaction patterns, describe the corresponding VoiceXML – MRCP mappings, and to present an example in the form of a complete VoiceXML application with associated MRCP flows. While reading this chapter, the reader may wish to refer back to Chapters 12 to 14 on the speech synthesiser, speech recogniser and recorder resources, as well as Chapter 16 on VoiceXML.

17.1 Interworking fundamentals

MRCP exposes a wide range of functions to the client including speech synthesis, audio playback, speech and DTMF recognition, audio recording, and speaker verification and identification. Different VoiceXML platforms may invoke MRCP to access media processing functions to various different degrees. For example, some VoiceXML platforms may depend wholly on MRCP for all their media processing requirements, while others may already incorporate the means to perform DTMF recognition and audio recording and therefore do not need access to the dtmfrecog and recorder resources. Note, however, that VoiceXML 2.1 does not provide speaker verification and identification functions[1] nor does it expose voice enrollment features. Indeed, even for those MRCP resources that a VoiceXML platform may invoke, not all features exposed by the protocol are required. For example, while the SET-PARAMS request method can be used to optimise the size of MRCP messages (i.e. by obviating the need to carry the same header field value in several messages within the same session), this message is not strictly required to implement standard VoiceXML functionality.

---

[1] Future versions of VoiceXML are expected to provide speaker verification and identification functions.

In this section we will focus on the speech synthesiser, speech recogniser and recorder resources, which provide a complete set of media processing functions required to support VoiceXML. In the following subsections, we break out the core VoiceXML interaction patterns into 'play prompts', 'play and recognise', and 'record' tasks. In each section, we review the VoiceXML elements that trigger the function, discuss the data representations sent over MRCP, describe the mapping of VoiceXML parameters to MRCP header fields, and summarise the return mappings from MRCP to VoiceXML. A complete VoiceXML example with corresponding MRCP flows is described in Section 17.2. Throughout the chapter, we assume the media and control sessions between the VoiceXML intepreter's MRCP client and the speech resources are already established.

17.1.1 Play prompts

During VoiceXML interpretation, prompts are queued for playback within the VoiceXML interpreter's prompt queue. The prompt queue is 'flushed' (i.e. played) at certain points within the execution – principally when the VoiceXML interpreter reaches a 'waiting state' at which point the prompts are played and the interpreter listens for input. Prompts are also flushed when the interpreter needs to fetch a resource (such as a VoiceXML document) and the `fetchaudio` attribute is specified – the prompts queued before the `fetchaudio` are played to completion and the `fetchaudio` is subsequently played until the fetch completes. A similar mechanism applies for `transferaudio` (the audio that plays while a call transfer attempt is underway).

In many VoiceXML applications, it is often necessary to trigger the playback of queued prompts without invoking a recogniser or recorder resource. Examples of such cases include:

- playback of prompts when barge-in is disabled;
- upon execution of a `<disconnect>` or `<exit>` element;
- execution of `<transfer>`;
- execution of `<record>`;
- upon execution of a document transition when `fetchaudio` is specified (i.e. resulting from `<goto>`, `<submit>`, `<link>`, `<choice>` or `<subdialog>`).

Play prompts is achieved by issuing a `SPEAK` request to the speech synthesiser resource. The VoiceXML interpreter provides inline SSML to the speech synthesiser resource in the message body of the `SPEAK` request. The VoiceXML interpreter is responsible for ensuring that the content to be rendered is valid SSML and, while VoiceXML itself leverages SSML for audio output, some pre-processing is still required. For example, the VoiceXML interpreter must enclose the SSML content extracted from the VoiceXML document in a `<speak>` element. Since the VoiceXML interpreter maintains its own prompt queue, it does not require the queuing functionality provided by MRCP's `SPEAK` queuing mechanism. Indeed, VoiceXML specifically requires that the prompt queue is only flushed at certain instances during the execution while MRCP will actually queue any received `SPEAK` request for immediate playback. When the VoiceXML prompt queue is large, a platform may choose to break the queue into smaller sections at playback time and send a series of `SPEAK` requests. It may also be appropriate to use multiple `SPEAK` requests if different parameters need to be applied to different prompts, e.g. different barge-in attributes or some prompts may specify a different default language requiring a different `Speech-Language` header field, etc.

In summary, the following steps are performed for 'prompt playback':

(i) During execution SSML is extracted from the VoiceXML document, pre-processed, and added to a queue in the VoiceXML interpreter. The main pre-processing steps are:

(a) enclose the SSML in a `<speak>` element;

(b) remove the `expr` attribute on each `<audio>` element, evaluate it and insert the result in a new `src` attribute;

(c) for each `<value>` element in the SSML content, evaluate the `expr` attribute content and insert the string result in place of the element;

(d) If any `<audio>` element includes the `maxage`, `maxstale`, `fetchtimeout`, or `fetchhint` attributes, remove the attributes and map their values to the corresponding MRCP header fields[2] (see Table 17.1).

(ii) The prompt queue is played via one or more `SPEAK` requests to the speech synthesiser resource. If prompts have different barge-in attributes, or `Content-Base`,[3] separate `SPEAK` requests must be issued. VoiceXML attributes and properties are mapped to the `SPEAK` request (or alternatively set for the session via `SET-PARAMS`) according to Table 17.1.

(iii) If the response to the `SPEAK` request specified a non-2*xx* status code, an `error.noresource` event[4] is thrown and the operation is aborted;

(iv) If `SPEECH-MARKER` events are received, the mark name and time (extracted from the `Speech-Marker` header field) is stored in the VoiceXML interpreter. In the case of recognition, the name of the last mark encountered is made available through the `markname` shadow variable property.

(v) The VoiceXML interpreter waits for the `SPEAK-COMPLETE` event to be received and maps the `Completion-Cause` result as described in Table 17.2. The timestamp reported in the `Speech-Marker` header field is used to calculate the time since the last mark was received and communicated to the application via VoiceXML's `marktime` shadow variable property.

(vi) If the VoiceXML interpreter needs to terminate the audio (for example, an ongoing fetch completes where `fetchaudio` was specified, or the user hangs up), a `STOP` request is issued to the speech synthesiser resource.

17.1.2 Play and recognise

Play and recognise functionality is triggered in VoiceXML primarily upon execution of a `<field>`, `<menu>`, or `<initial>` element. Additionally, play and recognise functions may be invoked for the `<record>` or `<transfer>` elements when grammars are activated (the primary purpose of the recognition here is to terminate the record or transfer action). A special case of 'play and recognise' occurs when no prompts are queued, in which case recognition without playback is performed.

The MRCP flows for 'play and recognise' involve both issuing a `SPEAK` request to the speech synthesiser and a `RECOGNIZE` request to the speech recogniser resource and differ slightly depending on whether barge-in is enabled or not. Note that the queuing mechanism for `RECOGNIZE` requests is not used by VoiceXML since, in general, subsequent recognition actions are dependent on the preceding one and cannot be queued. Inline grammars within the VoiceXML document are extracted

---

[2] Note that while VoiceXML provides more granular control of caching and fetching attributes (i.e. on a per-`<audio>` element), MRCP only allows these parameters to be specified on a per-`SPEAK` request (although note that this is mostly due to a limitation in the SSML language itself).

[3] Prompts may be queued from different VoiceXML pages with different base URIs.

[4] Platforms might choose to specifically map the `401 Method not allowed` status code to an `error.noauthorization` event.

Table 17.1 VoiceXML to MRCP parameter mappings (applicable to the SPEAK request)

VoiceXML attribute or property	MRCP parameter
bargein attribute on <prompt> or bargein property	Kill-On-Barge-In[a]
maxage attribute on <audio> or audiomaxage property	Cache-Control: max-age
maxstale attribute on <audio> or audiomaxstale property	Cache-Control: max-stale
fetchhint attribute on <audio> or audiofetchhint property	Audio-Fetch-Hint
fetchtimeout attribute on <audio> or fetchtimeout property	Fetch-Timeout
xml:lang attribute on <prompt> (or, if omitted, xml:lang attribute on <vxml>)	Speech-Language
xml:base attribute on <prompt> (or, if omitted, xml:base attribute on <vxml> or, if omitted, the content base of the VoiceXML document as determined by the VoiceXML document URI)	Content-Base

[a]This parameter only has an effect when a simultaneous recognition action is occurring – see next section.

Table 17.2 SPEAK-COMPLETE Completion-Cause return mappings

Completion-Cause	VoiceXML behaviour
000 normal	VoiceXML interpretation continues.
001 barge-in	VoiceXML interpretation continues (barge-in will be detected by START-OF-INPUT from the recogniser resource).
002 parse-failure	Throw error.noresource event.
003 uri-failure[a]	Throw error.badfetch event.
004 error	Throw error.noresource event.
005 language-unsupported	Throw error.unsupported.language event.
006 lexicon-load-failure	Throw error.badfetch event.

[a] Note: In the event that an <audio> URI cannot be played, no error is generated or reported by MRCP. Rather, fallback audio (if any) is played and synthesis continues as described in SSML and VoiceXML.

and pre-processed and the result is inserted directly into the message body of the MRCP request. If the fetchhint attribute on grammar (or the grammarfetchhint property) is set to prefetch, the grammar is dispatched to the recogniser resource via the DEFINE-GRAMMAR method. Otherwise the grammar may be sent as part of the RECOGNIZE request when the recognition action is triggered. Note that if an inline grammar needs to be used multiple times within the same session, its associated session: URI (derived from it's Content-ID when first supplied to the recogniser resource) may be used to avoid sending the grammar data to the recogniser resource on each subsequent RECOGNIZE request. Grammars which reference external documents are passed over to the recogniser resource in the form of a text/grammar-ref-list. Note that the document order of the grammars as they appear in VoiceXML must be preserved as they are passed over to the recogniser resource to maintain the correct grammar precedence (VoiceXML and MRCP rules stipulate that if an utterance matches more than one grammar, the first grammar in document order is given precedence). VoiceXML support of built-in grammars and universal grammars require no special support from MRCP, rather the VoiceXML interpreter is responsible for providing the appropriate grammars for recognition. Some features of VoiceXML (such as <menu>, <option> and <link>) also require that the VoiceXML interpreter generate appropriate grammars for recognition.

In summary, the following steps are performed for 'play and recognise':

(i) Inline grammars are extracted from the VoiceXML document and pre-processed. If the <grammar> element includes the maxage, maxstale, fetchtimeout, fetchhint, or weight attributes, they are removed and mapped according to Table 17.3.

(ii) If a fetchhint on prefetch applies to any of the active grammars, a DEFINE-GRAMMAR request is issued to the recogniser resource at grammar activation time. Otherwise, the grammars (both inline and URIs) are assembled in a multipart message body in VoiceXML document order. Only the grammar types specified by the VoiceXML inputmodes property are selected.

(iii) The prompts (if any) are dispatched for playback following the steps outlined in Section 17.1.1.

(iv) The next step depends on whether prompts were dispatched and barge-in is enabled or not. In either case, VoiceXML attributes and properties are mapped to the RECOGNIZE request (or alternatively set for the session via SET-PARAMS) according to Table 17.3. If no prompts were dispatched in the previous step, the RECOGNIZE request is issued. Otherwise:

(a) If barge-in is enabled, the RECOGNIZE request is issued to the speech recogniser resource at the same time as the SPEAK request. The RECOGNIZE request is specified with a Start-Input-Timers[5] value of false and the START-INPUT-TIMERS[6] request method is issued to the recogniser resource on receipt of the SPEAK-COMPLETE message.

(b) If barge-in is not enabled, the RECOGNIZE request is not sent until the SPEAK-COMPLETE event message is received and the Start-Input-Timers header field is specified with a value of true. The header field Clear-DTMF-Buffer is inserted with value true.

(v) If the response to the RECOGNIZE request specified a non-2xx status code, an error.noresource event[7] is thrown and the operation is aborted.

(vi) If the START-OF-INPUT[8] event is received and there is one or more outstanding SPEAK requests, the BARGE-IN-OCCURRED request is sent to the speech synthesis resource. The time stamp reported in the Speech-Marker header field is used to calculate the time since the last mark was received and communicated to the application via VoiceXML's marktime shadow variable property.

(vii) The VoiceXML interpreter waits for the RECOGNITION-COMPLETE event to be received and maps the Completion-Cause result as described in Table 17.4. The recognition results in the message body are parsed and mapped according to the rules described in Section 16.6.4. The Waveform-URI header field's value (if present) maps to the name$.recording, name$.recordingsize, and name$.recordingduration variables.

(viii) If the VoiceXML interpreter needs to terminate the recognition (for example, the user hangs up), a STOP request is issued to the recogniser resource.

---

[5] This header field is called Recognizer-Start-Timers in MRCPv1 (see Appendix A, located on the Web at http://www.daveburke.org/speechprocessing/).
[6] This method is called RECOGNITION-START-TIMERS in MRCPv1 (see Appendix A, located on the Web at http://www.daveburke.org/speechprocessing/).
[7] See footnote 4.
[8] This event is called START-OF-SPEECH in MRCPv1 (see Appendix A, located on the Web at http://www.daveburke.org/speechprocessing/).

Table 17.3 VoiceXML to MRCP parameter mappings (applicable to the `RECOGNIZE` request)

VoiceXML attribute or property	MRCP parameter
`timeout` attribute on `<prompt>` or `timeout` property	`No-Input-Timeout`
`bargeintype` attribute on `<prompt>` or `bargeintype` property	`Input-Type`
`maxage` attribute on `<grammar>` or `grammarmaxage` property	`Cache-Control: max-age`
`maxstale` attribute on `<grammar>` or `grammarmaxstale` property	`Cache-Control: max-stale`
`fetchhint` attribute on `<grammar>` or `grammarfetchhint` property	`Fetch-Hint`
`fetchtimeout` attribute on `<grammar>` or `fetchtimeout` property	`Fetch-Timeout`
`weight` attribute on `<grammar>`	`weight` parameter on the `text/grammar-ref-list`[a]
`xml:lang` attribute on `<grammar>` (or, if omitted, `xml:lang` attribute on `<vxml>`)	`Speech-Language`
`xml:base` attribute on `<grammar>` (or, if omitted, `xml:base` attribute on `<vxml>` or, if omitted, the content base of the VoiceXML document as determined by the VoiceXML document URI)	`Content-Base`
`recordutterance` property	`Save-Waveform`
`recordutterancetype` property	`Media-Type`
`confidencelevel` property	`Confidence-Threshold`
`sensitivity` property	`Sensitivity-Level`
`speedvsaccuracy` property	`Speed-Vs-Accuracy`
`completetimeout` property	`Speech-Complete-Timeout`
`incompletetimeout` property	`Speech-Incomplete-Timeout`
`maxspeechtimeout` property	`Recognition-Timeout`
`interdigittimeout` property	`DTMF-Interdigit-Timeout`
`termtimeout` property	`DTMF-Term-Timeout`
`termchar` property	`DTMF-Term-Char`
`maxnbest` property	`N-Best-List-Length`

[a] If a weight is specified on an inline grammar, that grammar must be sent to the recogniser resource via a `DEFINE-GRAMMAR` request and later referenced in the `RECOGNIZE` request via a `session:` URI in a `text/grammar-ref-list` where a `weight` parameter can be applied.

17.1.3 Record

Recording is triggered in VoiceXML upon execution of the `<record>` element. Before the recording begins, the queued audio is played to completion (i.e. the `RECORD` request is not sent to the recorder resource until after receipt of the `SPEAK-COMPLETE` event from the speech synthesiser resource). If no grammars are active then the steps described in Section 17.1.1 apply prior to recording. If grammars are active, then the steps described in Section 17.1.2 apply. Note that recognition can remain active in parallel with the recording. In the trivial case, this occurs if the `dtmfterm` attribute is present on

Table 17.4 RECOGNITION-COMPLETE Completion Cause return mappings

Completion-Cause	VoiceXML behaviour
000 normal	NLSML results extracted and
008 success-maxtime	VoiceXML interpretation continues.
001 nomatch	Throw nomatch event.
007 speech-too-early	
013 partial-match	
002 no-input-timeout	Throw noinput event.
003 hotword-maxtime	Throw maxspeechtimeout event
014 partial-match-maxtime	
015 no-match-maxtime	
004 grammar-load-failure	Throw error.badfetch event
005 grammar-compilation-failure	
009 uri-failure	
012 semantics-failure	
006 recognizer-error	Throw error.noresource event.
010 language-unsupported	Throw error.unsupported.language event.

the <record> element and is given a value of true (in this case any DTMF key terminates the recording). Parallel recognition and recording can also occur if a DTMF grammar is specified within the <record> element or a non-local grammar (e.g. one associated with a <link>) is active.

In summary, the following steps are performed for 'record':

(i) If the beep attribute is specified on the <record> element with a value of true, a short beep prompt must be added to the end of the prompt queue.

(ii) If any grammars are active, the steps described for 'prompt and recognise' in Section 17.1.2 are applied; otherwise the steps described for 'prompt playback' in Section 17.1.1 are applied.

(iii) The RECORD element is issued on receipt of the SPEAK-COMPLETE event (assuming there were prompts played). VoiceXML attributes and properties are mapped to the RECORD request (or alternatively set for the session via SET-PARAMS) according to Table 17.5. As an optimisation, the Capture-On-Speech header field may be specified with a value of true.

(iv) If the response to the RECORD request specifies a non-2xx status code, an error.noresource event[9] is thrown and the operation is aborted.

(v) If a parallel recognition action is ongoing and a recognition match is returned (via the RECOGNITION-COMPLETE event), the recording is stopped by issuing a STOP request to the recorder resource; otherwise step (vi) applies.

(vi) The VoiceXML interpreter waits for the RECORD-COMPLETE event to be received and maps the Completion-Cause result as described in Table 17.6. The recording results are obtained from the Record-URI header field and mapped to the name$.duration and name$.size shadow variables.

(vii) If the VoiceXML interpreter needs subsequently to <submit> the data to a document server, the recording must first be accessed from the Record-URI (this could involve a HTTP GET request, for example) to obtain the recording bytes.

---

[9] See footnote 4.

Table 17.5 VoiceXML to MRCP parameter mappings (applying to the RECORD request)

VoiceXML attribute or property	MRCP parameter
maxtime attribute on <record>	Max-Time
finalsilence attribute on <record>	Final-Silence
type attribute on <record>	Media-Type
timeout attribute on <prompt> or timeout property	No-Input-Timeout
sensitivity property	Sensitivity-Level

Table 17.6 RECORD-COMPLETE Completion-Cause return mappings

Completion-Cause	VoiceXML behaviour
000 success-silence	Recording results extracted and VoiceXML interpretation continues.
001 success-maxtime	Recording results extracted and VoiceXML interpretation continues. The name$.maxtime shadow variable is set to true.
002 noinput-timeout	Throw noinput event
003 uri-failure	Throw error.noresource event
004 error	

17.2 Application example

The application described here is a corporate auto-attendant, which allows callers to speak the name of the person they wish to talk to and automatically be transferred to that person without the hassle of looking up a phone directory or speaking to an operator. If the party cannot be contacted, the application allows the caller to record a message. Figure 17.1 illustrates the call flow.

17.2.1 VoiceXML scripts

The application consists of four VoiceXML documents, one SRGS file, and one PLS file. The files are summarised in Table 17.7.

The application root document, root.vxml, is illustrated in Listing 17.1 below. The purpose of this document is to declare an application-scoped ECMAScript variable to hold the contact information about the party the caller wishes to speak to. It also specifies default catch handlers for noinput, nomatch and error events.

The application entry point, index.vxml, is illustrated in Listing 17.2 below. The application commences by playing an audio 'jingle' and a welcome message. The field is then selected and executed, which results in the caller being prompted for the name of the person they wish to speak to. The grammar containing the names is specified in directory.grxml (described below). When the name is recognised, the field's filled action assigns the input item variable to the application-scoped variable declared in root.vxml for storage across the multiple VoiceXML files in the application.

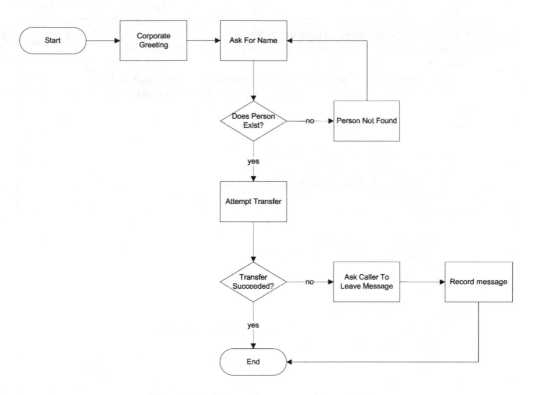

Figure 17.1 Call flow for auto-attendant application.

Table 17.7 Application files

File	Type	Purpose	Listing
root.vxml	VoiceXML	Application root. Stores the contact details of the desired party and specifies default catch handlers.	Listing 17.1
index.vxml	VoiceXML	Application entry point. Plays welcome message, collects name of party to contact and attempts call transfer.	Listing 17.2
directory.grxml	SRGS, SISR	Grammar for names with associated contact information.	Listing 17.3
names.pls	PLS	Lexicon for tuning name recognition.	Listing 17.4
record.vxml	VoiceXML	Records a message from the caller.	Listing 17.5
store_message.php	VoiceXML	Receives recorded message and stores; returns an empty VoiceXML document.	Listing 17.6

```xml
<?xml version="1.0" encoding="UTF-8"?>
<vxml version="2.1" xmlns="http://www.w3.org/2001/vxml">
    <!--
        An application-scoped variable holding the contact
        details of the person the caller wishes to speak to.
    -->
    <var name="person"/>

    <!-- Default catch handlers -->
    <catch event="noinput">
        I'm sorry, I didn't hear you.
        <reprompt/>
    </catch>

    <catch event="nomatch">
        I'm sorry, I didn't get that.
        <reprompt/>
    </catch>

    <catch event="error">
        We are experiencing technical difficulties.
        Please call back later.
        <exit/>
    </catch>
</vxml>
```

Listing 17.1 Application root document: root.vxml.

```xml
<?xml version="1.0" encoding="UTF-8"?>
<vxml version="2.1" xmlns="http://www.w3.org/2001/vxml"
    application="root.vxml" xml:lang="en-GB">
    <form>
        <!-- Welcome message -->
        <block>
            <prompt bargein="false">
                <audio src="audio/jingle.wav"/>
                Welcome to the Smart Company auto attendant.
```

Listing 17.2 Application entry point: index.vxml.

```
            </prompt>
        </block>
        <!-- Find person to contact -->
        <field name="person">
            <prompt bargein="true">
                Please say the name of the person you would like
                to speak to or say <emphasis>operator</emphasis> to
                speak to the operator.
            </prompt>
            <grammar src="grammar/directory.grxml"
                    type="application/srgs+xml"/>
            <filled>
                <assign name="application.person" expr="person"/>
            </filled>
        </field>

        <!-- Attempt transfer -->
        <transfer name="xfer"
                destexpr="'tel:' + application.person.number"
                transferaudio="audio/ringback.wav"
                connecttimeout="20s"
                type="consultation">
            <prompt>
                <lexicon uri="names.pls" type="application/pls+xml"/>
                Transferring you to
                <value expr="application.person.name"/>
            </prompt>
            <filled>
                <!-- If the person is busy or does not answer,
                    record message -->
                <if cond="xfer == 'busy'">
                    <prompt>The line is busy</prompt>
                    <goto next="record.vxml"/>
                <elseif cond="xfer == 'noanswer'"/>
                    <prompt>There was no answer</prompt>
                    <goto next="record.vxml"/>
                </if>
            </filled>
        </transfer>
    </form>
</vxml>
```

Listing 17.2   (Continued)

```xml
<?xml version="1.0" encoding="UTF-8"?>
<grammar version="1.0" xmlns="http://www.w3.org/2001/06/grammar"
        mode="voice" xml:lang="en-GB"
        tag-format="semantics/1.0" root="names">
    <lexicon uri="names.pls" type="application/pls+xml"/>
    <rule id="names">
        <tag>out.name=meta.current().text;</tag>
        <one-of>
            <item>The Operator<tag>out.number='1900';</tag></item>
            <item>Stephen Breslin<tag>out.number='1907';</tag></item>
            <item>Andrew Fuller<tag>out.number='1916';</tag></item>
            <item>James Bailey<tag>out.number='1914';</tag></item>
            <item>Amanda McDonnell<tag>out.number='1926';</tag></item>

            <!-- Add new names and numbers here -->

        </one-of>
    </rule>
</grammar>
```

Listing 17.3   Grammar encapsulating the directory database: directory.grxml.

The semantic interpretation result from the grammar consists of an ECMAScript object with two properties:

- name – the name of the person to connect to;
- number – the phone number of the person to connect to.

Once the field is filled, the FIA selects and executes the transfer. The destination for the transfer is obtained from the previously assigned application-scoped variable. A transfer type of consultation is used with the benefit that a port of the VoiceXML platform is not usurped after the caller and callee are connected while, at the same time, giving the opportunity to the caller to leave a message for the callee in the event that the connection cannot be established. If the transfer is successful, the connection.disconnect.transfer is thrown and the VoiceXML application simply exits. Otherwise, the transfer input item, xfer, is filled with the outcome of the transfer attempt. In the event that the callee is busy or does not answer, the VoiceXML application transitions to record.vxml.

   Listing 17.3 contains the grammar file directory.grxml. The structure of the grammar is very simple – it contains a single alternative with an entry for each name. SISR is used to annotate each name with a corresponding phone number (the number property of the rule variable). The raw text utterance is assigned to the name property of the rule variable. This approach allows for a convenient centralised directory database contained in a single file. Other approaches are of course possible. For example, the grammar could simply return the name and this value could be sent via the <submit> element to the web server, resulting in a server-side database lookup of the phone number. The resultant VoiceXML document could contain a <transfer> element to the appropriate destination.

```
<?xml version="1.0" encoding="UTF-8"?>
<lexicon version="1.0"
        xmlns="http://www.w3.org/2005/01/pronunciation-lexicon"
        alphabet="ipa"
        xml:lang="en-GB">
    <lexeme>
        <grapheme>McDonnell</grapheme>
        <alias>Mac Donnell</alias>
    </lexeme>
</lexicon>
```

Listing 17.4 Pronunciation lexicon file: names.pls.

The SRGS in Listing 17.3 specifies a pronunciation lexicon names.pls, which can be used to tune the recognition of names, for example by specifying aliases and multiple pronunciations for names. The PLS file is presented in Listing 17.4.

Listing 17.5 contains record.vxml, which is the file transitioned to when the callee cannot be contacted.

The caller is asked to leave a message for the person they wish to contact. The recording is terminated either by a maximum duration cutoff of 20 seconds, or a continuous detection of silence of 4 seconds, or by virtue of a hangup. The catch handler for the `connection.disconnect.hangup` submits the recording to the server using the relative URI cgi/store_message.php. Note that this handler is also called as a result of the caller hanging up or by the application triggering a hangup when the recording terminates due to `maxtime` or `finalsilence` settings. The store_message.php script contains some server-side programming logic (not shown) to extract the recorded message bytes in the `multipart/formdata` message sent in the HTTP POST request from the VoiceXML interpreter. After storing the message (e.g. to a file store or inside a database), the script returns a VoiceXML file that simply terminates – illustrated in Listing 17.6.

```
<?xml version="1.0" encoding="UTF-8"?>
<vxml version="2.1" xmlns="http://www.w3.org/2001/vxml"
      application="root.vxml">
    <form>
        <!-- Record message -->
        <record name="message" beep="true" maxtime="20s"
                type="audio/x-wav" finalsilence="4s">
            <prompt>
                <lexicon uri="names.pls" type="application/pls+xml"/>
                <value expr="application.person.name"/> is
```

Listing 17.5 VoiceXML file to record caller's message: record.vxml.

```
                   unavailable to take your call. Please leave
                   a message after the beep.
               </prompt>
               <filled>
                   Message recorded. Good bye.
                   <disconnect/>
               </filled>
           </record>

           <!-- Send the recording to the server when the user
                 hangs up -->
           <catch event="connection.disconnect.hangup">
               <if cond="message != undefined">
                   <submit next="cgi/store_message.php"
                             namelist="message application.person.name"
                             enctype="multipart/form-data"
                             method="post"/>
               </if>
           </catch>
       </form>
</vxml>
```

Listing 17.5 (Continued)

```
<?xml version="1.0" encoding="UTF-8"?>
<vxml version="2.1" xmlns="http://www.w3.org/2001/vxml"
       application="root.vxml">
    <form>
        <block>
            <!--
                 This it the VoiceXML document that is returned
                 after the voice message is stored server-side.
                 This document simply exits.
             -->
            <exit/>
        </block>
    </form>
</vxml>
```

Listing 17.6 VoiceXML document returned after submission of recorded message to store_message.php (PHP code that receives the recorded message omitted for simplicity).

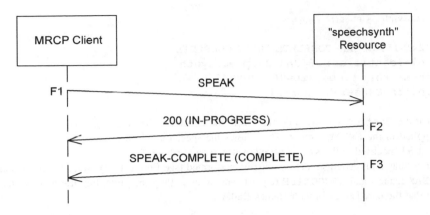

Figure 17.2 'Play Prompts' sequence for barge-in disabled prompts in Listing 17.2.

17.2.2 MRCP flows

In this section, we look at MRCP flows derived from the VoiceXML listings in Section 17.2.1. We start with the VoiceXML entry point, Listing 17.2, and assume that the initial URI is http://example.com/autoattend/index.vxml. Upon execution of the <block> element, its contents are simply added to the VoiceXML prompt queue. Note that barge-in is disabled for these prompts. Next, the FIA selects the <field> and executes it. The contained prompt (this time with barge-in enabled) is added to the prompt queue. Now that the VoiceXML execution has reached the waiting state, the prompt queue is ready to be flushed. Since the first set of prompts has barge-in disabled, a separate 'play prompts' action must be performed (see Figure 17.2).

F1 (client → speechsynth):

```
MRCP/2.0 399 SPEAK 10000
Channel-Identifier: 98f3bae2@speechsynth
Speech-Language: en-GB
Content-Base: http://example.com/autoattend/
Content-Type: application/ssml+xml
Content-Length: 202

<?xml version="1.0" encoding="UTF-8"?>
<speak version="1.0" xmlns="http://www.w3.org/2001/10/synthesis">
    <audio src="audio/jingle.wav"/>
    Welcome to the Smart Company auto attendant.
</speak>
```

F2 (speechsynth → client):

```
MRCP/2.0 119 10000 200 IN-PROGRESS
Channel-Identifier: 98f3bae2@speechsynth
Speech-Marker: timestamp=857206027059
```

F3 (speechsynth → client):

```
MRCP/2.0 157 SPEAK-COMPLETE 10000 COMPLETE
Channel-Identifier: 98f3bae2@speechsynth
Speech-Marker: timestamp=861500994355
Completion-Cause: 000 normal
```

Note that the `Content-Base` is derived from the absolute URI part of the VoiceXML application and supplied in the `SPEAK` request. This allows the speech synthesiser resource to fully resolve the relative URI specified in the `<audio>` element's `src` attribute.

Next, a 'play and recognise' action must be performed (see Figure 17.3). Since barge-in is enabled, the `SPEAK` request and `RECOGNIZE` requests are issued at the same time. For argument's sake, let us assume that the user barges in with 'James Bailey'.

F1 (client → speechsynth):

```
MRCP/2.0 454 SPEAK 20000
Channel-Identifier: 98f3bae2@speechsynth
Speech-Language: en-GB
Content-Base: http://example.com/autoattendant/
```

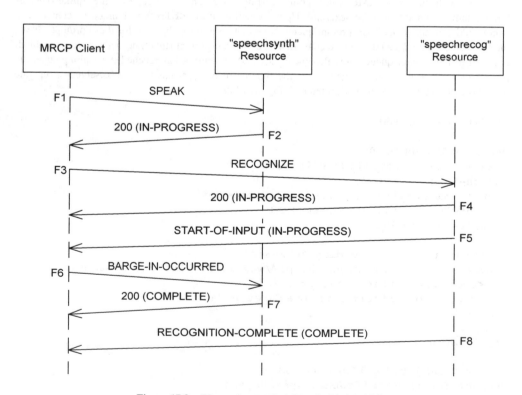

Figure 17.3 'Play and recognise' flow for Listing 17.2.

```
Content-Type: application/ssml+xml
Content-Length: 254

<?xml version="1.0" encoding="UTF-8"?>
<speak version="1.0" xmlns="http://www.w3.org/2001/10/synthesis">
    Please say the name of the person you would like
    to speak to or say <emphasis>operator</emphasis> to
    speak to the operator.
</speak>
```

F2 (speechsynth → client):

```
MRCP/2.0 119 20000 200 IN-PROGRESS
Channel-Identifier: 98f3bae2@speechsynth
Speech-Marker: timestamp=857206027059
```

F3 (client → speechrecog):

```
MRCP/2.0 215 RECOGNIZE 20001
Channel-Identifier: 23af1e13@speechrecog
Start-Input-Timers: false
Content-Type: text/grammar-ref-list
Content-Length: 55

<http://example.com/autoattend/grammar/directory.grxml>
```

F4 (speechrecog → client):

```
MRCP/2.0 79 20001 200 IN-PROGRESS
Channel-Identifier: 23af1e13@speechrecog
```

F5 (speechrecog → client):

```
MRCP/2.0 133 START-OF-INPUT 20001 IN-PROGRESS
Channel-Identifier: 23af1e13@speechrecog
Input-Type: speech
Proxy-Sync-Id: 2102
```

F6 (client → speechsynth):

```
MRCP/2.0 103 BARGE-IN-OCCURRED 20002
Channel-Identifier: 98f3bae2@speechsynth
Proxy-Sync-Id: 2102
```

F7 (speechsynth → client):

```
MRCP/2.0 147 20002 200 COMPLETE
Channel-Identifier: 98f3bae2@speechsynth
```

```
Speech-Marker: timestamp=861500994355
Active-Request-Id-List: 20000
```

F8 (speechrecog → client):

```
MRCP/2.0 549 RECOGNITION-COMPLETE 20001 COMPLETE
Channel-Identifier: 23af1e13@speechrecog
Completion-Cause: 000 success
Content-Type: application/nlsml+xml
Content-Length: 366
```

```
<?xml version="1.0" encoding="UTF-8"?>
<result
grammar="http://example.com/autoattend/grammar/directory.grxml"
xmlns="http://www.ietf.org/xml/ns/mrcpv2">
    <interpretation confidence="0.9">
        <instance>
            <name>James Bailey</name>
            <number>1914</number>
        </instance>
        <input>James Bailey</input>
    </interpretation>
</result>
```

Note that the START-INPUT-TIMERS request method is never sent due to the fact that barge-in occurred. The VoiceXML interpreter now maps the NLSML semantic result (an object with two properties: name and number) into the field's input item variable called person.

The next sequence involves a 'play prompts' action leading up to the VoiceXML call transfer. Since a transferaudio attribute is specified on the transfer element (see Listing 17.2), the prompts queued up to that point are flushed prior to playing the transferaudio. This in effect means that a second 'play prompts' is required specifically for the transferaudio. The VoiceXML interpreter can conveniently make use of the speech synthesiser's SPEAK queue in this case – the second SPEAK request for the transferaudio will initially be in the PENDING state. When the transferaudio is ready to be played, a SPEECH-MARKER event is generated to notify the VoiceXML interpreter that the request has transitioned to IN-PROGRESS and is being played. The VoiceXML interpreter initiates the call transfer attempt at this point. When the call transfer completes, either successfully or otherwise, the transferaudio must be stopped. Figure 17.4 illustrates the sequence diagram and the corresponding messages follow.

F1 (client → speechsynth):

```
MRCP/2.0 418 SPEAK 30000
Channel-Identifier: 98f3bae2@speechsynth
Speech-Language: en-GB
Content-Base: http://example.com/autoattendant/
Content-Type: application/ssml+xml
Content-Length: 218
```

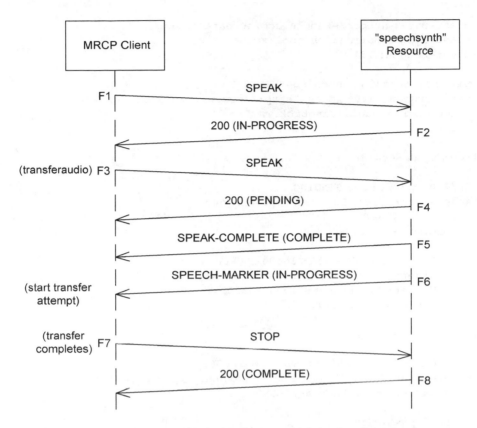

Figure 17.4   Two 'play prompts' actions leading up to and during the call transfer attempt.

```
<?xml version="1.0" encoding="UTF-8"?>
<speak version="1.0" xmlns="http://www.w3.org/2001/10/synthesis">
    <lexicon uri="names.pls" type="application/pls+xml"/>
    Transferring you to
    James Bailey
</speak>
```

F2 (speechsynth → client):

```
MRCP/2.0 119 30000 200 IN-PROGRESS
Channel-Identifier: 98f3bae2@speechsynth
Speech-Marker: timestamp=857206027059
```

F3 (client → speechsynth):

```
MRCP/2.0 354 SPEAK 30001
Channel-Identifier: 98f3bae2@speechsynth
Speech-Language: en-GB
```

```
Content-Base: http://example.com/autoattendant/
Content-Type: application/ssml+xml
Content-Length: 154

<?xml version="1.0" encoding="UTF-8"?>
<speak version="1.0" xmlns="http://www.w3.org/2001/10/synthesis">
    <audio src="audio/ringback.wav"/>
</speak>
```

F4 (speechsynth → client):

```
MRCP/2.0 75 30001 200 PENDING
Channel-Identifier: 98f3bae2@speechsynth
```

F5 (speechsynth → client):

```
MRCP/2.0 157 SPEAK-COMPLETE 30000 COMPLETE
Channel-Identifier: 98f3bae2@speechsynth
Speech-Marker: timestamp=861500994355
Completion-Cause: 000 normal
```

F6 (speechsynth → client):

```
MRCP/2.0 129 SPEECH-MARKER 30001 IN-PROGRESS
Channel-Identifier: 98f3bae2@speechsynth
Speech-Marker: timestamp=861500994355
```

F7 (client → speechsynth):

```
MRCP/2.0 68 STOP 30002
Channel-Identifier: 98f3bae2@speechsynth
```

F8 (speechsynth → client):

```
MRCP/2.0 147 30002 300 COMPLETE
Channel-Identifier: 98f3bae2@speechsynth
Speech-Marker: timestamp=865795961651
Active-Request-Id-List: 30001
```

For the final MRCP flow, we assume that the aforementioned call transfer completed with a status of noanswer. As a result, the record.vxml document will be fetched and executed (see Listing 17.5). This listing requires a 'play prompts' followed by a 'record' action. Since the beep attribute is specified on the <record> element, the VoiceXML interpreter must insert an audio recording of a beep – this is achieved in the example by specifying a URI to an internal web server in the VoiceXML interpreter that serves up the audio file for the beep. If the user terminates the recording with silence, a further 'play prompts' action is required to play the valediction message. In the following, we assume the caller hangs up after recording their message, thus requiring the VoiceXML interpreter to stop the recording and retrieve the audio up to that point. The audio is specified in the Record-URI header

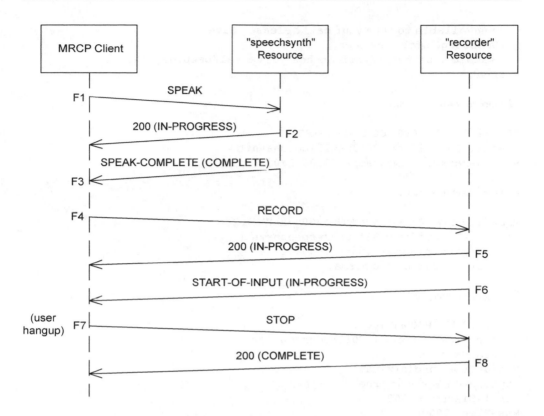

Figure 17.5 'Play prompts' action followed by 'record' action for Listing 17.5.

in the response to the STOP request (in this case, the recorder resource makes the recording available via its own internal Web server). Since, in Listing 17.6, the VoiceXML interpreter must HTTP POST the recording to the web server, the VoiceXML interpreter must first fetch the audio from the recorder resource via a HTTP GET prior to submitting the audio bytes. Figure 17.5 illustrates the flow and the corresponding messages follow.

F1 (client → speechsynth):

```
MRCP/2.0 554 SPEAK 40000
Channel-Identifier: 98f3bae2@speechsynth
Speech-Language: en-GB
Content-Base: http://example.com/autoattend/
Content-Type: application/ssml+xml
Content-Length: 357

<?xml version="1.0" encoding="UTF-8"?>
<speak version="1.0" xmlns="http://www.w3.org/2001/10/synthesis">
    <lexicon uri="names.pls" type="application/pls+xml"/>
    James Bailey is
```

```
    unavailable to take your call. Please leave
    a message after the beep.
    <audio src="http://client-host/internal/beep.wav"/>
</speak>
```

F2 (speechsynth → client):

```
MRCP/2.0 119 40000 200 IN-PROGRESS
Channel-Identifier: 98f3bae2@speechsynth
Speech-Marker: timestamp=857206027059
```

F3 (speechsynth → client):

```
MRCP/2.0 157 SPEAK-COMPLETE 40000 COMPLETE
Channel-Identifier: 98f3bae2@speechsynth
Speech-Marker: timestamp=861500994355
Completion-Cause: 000 normal
```

F4 (client → recorder):

```
MRCP/2.0 170 RECORD 40001
Channel-Identifier: 11F018BE6@recorder
Record-URI:
Media-Type: audio/x-wav
Capture-On-Speech: true
Final-Silence: 4000
Max-Time: 20000
```

F5 (recorder → client):

```
MRCP/2.0 77 40001 200 IN-PROGRESS
Channel-Identifier: 11F018BE6@recorder
```

F6 (recorder → client):

```
MRCP/2.0 88 START-OF-INPUT 40001 IN-PROGRESS
Channel-Identifier: 11F018BE6@recorder
```

F7 (client → recorder):

```
MRCP/2.0 66 STOP 40002
Channel-Identifier: 11F018BE6@recorder
```

F8 (recorder → client):

```
MRCP/2.0 184 40002 200 COMPLETE
Channel-Identifier: 11F018BE6@recorder
```

```
Active-Request-Id-List: 40001
Record-URI: <http://server-host/audio/file04.wav>;
    size=40000;duration=5000
```

17.3 Summary

In this chapter, we look at how VoiceXML and MRCP can interwork by first reviewing some fundamental interaction patterns employed by VoiceXML including 'play prompts', 'play and recognise', and 'record'. For each pattern, we summarised some straightforward mapping rules between VoiceXML and MRCP. Finally, we presented a complete VoiceXML application and studied the MRCP flows derived from it.

The material presented here serves both to illustrate how MRCP can be employed to provide advanced speech processing capabilities for real applications and also to elucidate some of the inner workings of modern IVR platforms. MRCP is a powerful and flexible protocol and may be employed in numerous different ways. Although vendors' IVR platform implementations will inevitably differ from each other in terms of how they use MRCP, we anticipate that many will employ similar approaches to those described in this chapter. While the influence of VoiceXML on the MRCP protocol is quite apparent, MRCP today goes considerably further than VoiceXML in terms of functionality provided and should serve as a good basis for new versions of VoiceXML as they are developed.

References

[1] The Definitive Guide to the IVR Marketplace: North America and EMEA. Understanding the IVR in developed markets. DMTC1041, *DataMonitor*, April 2005.

[2] J. Rosenberg, H. Schulzrinne, G. Camarillo, A. Johnston, J. Peterson, R. Sparks, M. Handley, and E. Schooler. *SIP: Session Initiation Protocol*. RFC 3261, Internet Engineering Task Force, June 2002.

[3] S. Shanmugham and D. Burnett. *Media Resource Control Protocol Version 2 (MRCPv2)*. Internet-Draft draft-ietf-speechsc-mrcpv2-11, Internet Engineering Task Force, September 2006 (work in progress).

[4] S. McGlashan, D. C. Burnett, J. Carter, P. Danielson, J. Ferrans, A. Hunt, B. Lucas, B. Porter, K. Rehor, and S. Tryphonas. *VoiceXML Extensible Markup Language (VoiceXML)* Version 2.0. W3C Recommendation, World Wide Web Consortium, March 2004.

[5] M. Oshry, R. J. Auburn, P. Baggia, M. Bodell, D. Burke, D. Burnett, E. Candell, J. Carter, S. McGlashan, A. Lee, B. Porter, and K. Rehor. *VoiceXML Extensible Markup Language (VoiceXML)* Version 2.1. W3C Candidate Recommendation, World Wide Web Consortium, September 2006.

[6] A. Hunt and S. McGlashan. *Speech Recognition Grammar Specification* Version 1.0. W3C Recommendation, World Wide Web Consortium, March 2004.

[7] L. Van Tichelen and D. Burke. *Semantic Interpretation for Speech Recognition*. W3C Candidate Recommendation, World Wide Web Consortium, January 2006.

[8] D. C. Burnett, M. R. Walker, and A. Hunt. *Speech Synthesis Markup Language (SSML)* Version 1.0. W3C Recommendation, World Wide Web Consortium, September 2004.

[9] P. Baggia. *Pronunciation Lexicon Specification (PLS)* Version 1.0. W3C Working Draft, World Wide Web Consortium, January 2006.

[10] D. Oran. *Requirements for Distributed Control of Automatic Speech Recognition (ASR), Speaker Identification/Speaker Verification (SI/SV), and Text-to-Speech (TTS) Resources*. RFC 4313, Internet Engineering Task Force, December 2005.

[11] H. Schulzrinne, S. Casner, R. Frederick, and V. Jacobson. *RTP: A Transport Protocol for Real-Time Applications*. RFC 3550, Internet Engineering Task Force, July 2003.

[12] R. Fielding, J. Gettys, J. Mogul, H. Frystyk, L. Masinter, P. Leach, and T. Berners-Lee. *Hypertext Transfer Protocol–HTTP/1.1*. RFC 2616, Internet Engineering Task Force, June 1999.

[13] S. Shanmugham, P. Monaco, and B. Eberman. *A Media Resource Control Protocol (MRCP) Developed by Cisco, Nuance, and Speechworks*. RFC 4463, Internet Engineering Task Force, April 2006.

[14] M. Handley and V. Jacobson. *SDP: Session Description Protocol*. RFC 2327, Internet Engineering Task Force, April 1998.

[15] D. Burke, M. Scott, J. Haynic, R. Auburn, and S. McGlashan. *SIP Interface to VoiceXML Media Services*. Internet-Draft draft-burke-vxml-02, Internet Engineering Task Force, November 2006 (work in progress).

[16] *Gateway Control Protocol: Version 2*. ITU-T Recommendation H.248, International Telecommunication Union, May 2002.

[17] H. Schulzrinne and S. Casner. *RTP Profile for Audio and Video Conferences with Minimal Control*. RFC 3551, Internet Engineering Task Force, July 2003.

[18] A. B. Roach. *Session Initiation Protocol (SIP)-Specific Event Notification*. RFC 3265, Internet Engineering Task Force, June 2002.

[19] R. Sparks. *The Session Initiation Protocol (SIP) Refer Method*. RFC 3515, Internet Engineering Task Force, April 2003.

[20] B. Campbell, J. Rosenberg, H. Schulzrinne, C. Huitema, and D. Gurle. *Session Initiation Protocol (SIP) Extension for Instant Messaging*. RFC 3428, Internet Engineering Task Force, December 2002.

[21] J. Rosenberg and H. Schulzrinne. *Reliability of Provisional Responses in the Session Initiation Protocol (SIP)*. RFC 3262, Internet Engineering Task Force, June 2002.

[22] T. Berners-Lee, R. Fielding, and L. Masinter. *Uniform Resource Identifiers (URI): General Syntax*. RFC 2396, Internet Engineering Task Force, August 1998.

[23] A. Niemi. *Session Initiation Protocol (SIP) Extension for Event State Publication*. RFC 3903, Internet Engineering Task Force, October 2004.

[24] J. Rosenberg. *The Session Initiation Protocol (SIP) UPDATE Method*. RFC 3311, Internet Engineering Task Force, September 2002.

[25] S. Donovan. *The SIP INFO Method*. RFC 2976, Internet Engineering Task Force, October 2000.

[26] G. Camarillo, G. Eriksson, J. Holler, and H. Schulzrinne. *Grouping of Media Lines in the Session Description Protocol (SDP)*. RFC 3388, Internet Engineering Task Force, December 2002.

[27] J. Rosenberg and H. Schulzrinne. *An Offer/Answer Model with the Session Description Protocol (SDP)*. RFC 3264, Internet Engineering Task Force, June 2002.

[28] A. B. Johnston. *SIP: Understanding the Session Initiation Protocol*. Second edition, Artech House, 2004.

[29] H. Sinnreich and A. B. Johnston. *Internet Communications Using SIP*. John Wiley & Sons, 2001.

[30] M. Handley, H. Schulzrinne, E. Schooler, and J. Rosenberg. *SIP: Session Initiation Protocol*. RFC 2543, Internet Engineering Task Force, March 1999.

[31] D. Yon and G. Camarillo. *TCP-Based Media Transport in the Session Description Protocol (SDP)*. RFC 4145, Internet Engineering Task Force, September 2005.

[32] J. Lennox. *Connection-Oriented Media Transport over the Transport Layer Security (TLS) Protocol in the Session Description Protocol (SDP)*. RFC 4572, Internet Engineering Task Force, July 2006.

[33] D. Fallside and P. Walmsley. *XML Schema Part 0*: Primer Second edition. W3C Recommendation, World Wide Web Consortium, October 2004.

[34] J. Clarke. *RELAX NG Specification*, Committee Specification, Oasis, December 2001.

[35] D. Mills. *Network Time Protocol (Version 3) Specification, Implementation and Analysis*. RFC 1305, Internet Engineering Task Force, March 1992.

[36] *Dual Rate Speech Coder For Multimedia Communications Transmitting at 5.3 and 6.3 kbit/s*. ITU-T Recommendation G.723.1, International Telecommunication Union, March 1996.

[37] *Coding of Speech at 8 kbit/s Using Conjugate-Structure Algebraic-Code-Excited Linear-Prediction (ACELP)*. ITU-T Recommendation G.729, International Telecommunication Union, March 1996.

[38] *Pulse Code Modulation (PCM) of Voice Frequencies*. ITU-T Recommendation G.711, International Telecommunication Union, 1988.

[39] W. C. Chu. *Speech Coding Algorithms: Foundations and Evolution of Standardized Coders*. Wiley–Interscience, 2003.

[40] *AMR Speech Codec; Transcoding Functions (Release 6)*. 3GPP TS 26.090, 3rd Generation Partnership Project, December 2004.

[41] *Adaptive Multi-rate – Wideband (AMR-WB) speech codec; Transcoding Functions (Release 6)*. 3GPP TS 26.190, 3rd Generation Partnership Project, July 2005.

[42] J. Sjoberg, M. Westerlund, A. Lakaniemi, and Q. Xie. *Real-Time Transport Protocol (RTP) Payload Format and File Storage Format for the Adaptive Multi-Rate (AMR) and Adaptive Multi-Rate Wideband (AMR-WB) Audio Codecs*. RFC 3267, Internet Engineering Task Force, June 2002.

[43] S. Andersen, A. Duric, H. Astrom, R. Hagan, W. Kleijn, and J. Linden. *Internet Low Bit Rate Codec (iLBC)*. RFC 3951, Internet Engineering Task Force, December 2004.

[44] H. Schulzrinne and S. Petrack. *RTP Payload for DTMF Digits, Telephony Tones and Telephony Signals*. RFC 2833, Internet Engineering Task Force, May 2000.

[45] J. Rosenberg, H. Schulzrinne, and P. Kyzivat. *Indicating User Agent Capabilities in the Session Initiation Protocol (SIP)*. RFC 3840, Internet Engineering Task Force, August 2004.

[46] J. Rosenberg, H. Schulzrinne, and P. Kyzivat. *Caller Preferences for the Session Initiation Protocol (SIP)*. RFC 3841, Internet Engineering Task Force, August 2004.

[47] T. Melanchuk and C. Boulton. *Locating Media Resource Control Protocol Version 2 (MRCPv2) Servers.* Internet-Draft draft-melanchuk-speechsc-serverloc-00, Internet Engineering Task Force, January 2006 (work in progress).

[48] B. Ramsdell. *S/MIME Version 3 Message Specification.* RFC 2633, Internet Engineering Task Force, June 1999.

[49] M. Baugher, D. McGrew, M. Naslund, E. Carrara, and K. Norrman. *The Secure Real-time Transport Protocol (SRTP).* RFC 3711, Internet Engineering Task Force, March 2004.

[50] F. Andreasen, M. Baugher, and D. Wing. *Session Description Protocol (SDP) Security Descriptions for Media Streams.* RFC 4568, Internet Engineering Task Force, July 2006.

[51] D. Kristol and L. Montulli. *HTTP State Management Mechanism.* RFC 2109, Internet Engineering Task Force, February 1997.

[52] D. Kristol and L. Montulli. *HTTP State Management Mechanism.* RFC 2965, Internet Engineering Task Force, October 2000.

[53] D. C. Burnett, P. Baggia, J. Barnett, A. Buyle, E. Eide, and L. Van Tichelen. *SSML 1.0 Say-as Attribute Values.* W3C Working Group Note, World Wide Web Consortium, May 2005.

[54] H. Schulzrinne, A. Rao, and R. Lanphier. *Real Time Streaming Protocol.* RFC 2326, Internet Engineering Task Force, April 1998.

[55] H. Alvestrand. Tags for the Identification of Languages. RFC 3066, Internet Engineering Task Force, January 2001.

[56] *Standard ECMA-327*, third edition Compact Profile. European Computer Manufacturer's Association, June 2001.

[57] *Standard ECMA-262*, third edition. European Computer Manufacturer's Association, December 1999.

[58] D. A. Dahl. *Natural Language Semantics Markup Language for the Speech Interface Framework.* W3C Working Draft, World Wide Web Consortium, November 2000.

[59] M. Johnston, W. Chou, D. A. Dahl, G. McCobb, and D. Raggett. *EMMA: Extensible MultiModal Annotation Markup Language,* W3C Working Draft, World Wide Web Consortium, September 2005.

[60] J. R. Deller, J. H. L. Hansen, and J. G. Proakis. *Discrete-Time Processing of Speech Signals.* Second edition, IEEE Press, Wiley–Interscience, 2000.

[61] L. Rabiner and B-H Juang. *Fundamentals of Speech Recognition.* Prentice Hall, 1993.

[62] J. Marsh, D. Veillard, and N. Walsh. *xml:id Version 1.0.* W3C Recommendation, World Wide Web Consortium, September 2005.

[63] D. Beckett and B. McBride. *RDF/XML Syntax Specification (Revised).* W3C Recommendation, World Wide Web Consortium, February 2004.

[64] *Dublin Core Metadata Initiative.* See http://dublincore.org.

[65] I. Jacobs and N. Walsh. *Architecture of the World Wide Web,* Volume 1. W3C Recommendation, World Wide Web Consortium, December 2004.

[66] *Codes for the Representation of Currencies and Funds.* ISO 4217:2001, International Organization for Standardization, 2001.

[67] A. Vaha-Sipila. *URLs for Telephone Calls.* RFC 2806, Internet Engineering Task Force, April 2000.

[68] B. P. Bogart, M. J. R. Healy, and J. W. Tukey. The quefrency alanysis of time series for echoes: Cepstrum, pseudo-autocovariance, cross-cepstrum and saphe cracking. In M. Rosenblatt (ed.), *Proceedings of the Symposium on Time Series Analysis.* John Wiley & Sons, pp. 209–243, 1963.

[69] L. R. Rabiner and M. R. Sambur. An algorithm for determining the endpoints of isolated utterances. *Bell System Technical Journal,* **23**, pp. 552–557, December 1975.

[70] T. Dutoit. *An Introduction to Text-To-Speech Synthesis.* Kluwer Academic Publishers, 1997.

[71] J. P. Campbell. Speaker recognition: A Tutorial. *Proceedings of the IEEE,* **85**(9) pp. 1437–1462, September 1997.

[72] F. Bimbot *et al.* A tutorial on text-independent speaker verification. EURASIP *Journal on Applied Signal Processing,* **4**, pp. 430–451, 2004.

[73] J. Makhoul and R. Schwartz. State of the art in continuous speech recognition. *Proceedings of the National Academy of Sciences,* **92**, pp. 9956–9963, October 1995.

[74] J. Schroeter. Text-to-Speech (TTS) Synthesis, in R. C. Dorf (ed.), *Circuits, Signals, and Speech and Image Processing, The Electrical Engineering Handbook,* third edition, December 2005.

[75] A. Hunt and A. Black. Unit selection in a concatenative speech synthesis system using a large speech database. *Proc. ICCASSP '96,* pp. 373–376, 1996.

[76] E. Moulines and F. Charpentier. Pitch-synchronous waveform processing techniques for text-to-speech synthesis using diphones. *Speech Communication,* **9**, pp. 453–467, 1990.

[77] Y. Stylianou. Applying the harmonic plus noise model in concatenative speech synthesis. *IEEE Transactions on Speech and Audio Processing,* **9**, pp. 21–29, 2001.

[78] L. Masinter. *Returning Values from Forms: Multipart/Form-data*. RFC 2388, Internet Engineering Task Force, August 1998.

[79] T. Bray, J. Paoli, C. M. Sperberg-McQueen, E. Maler, and F. Yergeau. *Extensible Markup Language (XML) 1.0*, third edition. W3C Recommendation, World Wide Web Consortium, February 2004.

[80] T. Bray, D. Hollander, and A. Layman. *Namespaces in XML*. W3C Recommendation, World Wide Web Consortium, January 1999.

Acronyms

AMR – Adaptive Multi-Rate
ASR – Automatic Speech Recognition
B2BUA – Back-to-Back User Agent
CELP – Code Excited Linear Prediction
DSR – Distributed Speech Recognition
DTD – Document Type Definition
DTMF – Dual Tone Multi-Frequency
ECMA – European Computer Manufacturer's Association
EMMA – Extensible MultiModal Annotation
ETSI – European Telecommunications Standards Institute
FIA – Form Interpretation Algorithm
GMM – Gaussian Mixture Model
GSM – Global System for Mobile Communications
HMM – Hidden Markov Model
HNM – Harmonic Plus Noise Model
HTTP – Hyper-text Transport Protocol
IANA – Internet Assigned Numbers Authority
IETF – Internet Engineering Task Force
IMS – IP Multimedia Subsystem
IP – Internet Protocol
IPA – International Phonetic Alphabet
IP Sec – IP Security
ISDN – Integrated Services Digital Network
ITU – International Telecommunications Union
IVR – Interactive Voice Response
LPC – Linear Predictive Coding
MFCC – Mel Frequency Cepstral Coefficients
MIME – Multipurpose Internet Mail Extensions
MRB – Media Resource Broker
MRCP – Media Resource Control Protocol
MTU – Maximum Transmission Unit
NLSML – Natural Language Semantics Markup Language
NLU – Natural Language Understanding
NTP – Network Time Protocol
PBX – Public Branch Exchange
PCM – Pulse Code Modulation

Speech Processing for IP Networks   Dave Burke
© 2007 John Wiley & Sons, Ltd

PLS – Pronunciation Lexicon Specification
PSTN – Public Switched Telephony Network
RFC – Requests For Comments
RTP – Real-Time Protocol
RTCP – RTP Control Protocol
RTSP – Real-Time Streaming Protocol
SCTP – Stream Control Transport Protocol
SDP – Session Description Protocol
SIP – Session Initiation Protocol
SISR – Semantic Interpretation for Speech Recognition
SLM – Statistical Language Model
SRGS – Speech Recognition Grammar Specification
SRTCP – Secure RTCP
SRTP – Secure RTP
SSL – Secure Sockets Layer
SSML – Speech Synthesis Markup Language
TCP – Transmission Control Protocol
TD-PSOLA – Time Domain Pitch-Synchronous Overlap Add
3GPP – Third Generation Partnership Project
TLS – Transaction Layer Security
TTS – Text-to-Speech
UAC – User Agent Client
UAS – User Agent Server
UDP – User Datagram Protocol
URI – Uniform Resource Indicator
VAD – Voice Activity Detection
VoiceXML – Voice Extensible Markup Language
WWW – World Wide Web
W3C – World Wide Web Consortium
XML – Extensible Markup Language
XMLNS – XML Namespace

Index

Analysis-by-synthesis   31, 99, 101
Automatic speech recognition (ASR) – *see* Speech
  recognition

Barge-in   16
  communicating to speech synthesiser resource
    193–194, 204
  detecting   51, 233, 255, 277
  VoiceXML   303, 305, 323

Cepstrum   16–19
Code excited linear prediction (CELP)   99, 100, 101
Codec   95–101

Diphone   34, 35
Dual-tone multifrequency (DTMF)   106–107

ECMAScript
  use in SISR   152, 153–157
  use in VoiceXML   289, 306–309
Endpoint detection   16, 23, 42, 242, 243, 282

Formant   10, 31

Gaussian mixture model (GMM), 28–30
Grammars
  context free   143
  finite state   23, 24, 26, 143
  n-gram Markov grammar   25–26
  perplexity   144
  recursive   143
  regular – *see* Grammars, finite state
  Statistical Language Model (SLM)   25, 149
  see also Speech Recognition Grammar
    Specification

Hidden Markov Model (HMM)   15, 19–21
  Baum–Welch algorithm   21, 23, 29
  beam search   22, 23, 35
  continuous observation   21
  discrete observation   20, 21, 24
  Expectation–Maximisation (EM) algorithm   29, 30
  Forward–Backward algorithm – *see* Baum–Welch
    algorithm
  lattice   21–23, 25
  observation probability   20, 21, 23, 24, 25
  transition probability   20, 21, 23, 24
  Viterbi algorithm   21, 22, 23, 25
Hotword recognition   208
  in MRCP   213, 216, 235, 237, 238, 239
  in VoiceXML   297, 303, 316
Human speech production   9–13, 31, 97
Hyper Text Transfer Protocol (HTTP)   xiv, 6
  digest authentication   75
  HTTPS   53, 257
  its influence on
    MRCP   xiii, 47, 49, 115, 121, 122, 123
    SIP   57, 60, 62, 75

International Phonetic Alphabet (IPA)   12–13, 134,
    135, 177
Internet Assigned Numbers Authority (IANA)   5, 49,
    66, 122, 241, 258, 281
Internet Engineering Task Force (IETF)   5–6
  Internet draft   5
  Internet Engineering Steering Group (IESG)   5
  Request For Comments (RFC)   5, 6
  Speech Services Control (SpeechSC) Working
    Group   xiv, 7, 8
IP security (IPSec)   75, 108, 126